Solaris® for Managers and Administrators

Third Edition

Curt Freeland, Dwight McKay, and
Kent Parkinson

 OnWord Press
Thomson Learning™

Africa • Australia • Canada • Denmark • Japan • Mexico • New Zealand
Philippines • Puerto Rico • Singapore • United Kingdom • United States

NOTICE TO THE READER

Publisher does not warrant or guarantee any of the products described herein or perform any independent analysis in connection with any of the product information contained herein. Publisher does not assume, and expressly disclaims, any obligation to obtain and include information other than that provided to it by the manufacturer.

The reader is expressly warned to consider and adopt all safety precautions that might be indicated by the activities herein and to avoid all potential hazards. By following the instructions contained herein, the reader willingly assumes all risks in connection with such instructions.

The publisher makes no representation or warranties of any kind, including but not limited to, the warranties of fitness for particular purpose or merchantability, nor are any such representations implied with respect to the material set forth herein, and the publisher takes no responsibility with respect to such material. The publisher shall not be liable for any special, consequential, or exemplary damages resulting, in whole or part, from the readers' use of, or reliance upon, this material.

Trademarks

OnWord Press Staff

Publisher: Alar Elken

Executive Editor: Sandy Clark

Acquisitions Editor: James Gish

Managing Editor: Carol Leyba

Development Editor: Daril Bentley

Editorial Assistant: Fionnuala McAvey

Executive Marketing Manager: Maura Theriault

Executive Production Manager: Mary Ellen Black

Production and Art & Design Coordinator: Leyba Associates

Manufacturing Director: Andrew Crouth

Technology Project Manager: Tom Smith

Cover Design by Cammi Noah

Copyright © 2001 by OnWord Press
SAN 694-0269
Third Edition, 2000

10 9 8 7 6 5 4 3 2 1

Printed in Canada

Library of Congress Cataloging-in-Publication Data
Freeland, Curt.
    Solaris for managers and administrators / Curt Freeland, Dwight McKay, and Kent Parkinson. —3rd ed.
      p.   cm.
    Includes index
    ISBN 0-7668-2137-4
    1. Operating systems (Computers) 2, Solaris (Computer file) I. Freeland, Curt. II. McKay, Dwight. III. Parkinson, Kent.

For more information, contact
OnWord Press   An imprint of Thomson Learning
Box 15-015   Albany, New York USA 12212-15015

You can request permission to use material from this text through the following phone and fax numbers.

Phone: 1-800-730-2214; Fax: 1-800-730-2215; or visit our Web site at www.thomsonrights.com

About the Authors

Curt Freeland is a member of the faculty and Director of Facilities for the Computer Science and Engineering Department at the University of Notre Dame. In addition to serving as the departmental systems administrator, Curt teaches senior-level courses in systems administration and in network management for the department. Curt has been working with UNIX systems for more than 20 years. He received a BSE degree in Systems Engineering from Purdue University in 1985, where he worked as Manager of Systems Engineering for the Purdue University Engineering Computer Network for 17 years.

Dwight McKay is Systems and Network Manager for the Markey Center for Structural Biology at Purdue University. Dwight has over 16 years of experience on a wide variety of UNIX systems, as well as VMS, Macintosh, and Windows. His role at the Markey Center involves supporting scientific computing needs, very large data sets, and parallel computation using clustered UNIX systems. Dwight holds a degree from Princeton University.

Kent Parkinson is a principal of Parkinson and Associates, a UNIX systems consulting and training firm located in West Lafayette, Indiana. A former faculty member of Purdue University, Kent has worked with UNIX systems more than 20 years, and with Sun/Solaris operating environments since their respective introductions.

Acknowledgments

The authors would like to thank Carol Leyba, Daril Bentley, James Gish, and the rest of the staff at Thomson Learning/OnWord Press for their patience, encouragement, and guidance throughout this project. We would also like to thank all of the people who read and commented on our manuscripts. This edition of the book is much better thanks to your contributions and suggestions!

—*Curt Freeland, Dwight McKay, and Kent Parkinson*

Thanks to Dwight McKay and Kent Parkinson for their continued friendship, patience, and for once again entrusting them with the production of the book. The success of the first two editions made the production of the third edition much easier to handle!

Thanks to Dr. Steven C. Bass and the rest of the CSE department faculty for their support and encouragement, and for offering me the opportunity to teach at the University of Notre Dame. Thanks also to Dr. Jeffrey M. Squyres, Rich Sudlow, Terry McCoy, Tom Hanstra, Paul Go, Milind Saraph, Johanes Suhardjo, Dale Southard, and the staff of Notre Dame's Office of Information Technology for their assistance. They never (well, rarely ever) complained as I broke things, asked questions, and made unreasonable demands of their time and expertise. I would also like to thank the students of CSE444 for their comments and suggestions on the material in this book. My teaching assistant, Mike (Maddog) McNally not only helped manage my courses, but has also been very helpful with his critiques and comments on courses.

Most of all, to my parents, thank you for teaching me that any job worth doing is worth doing well.

Last, but certainly not least, thanks to my wife (Sally), kids (Shaun and Erin), and the rest of my family for putting up with me during the revision of the manuscript.

—With love, and gratitude, Curt Freeland

Thanks to Curt Freeland and Kent Parkinson for their friendship and support. It is a privilege and a pleasure to work with you both. Thanks also to Dr. Janet Smith and the faculty and staff of the Markey Center for Structural Biology at Purdue University for their encouragement and support of my writing.

But most of all, thanks to my wife (Mary) and my daughters (Catherine and Susan). You've helped me meet the deadlines, given me the time to write when I needed it, and hugged me when I was tired. I could not have finished this without you.

—With deep appreciation, Dwight McKay

First and foremost, we would like to dedicate this book to our readers. We hope that the material in this book is helpful to you as you pursue your goals. Much of the material we present is based on our years of learning through trial and error. We hope that you can benefit from the many errors we committed, as we were learning about UNIX (in general) and Solaris.

We also dedicate this to all of our mentors, with our sincere gratitude and appreciation.

Contents

CHAPTER 4
Fundamental System Administration Tools 69

PART II: SYSTEM MANAGEMENT 109

CHAPTER 5
Refining System Boot and Shutdown Procedures 111

CHAPTER 6
Creating, Deleting, and Managing User Accounts 143

CHAPTER 7
Managing System Security 173

CHAPTER 11
Disk and File System Maintenance 295

CHAPTER 12
Managing System Software 329

Introduction

Welcome to the third edition of *Solaris for Managers and Administrators*. This book was written by systems administration professionals with over 50 years of combined experience in managing corporate and academic computing facilities. Our intent was to approach the topics we cover from a practitioner's perspective.

As time marches on, so does technology. With this edition of the book, we have attempted to include coverage of many of the promising new technologies we believe will have an impact on systems administrators over the upcoming years. As we reviewed and updated the manuscript, we also made every attempt to fix any inaccuracies in the first two editions of the book.

Who Should Read This Book

Solaris for Managers and Administrators was written for beginning Solaris systems administrators and managers who have general UNIX knowledge but little or no Sun systems administration experience. The purpose of this book is to deliver practical information regarding the installation and maintenance of Solaris systems situated in local area networks.

Throughout the book, helpful tips and examples are used to explain and illustrate important concepts. Many examples used in the book derive from the authors' experience as systems administrators. Polling experienced and inexperienced systems administrators to identify the tools they used, and how they approach daily systems administration tasks, derived other examples presented in this book. Much of the material in the book has been taught as course material for CSE444 – "An Introduction to System Administration" and CSE458 "Network Management" at the University of Notre Dame since the fall of 1996.

This text is intended as a handy reference for beginning and experienced Solaris systems administrators. The authors hope that readers enjoy learning about the topic of Solaris systems administration as much as much as we have enjoyed producing the book. If you have any comments, questions, or recommendations about the book, please contact the authors at the Delmar Publishers address found on the copyright page of this book.

Book Organization

Part I, "Fundamentals," covers the concepts of the Solaris distributed environment, Solaris terminology, preparing for a Solaris installation, and basic administration tools.

Part II, "System Management," consisting of the next 12 chapters, is focused on fundamental systems administration tasks such as boot and shutdown procedures, managing user accounts, system security maintenance, and working with device names. Separate chapters are devoted to disk subsystem hardware and software, disk and file system maintenance, system software management, adding terminals and modems, managing printers, system backups and disaster recovery, and automating routine administrative tasks.

Part III, "Network Services," the final six chapters, covers the basics of network hardware, the use of network utilities such as the Network

File System (NFS) and Network Information System (NIS+), diskless and dataless client administration, automating NFS with *automount*, and network name services. The final chapter consists of an overview on providing World Wide Web services on the Solaris platform.

The book concludes with an appendix that summarizes material presented in the release notes and hardware compatibility documentation that ships with each release of Solaris, followed by a detailed index.

Typographical Conventions

Italic font in regular text is used to distinguish command names, daemons, command options, file names, directory and path names, and similar items. Italic is also used to highlight terms and for emphasis. The following are examples.

- Use the *chmod* command to execute the changes. Like *chown*, *chmod* has an *-r* option for handling entire directories.

- The */usr/sbin/sys-unconfig* command allows the administrator to change the network information of the host.

- Infrequently required services can be started on demand by a special daemon known as *inetd*.

- These logical drives became known as *partitions*.

- The files are located in the */sbin* directory, and script directories are located in the */etc* directory.

- The *etc/inittab* file contains a series of directives for the *init* process.

The following is an example of the monospaced font used for examples of command statements and computer/operating system responses.

```
glenn% /etc/mount
/ on /dev/dsk/c0t3d0s0 read/write/setuid on Tue Aug 9 05:58:19 1997
/usr on /dev/dsk/c0t3d0s6 read/write/setuid on Tue Aug 9 05:58:19 1997
/proc on /proc read/write/setuid on Tue Aug 9 05:58:19 1997
/dev/fd on fd read/write/setuid on Tue Aug 9 05:58:19 1997
```

Named keys on the keyboard are enclosed in angle brackets, such as <Enter>, <Shift>, and <Ctrl>.

Key sequences, or instructions to press a key immediately followed by another key, are linked with a plus sign, such as <Ctrl>+G.

Throughout, you will see Notes, Tips, and Warnings. These are described in the examples that follow.

➥   *NOTE: Information on features and tasks that is not immediately obvious or intuitive appears in notes.*

✓   *TIP: Tips on command usage, shortcuts, and other information aimed at saving you time appear like this.*

✗   *WARNING: The warnings appearing in this book are intended to help you avoid committing yourself to results you may not have intended, and to avoid losing data or other unhappy consequences.*

PART 1
Fundamentals

CHAPTER 1
An Introduction
to Distributed Computing

When the first electronic computer was introduced, few people saw much use for this new technology. In fact, many people were afraid of the early computers. Much of this apprehension resulted from a lack of knowledge about what computers could and could not do.

The first computers were slow, difficult to use, extremely large, outrageously expensive, and nearly impossible to maintain. A corporation was "lucky" if it could afford to purchase and operate one computer. All corporate computing was handled at one central computing facility by computer specialists.

As technology marched forward, computers became faster, smaller, and more easily maintained. Suddenly it became feasible for corporations to operate many computers instead of just one. Corporate decision makers were then faced with unanticipated questions such as maintaining a single, large computing center versus more diverse and distributed computing resources.

A Brief History of the Microcomputer

Over the past decade, corporate computing has undergone many drastic changes. (In this chapter, the terms *corporation* and *corporate* are interchangeable with *organization* and *organizational*, and include any public or private sector entity using computing power in daily operations.) As the complexity of data processing jobs increased, typical corporate mainframe computers could not keep pace with computing demands.

These older *mainframe* computers employed discrete logic circuits (chips) to implement the computer hardware. Many computer manufacturers were having problems making these discrete components operate at the speeds required to produce faster computers.

At about the same time, chip manufacturers were developing single-chip computers that could outperform many of the discrete logic mainframes on the market. These single-chip computers were dubbed *microprocessors* or *microcomputers*. Many computer manufacturers began developing a new type of personal desktop computer system based on the microprocessor chips. These high-performance systems soon became known as *workstations*.

Traditional Mainframe Computing

The workstation computing revolution presented corporate computing with the following quandary: should the corporation rely on one large computer system that was not keeping up with computing demands, or should it "downsize" to the new workstations and distribute computing power among many smaller systems?

Corporations had to examine computing needs and how these needs could be addressed by the systems available on the market. Decisions were often reduced to a choice between the mainframe computing model and the distributed computing model.

Mainframe Model

What is the mainframe computing model? *Mainframe computing* is the term used when the corporation employs one large (mainframe) computer system for all corporate computing, an example of which is shown in the following illustration. This mainframe system provides computing power for everything from payroll to materials management and production management.

Typical corporate mainframe computing environment.

Access to the mainframe typically occurs via an ASCII terminal. These terminal connections may be local (in the same building or complex as the mainframe) or remote (via telephone lines and modems). Regardless of the type of connection, the data rate between the mainframe and the terminal is usually slow. Viewing a large document or accessing information from a database may take several minutes via ASCII terminal connections.

Hybrid Mainframe Model

A typical extension of the mainframe computing environment produced a tree-shaped computing environment. In this model, the corporation still relied on a mainframe computer (at the root of the tree). Smaller minicomputers were distributed throughout branch offices. These systems were tightly coupled to the mainframe, and in many cases were completely controlled by the mainframe.

Even smaller microcomputers were connected to the branch minicomputers. The microcomputers were used to control machinery and collect data. This information was shipped to the minicomputer for processing. The results of these computations were shipped off to the corporate mainframe for storage and/or further processing.

This version of mainframe computing typically used telephone lines and modems for the interconnection of system elements. Similar to the traditional mainframe computing model, the hybrid mainframe model was characterized by low data rates and lengthy delays in data access. An example of a hybrid mainframe computing environment is shown in the following illustration.

As minicomputers and microcomputers became more powerful, corporations had to evaluate their computing objectives. Among other things, corporations were forced to weigh the advantages and disadvantages of respective computing models, and make decisions about future directions.

A hybrid mainframe computing environment.

Advantages of Mainframe Computing

The mainframe computing model presents certain advantages over alternative models. One of its major advantages is the centralization of corporate data. In some instances, it is desirable to store all corporate data in one place. In other instances, reliance on a single central repository of information became a liability to the corporation.

As an example of mainframe computing advantages, consider the credit card division of a large banking operation. If all credit card customer data are maintained on a single system, corporate headquarters can easily determine the "value" and status of any and all customers. If the same data were distributed among many geographically diverse computing centers, bank management would not be able to track the "value" of customers as easily.

Another advantage of mainframe computing is that the corporation does not have to worry about maintaining large numbers of geographically separated computers. A single computer at a single site presents a relatively simple maintenance task for the corporate computing division.

Finally, certain applications simply require enormous amounts of computing power to accomplish basic objectives. These applications are best served with a mainframe system. Examples here might include the operation of international airline ticket offices and the trading offices of global stock exchanges.

Disadvantages of Mainframe Computing

Although the centralization of data allows corporate offices to monitor everyday operations, it could also cause substantial operating expenses and delays. These expenses and delays often appear in conjunction with the distribution of corporate data to branch offices or other entities requiring data access.

Revisiting the banking credit card division example previously discussed, what happens if customers are serviced by geographically distributed branch offices? If a branch office requires information pertaining to customers serviced by a different branch, tens of gigabytes of information may have to be exchanged between the two branches. The use of a typical ASCII terminal connection over a phone line would require days to access the information required to conduct the branch office's daily business.

Another disadvantage of mainframe computing is the "single point of failure" problem. If the corporation owns a single, large computer system and the system fails, all corporate data processing capability is off-line. Although many manufacturers provide 24-hour service contracts, problem resolution often requires minutes (or hours) to diagnose the problem and replace the failed module.

Finally, a typical mainframe computer is a very large, very expensive system. As the mainframe becomes obsolete, the corporation must plan well in advance to provide the capital for upgrading the system. Because mainframe computers from different manufacturers rarely run the same operating systems or the same versions of language compilers and other applications, the corporation must also plan for the crossover between the old and new systems.

Distributed Computing Model

The distributed computing environment was originally an off-shoot of the modified mainframe computing environment. As the minicomputers and microcomputers positioned at the leaves of the tree became more powerful, they assumed some of the data storage and processing responsibilities. If the mainframe required access to the information on these leaves, it had to request that information from the workstations.

In addition to more powerful processing elements, interconnection technology underwent significant advances. In the distributed computing model, processing elements are typically tied together via *local area* or *wide area* networks. These connections are orders of magnitude faster than a mainframe's ASCII terminal connections. Documents that once required several minutes to access could now be accessed in less than a minute.

These developments eventually led to the distribution of corporate data among the multitude of processing elements owned by the corporation. No single processor accessed all corporate data, but each processor had access to the data required to maintain relevant activities for a particular segment of the corporation. Such data became almost instantly available to other processing elements via local and wide area data networks. An example of a distributed computing environment is shown in the following illustration.

A distributed computing environment.

In today's distributed computing environment, the data processing chores are distributed among many systems. These systems are typically high-speed workstations linked via a local area or wide area network. Each of these processing elements has a specific task assigned as its primary function. However, each of these workstations is typically a general purpose computer that can also provide processing capabilities beyond its primary task.

Advantages of Distributed Computing

One of the principal advantages of the distributed computing model is geographic distribution. Systems are distributed by geographic region as required to best serve corporate needs. By linking these systems via high-speed networks, the branch offices have instant access to the data required for everyday operations. In addition, the corporate office computers can still access the branch office data via high-speed networks with minimal delays.

Another advantage of the distributed computing model is that each component of the system is typically less expensive than its equivalent in the mainframe computer model. Consequently, the corporation can often purchase several computer systems for the same capital investment required to purchase one mainframe computer.

As the workstations in the distributed computing model become obsolete, the capital burden of upgrading to new systems can be spread over a longer period of time by upgrading the critical systems first, and the less critical systems at a later date. In addition, because many of the workstations run the same operating system, the corporation should not have as many problems porting software to the new systems.

Another major advantage of the distributed computing model is the "many points of failure" feature. Because no single computer stores all corporate data, no single system failure should be able to shut down corporate data processing. Instead, if one system fails, it is often possible to have another system automatically assume the computing responsibilities of the failed system. This feature allows the corporation to maintain operations and provide maximum access to critical corporate data at all times.

Disadvantages of Distributed Computing

Although distributed computing offers many advantages, it also offers a few disadvantages to the corporate data processing center. Because workstations are widely distributed over a broad geographic area, maintenance can become a problem for the corporation. Spatial dispersion of computing equipment may force the corporation to add personnel in order to maintain these systems.

File system backups, installation of new software, system security, and installation of user accounts can also become complicated in the distributed computing environment. When the corporation operates only one computer, the tasks of monitoring the system, allowing access, and installing software are contained within a single location.

In the event of a network failure, corporate office computer systems may not be able to access certain essential data for indeterminate periods of time. This is not an acceptable mode of operation in some applications (e.g., stock trading, control of the space shuttle, and banking transactions).

Evolution of System Software

As minicomputer and microcomputer hardware evolved, so did system software. Dozens of companies produced workstations. Each company developed operating system software to control its computers. This diversity of hardware and software had to be factored into the corporate decision making process.

Several computer manufacturers offered system software based on traditional mainframe operating systems. Corporate computing centers could easily migrate their applications to such operating systems. Unfortunately, some of these operating systems were incapable of extension to support newer models of system hardware.

At about the same time minicomputers were introduced to the market, a new operating system that would be portable across several hardware platforms was emerging. The UNIX operating system, in conjunction with workstation technology, changed the direction of corporate computing.

A Brief History of UNIX

The UNIX operating system was originally developed at American Telephone and Telegraph's Bell Laboratories in the late 1960s and early 1970s. Originally intended as a research tool for use within Bell Labs, word of the power and flexibility of UNIX spread quickly. UNIX quickly became the software of choice in many research and higher education institutions.

UNIX Branches Out

As the UNIX operating system offered by AT&T evolved and matured, it became known as System V (five) UNIX. As the developer of UNIX, AT&T licensed other entities to produce their own versions of the UNIX operating system. One of the more popular of these licensed UNIX variants was developed by The University of California at Berkeley's Computer Science Research Group. The Berkeley UNIX variant was dubbed Berkeley Software Distribution (BSD) UNIX.

The BSD version of UNIX rapidly incorporated networking, multiprocessing, and other innovations that sometimes led to instability. In an academic environment these temporary instabilities were not considered major problems, and researchers embraced the quickly evolving BSD UNIX environment. In contrast, corporate computing centers were wary of converting to operating systems with a history of instability.

Unlike BSD UNIX, AT&T's System V UNIX offered stability and standardization. New capabilities were introduced at a slower rate, often after evaluating the results of introducing the same capabilities in the BSD UNIX releases. Corporate computing centers tended to favor the stability of AT&T's version of UNIX over that of BSD UNIX. The development of the family of UNIX systems is shown in the following illustration.

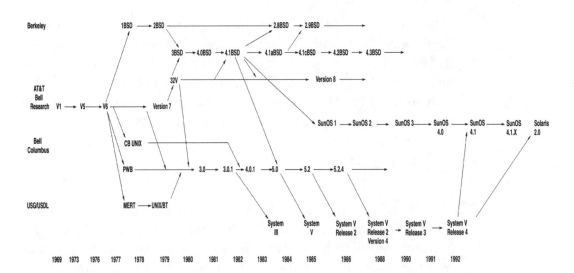

Genealogy of the UNIX operating system.

A New Player: Linux

One of the more recent additions to the UNIX family tree is a variant named Linux. Linux was developed in 1991 by Linus B. Torvalds, a graduate student at the University of Helsinki. Linux originated as a set of extensions to a miniature version of UNIX dubbed Minix. The Linux operating system was originally developed to run on systems that used the Intel x86 architecture, but recent versions will also run on Sun's SPARC architecture as well.

Linux is a POSIX (Portable Operating System Interface) compliant UNIX operating system. Because Linux offers both Berkeley and System V functionality, it has become extremely popular within the educational market. Many system vendors now offer a version of Linux on their systems. Indeed, there are several versions of Linux available for some architectures. For more information on Linux, visit *http://www.linux.org*.

A Brief History of Sun Microsystems

Sun Microsystems Computer Corporation was one of the manufacturers at the forefront of the distributed computing revolution. The company's first workstations employed microprocessors from the Motorola MC68000 chip family. Sun Microsystems founders believed that a UNIX operating system would be the most desirable option for their new line of workstations. The early Sun systems ran an operating system called SunOS (Sun Operating System, pronounced Sun Oh-ess). SunOS was based on the BSD UNIX distribution.

Enter the SPARC

Although the systems based on microprocessor chips were faster than many mainframes, there was still room for improvement. Sun promptly began developing its own microprocessor. In the 1980s Sun introduced a Reduced Instruction Set Computer (RISC) chip called the Scalable Processor ARChitecture (SPARC) processor. The first implementation of the SPARC chip ran at twice the speed of the fastest MC68000-based systems Sun was producing at the time.

The SPARC processor chip allowed Sun to produce very powerful, inexpensive, desktop workstations. The SPARC systems also ran the SunOS operating system, thus preserving customers' software development investments. Many other workstation manufacturers were delivering operating systems based on AT&T's System V release 3 operating system standard (also known as the System V Interface Description or SVID). Sun's customer base had to decide between the Sun BSD based operating system and competitors' System V based offerings.

Sun Develops a New Operating System

In the late 1980s Sun announced plans to develop a new operating system based on the AT&T System V release 4 UNIX. The new operating system is called *Solaris*. In order to make the progression of operating system release numbers easy to remember, Sun's original operating system is still known as SunOS version 4.X. The new Solaris operating system was referred to as SunOS 5.X. In 1999, Sun released a new version of the Solaris operating system named Solaris 7 (SunOS 5.7), and then in early 2000 Sun introduced Solaris 8 (SunOS 5.8). Because Solaris is a System V based operating system, it is quite different from the BSD-based SunOS operating system.

The Solaris operating system presented two problems for Sun customers. The first problem was that many applications available under BSD UNIX were not available under System V UNIX. Software vendors had to expend great effort to recode portions of respective applications in order to make them operate under Solaris.

The second problem presented by Solaris to customers was that many BSD-fluent system administrators had no experience with a System V operating system. Although most user-level commands are similar under SunOS and Solaris, the system administration commands are very different. Many of the "old reliable" commands from BSD-based UNIX are missing from Solaris. Next, new commands with more functionality replaced these BSD commands. The architecture of the operating system is also very different under Solaris. Access to and management of system devices under Solaris is foreign to BSD system administrators.

Summary

Sun Microsystems is one of the market leaders in the downsizing/ rightsizing of corporate computing environments. Recently, Sun introduced Solaris, a new operating system for its workstations. Although the new operating system has many similarities to the SunOS operating system, there are also many differences between the two offerings.

The primary goal of this book is to deliver practical information regarding the installation and maintenance of Solaris systems. This chapter presented a brief history of the UNIX operating system, and examined selected advantages and disadvantages of the mainframe and workstation computing models. Subsequent chapters of the book explore the Solaris operating system in detail.

CHAPTER 2
Solaris Concepts and Terminology

Chapter 1 clarified the differences between a mainframe and a distributed computing environment. The remainder of this book deals with the Solaris distributed computing environment. This chapter focuses on concepts and terminology related to installation methods and distributed environments. Installation methods encompass configuration and the role of servers, clients, and diskless clients.

Solaris Operating Environment

The Solaris Operating Environment is based on System V Release 4 (SVR4), which is an AT&T UNIX variant. However, the first release of Solaris (Solaris 1) was based on the Berkeley Software Distribution (BSD), commonly known as Berkeley UNIX.

Sun has twice changed the numbering system for Solaris. Originally, releases were numbered based on the underlying SunOS kernel. Later, Sun renumbered Solaris, starting with Solaris 1.0.1, which was the same as SunOS 4.1.2 under the earlier numbering system. After Solaris version 2.6, Sun once again changed the numbering scheme, making the next release Solaris 7 (SunOS 5.7), and the current release Solaris 8 (SunOS 5.8).

SunOS

SunOS is the heart of the Solaris operating environment. Like all operating systems, SunOS is a collection of software that manages system resources and schedules system operations. Solaris 2 and later versions have a kernel based on AT&T's System V Release 4, with numerous additions provided by Sun to support modern multi-processor hardware and programming techniques such as multi-threading.

CDE

The Common Desktop Environment (CDE) is a collection of user productivity tools similar to those found in the older Open-Windows DeskSet. However, unlike the OpenWindows DeskSet, CDE tools are available on many different vendors' workstations and operating environments. The Solaris version of CDE provides users a common look and set of tools on their desktops in work environments that may include several different vendors' workstations. CDE is used throughout this book where graphical tools are shown.

Another feature of CDE is the Motif GUI. Motif differs from OpenLook in that it includes 3D-style window borders and buttons along with graphical controls to adjust environmental parameters such as screen saving, color schemes, and menus. The "dashboard," a thin-strip window at the bottom of the display shown in the following illustration, provides easily modified pop-up menus and common functions such as a clock, calendar, printer control, and e-mail arrival notification. The dashboard also contains active icons for displaying the current time and system CPU and disk activity.

CDE window and selected CDE tools.

CDE includes a suite of productivity tools, including the HotJava web browser, a mail reader, calendar, note cards, and address book.

OpenWindows

OpenWindows, often referred to as Openwin, is Sun's older graphical user interface. OpenWindows, shown in the following illustration, is based on AT&T's OPEN LOOK graphical user interface functional specification. Some long-time Solaris users prefer the consistent, clean look of OpenWindows. When you log into a Solaris system, you have the option of switching between CDE and OpenWindows as needed.

OpenWindows graphical user interface.

DeskSet

OpenWindows DeskSet, shown in the following illustration, is a collection of productivity applications, including the Mail Tool and Calendar Manager. The Mail Tool is used to compose, send, and receive electronic mail, and the Calendar Manager to make appointments and schedule resources.

OpenWindows DeskSet tools.

Working with Hosts

Host is another word for a *computer*. The term *host* comes from the notion that the computer hardware hosts the operating system and applications. The hardware is like a house in which the "guests" (programs) live.

In order to distinguish one computer from another, hosts are assigned names, or *hostnames*. To continue the house metaphor, the hostname would be the system's address. In the Solaris operating environment, hostnames are assigned by the system administrator.

Assigning Hostnames

Hostnames usually follow a uniform naming convention. Names do not need to be stuffy or boring. The rocket motor division of an aerospace company, for example, may decide to name its hosts after astronauts, whereas its aircraft division would use names of famous test pilots. An advantage of the naming convention is to make it easier to keep track of which hosts belong to which organizations.

Another advantage of a naming scheme is that it makes it easy for system users to remember the name when reporting problems or describing tasks. Furthermore, system users favor environments with organized hostnames because such systems appear friendlier. A typical host naming convention is shown in the illustration that follows.

✓   *TIP: The* uname *command returns system information about the current host. For example, when the command* uname -n *is entered at the system prompt, the system returns the hostname.*

192.168.0.0.network

A typical host naming convention.

shepard   grissom   glenn   mercury

The previous illustration shows four hosts connected to a network. The host named *mercury* is a large-capacity computer that provides network services to other computers connected to the network. These are frequently referred to as *datacenters* or *servers*. The hosts named *shepard, grissom*, and *glenn* are desktop workstations that use the network services provided by *mercury*. Desktop workstations are often referred to as *clients*. The roles of clients and servers in a network are discussed in the section of the same name later in this chapter.

Assigning Network Addresses

Along with the host name, the system administrator assigns each host an Internet address. The naming service maps the host name to the assigned Internet address. Each Ethernet interface on a host also has an Ethernet address, built into the Ethernet hardware. The address resolution protocol (ARP) maps Internet addresses to Ethernet addresses.

Internet Address

The Internet address, similar to a telephone number, enables hosts to communicate with one another. For example, in the case of a long distance telephone call, the caller dials the area code, exchange number, and line number in order to communicate with a specific telephone location. In the same way, a host's Internet address describes where a host is on the Internet, which in turn allows network traffic to be directed to the host.

In the previous illustration, the hosts are connected to the Internet network number 192.168.0.0. (See Chapter 17, "Network Configuration and Management," for a detailed discussion of Internet addressing.) The Internet address assigned to a host can and often does change over the host's lifetime.

Host Ethernet Address

An Ethernet address functions like a Social Security number in that it is a unique and permanent hardware address assigned by the hardware manufacturer. Hosts are identified via such addresses on the Ethernet, which in turn enables them to communicate.

➡️   *NOTE: An Ethernet address uniquely identifies hosts on a network regardless of hostnames, whereas an Internet address identifies a host's location on the Internet.*

UNIX Operating System Design

Like all UNIX systems, Solaris uses a hierarchical operating system design. Each layer in the hierarchy is responsible for specific system functions and is capable of communicating with adjacent layers. The illustration that follows depicts a UNIX hierarchical operating system as a series of concentric circles.

UNIX hierarchical operating system.

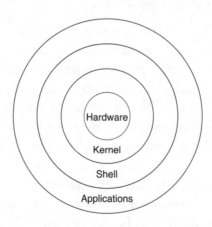

Hardware Core

The core of all computers is hardware. For hardware components such as CPUs, disks, and monitors to work as a team, they require management. The *kernel* is a collection of software components whose primary functions are the control and management of the system hardware. When information is written to a disk drive, the kernel manages the hardware components to make the write occur. Hardware components are directed by the kernel through software components known as *drivers*.

➤   *NOTE: The term* kernel *is used to indicate a part of the operating system found at the core of the UNIX environment.*

Shell Layer

The *shell layer* is the primary interface between user and computer. This layer processes command line instructions issued by a system user and passes them on to the kernel for further processing.

➤   *NOTE: Solaris provides the Bourne shell (sh), C- shell (csh), and Korn shell (ksh) versions. The Bourne shell is the default for the root or primary administration user.*

Applications Layer

The *applications layer* holds word processors, spreadsheets, and databases. When the user commands the shell, the kernel starts the application, which in turn calls up the applications layer. This layer interacts with both user and system hardware (via the kernel) to accomplish the desired task.

Modularity

One of the major advantages of the modular design of the UNIX operating system is portability. In order to port a UNIX operating system to a new hardware platform, only the kernel needs to be changed. The shell and applications layers remain unaffected. This greatly reduces system development time as new hardware technology becomes available. More important, however, is that end users are essentially unaffected by changes in hardware. Applications run on one hardware platform will perform identically on another.

System Architectures

Architecture describes the design and interaction of the various components of a computer system. There are two types of architecture administrators must be aware of when configuring and maintaining Solaris systems: application and kernel.

Application Architecture

Application architecture is the term used to define the applications and commands that run on a CPU. In the mid-1980s, Sun's dominant architecture was the Sun3, based on the Motorola 68000 series CPU chip set. In the late 1980s, the dominant architecture became Sun4, based on Sun's Scalable Processor ARChitecture (SPARC) chip set. In the mid 1990s, Sun introduced the 64-bit version of the SPARC chip, called UltraSPARC. UltraSPARC-based systems are available in sizes from desktop workstations up to multi-processor servers. Table 2-1, which follows, describes the differences between the Sun application architectures.

➥   NOTE: *Application binaries built for the Sun3 architecture are not com-*
       patible with Sun4 architecture. (See the tables in the next section.)

Kernel Architecture

Kernel architecture defines the kernel (operating system) version
and kernel-specific binaries that run on a CPU. Table 2-2, which
follows, lists several Sun systems, CPU models, kernel architectures, and applications architectures.

Table 2-1: Differences Between Sun Application Architectures

Application Architecture	CPU Architecture
Sun3	Motorola 68xxx
Sun4	SPARC
Sun4u	UltraSPARC

Table 2-2: Sun Systems Supporting Solaris

System Name	CPU Model	Kernel Architecture	Application Architecture	Last Supported Solaris Version
SPARCclassic	4/15	sun4m	sun4	7
SPARCstation LX	4/30	sun4m	sun4	7
SPARCstation 10	SS10	sun4m	sun4	7
Sun – 4/100	4/110	sun4	sun4	2.4
Sun – 4/200	4/2xx	sun4	sun4	2.4
SPARCserver 300	4/3xx	sun4	sun4	2.6
SPARCserver 400	4/4xx	sun4	sun4	2.6
SPARCserver 600MP	SS6xxMP	sun4m	sun4	2.5
SPARCserver 1000	SS1000	sun4d	sun4	8

System Name	CPU Model	Kernel Architecture	Application Architecture	Last Supported Solaris Version
SPARCserver 2000	SS2000	sun4d	sun4	8
SPARC Engine 1E	1E	sun4e	sun4e	2.4
Ultra 1, 5, 10, 30, 60, 80 & Enterprise 250	Ultra-1	sun4u	sun4u	None
Enterprise 250	Ultra-250	sun4u	sun4u	None
Enterprise 450	Ultra-4	sun4u	sun4u	None
Enterprise 3000, 4000, 5000, 6000, 10000	Ultra-Enterprise	sun4u	sun4u	None

⊷   *NOTE: The right-most column in the previous table, labeled "Last Supported Solaris Version," lists the last version of Solaris to support a given system type.*

When purchasing applications for a Sun system, check the host application architecture. When configuring the host kernel (O/S), check the host kernel architecture type.

✓   *TIP: Use the* uname *command, as follows, to obtain information on the host's kernel architecture.*

```
% uname -m
sun4c
```

✓   *TIP: Similar to the* uname *command, the* showrev *command can be used to display system information, as follows.*

```
# showrev
Hostname: mercury
Hostid: 240015cf
Release: 5.5
Kernel architecture: sun4m
Application architecture: sparc
Hardware provider: Sun_Microsystems
Domain: bigcorp
Kernel version: SunOS 5.5 Generic 103093-07
    September 1996
```

Roles of Servers and Clients

Servers and clients are the two types of hosts in a distributed computing environment. A server is a process or program that provides services to hosts on a network. If a host runs one or more server processes, it can also be referred to as a server. A client is both a "process" that uses services provided by the server and a host that runs a client process.

Process here refers to the program currently running. It is possible to have multiple instances of a single program, each separately serving different clients. The mail transport agent, *sendmail*, is a good example.

A wide variety of server and client processes can be operating in a network environment. For example, file servers share disk storage with hosts, application servers, applications, boot servers, and boot services. Print servers attach printers and provide network printing services. A typical server/client configuration is shown in the following illustration.

Typical network server/client configuration.

The previous illustration demonstrates that the host *grissom* (a network client) is able to use resources (i.e., printer and disk) attached to the host *mercury* (a network server). The *mercury* host is *sharing* its resources with *grissom*. A more detailed discussion of resource sharing appears in the chapters on the network file system (NFS), network name services (NIS+, DNS), and managing printers.

Host Configurations

There are two standard host configurations in the Solaris environment: *standalone* and *diskless*. These terms seem straightforward, but can be confusing. For example, a diskless client can have a local disk. The important point to remember is that the host's configuration refers to how it boots.

In order for a host to boot to the multi-user state, it must be able to locate three areas on the disk: *root* (/), *swap,* and *usr* (pronounced "user"). The root contains the boot code, system files, and directories necessary for host operations. The *usr* area holds system software, executable commands, and system libraries.

Finally, *swap* is used for virtual memory, which extends the host's total available memory. This area allows currently unused portions of a program (or its data) to be moved out of memory and on to disk, thereby freeing up space to run more programs. All programs and data in the memory are divided into segments known as *pages*. The virtual memory component of the operating system kernel moves unused pages out of memory and onto *swap*, and then moves the pages out of *swap* and back into memory when needed again.

Standalone

Standalone hosts boot up independently of network services. When booted, they obtain *root*, *swap*, and *usr* areas from a local disk. The term *standalone* refers to a host *not* connected to a network. "Networked standalone" refers to a standalone host *connected* to a network. One advantage of networked standalone hosts is that in the event of a network shutdown they can continue to operate (although use of network services may not be possible). A standalone host configuration is shown in the following illustration.

A standalone host configuration.

Host Disk

Diskless Client

Diskless clients are wholly dependent on network services to boot and operate. A diskless client's *root, swap,* and *usr* are supplied by a client server on the network. Advantages of the diskless client configuration include reduced equipment cost (because no local disk is required) and centralized services. The major disadvantage of using diskless clients is network load. When a diskless client boots, pages (writes) to the *swap* area, or executes a library routine, all of the I/O must go through the network.

As noted previously, every host must have both a unique *root* and *swap* area in order to function properly. A host configured as a boot server has two file systems on its disk that store *root* and *swap* for diskless clients (see previous illustration).

The */export* file system on the server contains a directory known as *root*. The */export/root* directory contains a subdirectory for each diskless client supported by the server. For example, */export/root/ glenn* located on *mercury*'s disk holds the root for the diskless client *glenn*.

The diskless client *swap* is used in a fashion similar to the *root* area. Special "swap files," one per supported diskless client, are located under the */export/swap* file system. For example, the */export/ swap/ glenn* file is the *swap* area for the diskless client named *glenn*. The illustration that follows shows a diskless client and boot server configuration.

✓   TIP: *A quick method of identifying diskless clients supported by a host is to change directory to* /export/swap *(assuming it exists) and list the files, as per the following.*

```
cd /export/swap
ls - l
-rw-----T 1 root 285036 Jan 6 1992 glenn
```

Diskless client and boot server configuration.

Software Terminology

Relationship among a software configuration, software cluster, and software packages.

Solaris bundled and unbundled software are distributed as software packages. These packages consist of files that perform diverse functions, such as on-line manual pages or OpenWindows demonstrations. There are more than 80 software packages in Solaris. The format of the Sun software package name is *SUN-Wxxx*. For example, the package name for the on-line manual pages is *SUNWman* and the package name for the OpenWindows demos is *SUNWowdem*.

The software packages in Solaris conform to a software standard known as the applications binary interface (ABI). ABI-compliant software is easily installed or removed with standard systems administration utilities. These utilities are discussed in Chapter 12, "Managing System Software."

Related software packages are grouped into software clusters. Cluster names use the *SUNWC* prefix. For example, the Open-Windows version 3 cluster contains 13 software packages related to OpenWindows version 3. The following illustration shows the relationship among the concepts of software configuration, cluster, and package.

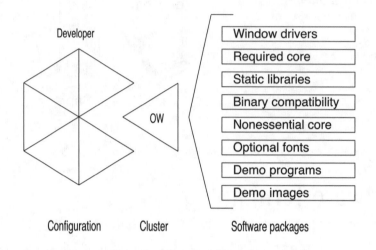

Working with Software Configurations

SunInstall further groups software packages and clusters into categories known as software configurations. The four configurations in Solaris 8 are *End user*, *Developer*, *Entire*, and *Entire plus OEM*. Each configuration supports a different level of system sophistication. The illustration that follows shows the relationship of the four configurations.

➥   *NOTE: The configuration space requirements listed in the following sections are approximations based on Solaris 8. The actual size of a configuration depends on system architecture, release level of the operating system, and any user customizations performed at the time of installation. Space for user file systems, mail, web or print server storage, and so on are not included in these figures.*

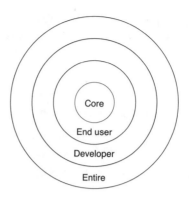

Relationship of the four software configurations.

End User

End user contains core configuration software and additional software typically used by end users, including CDE, OpenWindows, and the end user version of AnswerBook. It does not contain the on-line manual pages. *End user* requires approximately 1.25 Gb of disk space.

Developer

Developer contains both the core and end user configuration software, and additional software typically used by systems and software developers. It also includes on-line manual pages and the full implementation of CDE, OpenWindows, and compiler tools. It does not contain compilers or debugging tools. In Solaris 1 and earlier releases of Sun operating systems, compilers were bundled with the release media. Compilers and debuggers are unbundled in Solaris and purchased and installed separately. Developer requires approximately 1.7 Gb of disk space.

Entire and Entire plus OEM

Entire contains the entire Solaris release and requires approximately 2.2 Gb of disk space. The "plus OEM" version includes additional software provided by the Sun clone manufacturers. Its size varies depending on the specific additional software offered by the clone vendor. For Sun systems, the *Entire plus OEM* package requires approximately 2.3 Gb of disk space. Table 2-3, which follows, presents disk space requirements for Solaris 2.8 software configurations.

➤➤   *NOTE: It is important to note that software configurations can be modified. For example, the on-line manual pages cluster could be added to the* End user *configuration, or removed from the* Developer *configuration. These modifications can be made when the system is being constructed or afterward.*

Table 2-3: Disk Space Requirements for Solaris 2.8 Software Configurations

Software Configuration	Disk Space Requirement (Gb)
End user	1.25
Developer	1.7
Entire	2.2
Entire plus OEM	2.3

Obtaining System Access

On a Solaris system, every would-be user must gain access to it through a *log-in* procedure. The log-in process is a security mechanism that helps ensure that only authorized users gain access to system resources and information.

User Account

Before a user can log in, the system administrator must create a user account. The account contains a variety of information about the user, including areas of the system the user can access. In order to log in to a system, users need a name and a password.

The creation of user accounts is addressed in Chapter 6, "Managing User Accounts."

Log-in Conventions

The log-in name is a lower-case alphanumeric string up to eight characters long that identifies the user. Sites often employ users' last names. For example, Kathy Smith's log-in might be *smith*. Should more than one person share the same last name, the last name and first initial serves the purpose. For example, Mark Smith would be assigned *smithm*.

The simpler the log-in convention, the easier it is for system administrators to track and manage users. Simple conventions also facilitate user communication through utilities such as electronic mail.

> •◆   *NOTE: Experience has shown that a simple log-in naming convention will reduce user account management time as well as security problems. This is especially true in large or changing system environments.*

Assigning Passwords

System users must also have a password. The log-in name identifies a user to the system; the password ensures the system that the user is who she claims to be. By default, passwords must be at least six characters in length. In addition, passwords should never use the proper spelling of words, nor have anything to do with the user that could be easily guessed. Choosing secure passwords is discussed in Chapter 6, "Managing User Accounts."

Special Systems Account

A special systems account frequently used in system management is known as *root*. The *root* user account must be carefully guarded because it has unlimited access privileges. In the wrong hands, *root* access allows a user to cause extensive system damage and data corruption.

Logging In

To log in to the system on a terminal (in this case, as *root*), type the log-in name at the log-in prompt and press <Return>. (See the following illustration.) The system then prompts for a password. Type in the appropriate password and press <Return>. Note that the password is not echoed to the terminal screen. This feature prevents a bystander from seeing the password. If the information provided to the system is correct, a command prompt appears on the screen. The command prompt is the shell interpreter that passes commands for execution to the kernel. To log off the system, enter the *exit* command at the shell prompt and press <Return>.

Logging in and logging off a terminal.

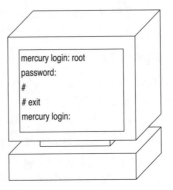

```
mercury login: root
password:
#
# exit
mercury login:
```

To log in on a workstation console, the user name is typed into the text box in the middle of the log-in window, shown in the following illustration.

Workstation log-in window.

The log-in window contains four buttons under the text box. The OK button brings up a window for entering the account password. The Start Over button resets the log-in window and workstation screen. The Options button opens a drop-down menu of choices that includes a text-only log-in option and a choice of OpenWindows or CDE as the graphical user interface to be started upon log-in. The Help button brings up a help screen.

The password window, shown in the illustration that follows, shows the same four buttons, and indicates the graphical environment that will be started in the box to the right of the text entry box. Note that the password text will not appear when typed, to avoid showing the password to an onlooker.

✗   *WARNING: Never leave a* root *log-in unattended! A clever system cracker could take advantage of the unattended* root *log-in to alter critical system files, thus allowing the cracker undetected* root *access. The integrity of an entire corporate computer system can be compromised due to lax* root *security.*

Password window.

Summary

This chapter covered the fundamental concepts and terminology used in the Solaris operating environment. An understanding of these terms, the divisions of the Solaris environment, and the conventions used for naming types of systems is important to following the discussion in subsequent chapters. Careful attention should be paid to the role of consistent and straightforward naming schemes as described here, with user accounts and host names. A good naming scheme forms an essential foundation to many system administration tasks. With these concepts in mind, the system administrator can more readily understand Sun's documentation, and move on to perform a system installation or perform system management duties.

# CHAPTER 3

Preparing for a Solaris Installation

One of your principal tasks as a systems administrator is planning for company software needs under Solaris and identifying a hardware configuration to support the software. This chapter focuses on basic pre-installation considerations to keep in mind while reviewing software applications.

Requirements may differ depending on whether you are working with a new system installation or an upgrade from the Solaris 1.X (SunOS) to the Solaris operating system. The results of this evaluation are then applied to your application and operating system demands to determine what they mean in terms of hardware configurations.

Evaluating Applications

If you are planning to upgrade a SunOS system to a Solaris system, you should be aware of several important differences between these operating systems. Some of the differences involve changing the names of commands or function calls that perform certain operations. Other changes involve portability issues such as how the compiler links or includes the library calls into the binaries. Because of these differences, you will need to examine the software currently in use on your system to differentiate mission-critical applications from non mission-critical applications.

Although many applications you have been using under SunOS are available under Solaris, others may not be available to run under the newer OS. This is the time to evaluate the packages to determine if you need to replace certain existing functionality, and if so, how replacement should be accomplished. In some cases, it may be possible to use existing SunOS binaries under Solaris. You will need to understand some of the internals of the applications before they can be used under Solaris.

When you are planning a new installation, some of the same compatibility and portability concerns may come into play. Even in this case, it is best for system administrators to understand the differences between the two operating systems so that the differences can be explained to users who have experience with SunOS or other Berkeley-derived UNIX operating systems.

Program Linking

Dynamically and statically linked binaries are found in SunOS applications. Linking is the process performed by the link editor (*/usr/ccs/bin/ld*) to resolve references to variables within program binaries. In early versions of SunOS, all binaries were statically linked. Under Solaris, the preferred method of linking binaries is dynamic. What are the differences between the two methods of linking a program binary?

A statically linked program links (or includes) all library functions at compile time. If changes are made to a library routine, the program must be recompiled to take advantage of those changes. A statically linked program will usually be quite large because it includes all library functions.

A dynamically linked program is one that loads the library functions at run time. This allows the program to include the latest version of all library functions upon invocation. No recompilation is required to enjoy the advantages of the updated library routines. Dynamically linked programs are usually smaller than their statically linked counterparts, because the library functions are not included in the on-disk binary image.

Native Mode Applications

In Solaris, applications may be run in native mode or binary compatibility mode. There are benefits and costs to both operating modes. For instance, because Solaris is based on industry standards, native mode applications enjoy application portability across all System V Release 4 platforms. Application code developed on one vendor's System V Release 4 system should compile and operate correctly on another vendor's System V Release 4 machine. Applications that contain native mode binaries will operate on Solaris systems, but will not operate on SunOS 4.x systems.

To be considered a native mode application, a program must be compiled under the Solaris environment. This typically means that the binaries are dynamically linked programs.

➥   *NOTE: As noted in Chapter 2, compilers are not included in the core Solaris operating system. The compilers must be purchased and installed separately. Keep this in mind for subsequent cataloging of applications software.*

✓   *TIP: Developers can save significant time porting applications to Solaris if the applications are developed under SunOS 4.1 using the System V environment, OpenWindows, and dynamic linking.*

Binary Compatibility Mode Applications

What happens if you do not have source code for the application, or if a critical application is not available for Solaris? Would such a situation prevent the corporation from converting to Solaris? Fortunately, in many cases, the answer is no.

Solaris provides a binary compatibility mode package (BCP) for existing SunOS 4.x programs, including SunView applications. It might be possible to run your mission-critical application in binary compatibility mode until a Solaris version is available.

••   *NOTE: The binary compatibility package is intended as a short-term transition aid, and should not be used as a long-term solution.*

In order to operate in compatibility mode, the application must strictly adhere to some simple guidelines. Table 3-1, which follows, summarizes the restrictions on running an application in compatibility mode.

Table 3-1: Binary Compatibility Mode Restrictions

Feature	Do Not Use	Must Use
Dynamic linking		X
Traps to operating system	X	
Application-specific IOCTLs	X	
Read/write directly to system files	X	
Access /dev/kmem or libkvm	X	
Use of undocumented SunOS interfaces/ features	X	

Applications run in compatibility mode must dynamically link the C library and other system library routines. Statically linked applications will not work in binary compatibility mode.

Another condition for proper operation in binary compatibility mode includes the absence of traps directly to the operating system, application-specific IOCTLs or drivers, and direct read or write access to the system files. Further conditions require that the application not attempt to access and interpret kernel data structures through the */dev/kmem* or *libkvm* facilities, nor should applications attempt to use any undocumented SunOS interface.

In the case of locally developed software or applications that include source code, it is relatively easy to determine if the code follows the rules of the table. In order to determine if third-party or binary-only applications will work in binary compatibility mode, it may be necessary to contact the original software supplier.

32-bit Versus 64-bit Applications

Beginning with Solaris 7, another consideration comes into play: 32-bit versus 64-bit applications. All Solaris releases prior to Solaris 7 supported 32-bit applications under a 32-bit kernel. Solaris 7 and Solaris 8 provide 64-bit operating system environments on the UltraSparc hardware (*sun4u*). Older hardware still only offers a 32-bit operating system environment (*sun4m*).

➡ *NOTE: By default, the UltraSparc 1 systems boot the 32-bit kernel. In order to allow these systems to run the 64-bit kernel, you must edit the file* /platform/platform_name/ boot.conf *and remove the comment character from the line* ALLOW_64BIT_KERNEL_ON_ Ultra-SPARC_ 1_CPU. *It may also be necessary to install a firm-*

*ware upgrade on these machines in order to install the Solaris 7 or
Solaris 8 Operating System software.*

In addition, the 64-bit environment makes it possible for users
and developers to create 64-bit applications. This feature may
cause problems for users and administrators though, as the 64-bit
applications will not run in a 32-bit environment! Users and
administrators must take this into account if the organization
supports a mixture of 32- and 64-bit hardware. 32-bit applications should operate properly in 32-bit mode within the 64-bit
kernel environment.

Cataloging Current Applications

Planning for application availability under Solaris involves cataloging existing and planned applications, and determining the
availability date of these applications under Solaris. The cataloging process requires that you examine your current systems to
identify the software in use. Once you develop a catalog of the
required applications, you need to determine when these applications will be available for use under Solaris, and how this will
affect your corporate computing plans.

In order to facilitate the cataloging process, you should develop a
form for the tabulation of local application information. This
form should allow you to collect as much information as possible
about the applications. Some of the more important information
about applications includes package size in megabytes (Mb), the
number of users who will require this application, and whether
the application runs in native or binary compatibility mode. A
sample form is shown in the following illustration.

Software Inventory Form

System Model: _____

System Configuration: Server ☐   Standalone ☐   Diskless ☐

System Hostname: _____

System IP Address: _____

Subnet Mask: _____

Domain: _____

Name Service: NIS ☐   NIS+ ☐   NONE (includes DNS) ☐

Name Server Hostname: _____

Name Server IP Address: _____

Time Zone: _____

Disk Layout

Device	Partition Size	Mount Point	Device	Partition Size	Mount Point
Disk 1	a		Disk 2	a	
	b			b	
	c			c	
	d			d	
	e			e	
	f			f	
	g			g	
	h			h	

Locally Developed Software

Application	Language	Native/BCP	When Available	#Users	Size

Third-Party Software

Application	Language	Native/BCP	When Available	#Users	Size

SunSoft Products

Application	Language	Native/BCP	When Available	#Users	Size

Sample software inventory form.

Evaluating Hardware Requirements

Once your software requirements are identified, you need to determine hardware dependencies. For instance, do you have enough disk space to support the operating system, users' files, and necessary application software?

You also need to consider special hardware requirements while the Solaris installation plan is developed. Do you rely on third-party or other special hardware to operate your enterprise? If so, is a special device driver required for the operation of this hardware? Which systems have this hardware? Do you know if software is available to allow the operation of this hardware under Solaris?

Cataloging Hardware Requirements

When you consider hardware requirements, you need to pay close attention to any nonstandard hardware on your system. As with the software, all hardware found on the system is cataloged. Again, you could develop a form to facilitate the cataloging process. A sample form for cataloging hardware is shown in the following illustration.

Hardware Inventory Form

System Name: _____   IP Address: _____   System Memory: _____ MB

Disk 1	Controller:		Unit:	Disk 2	Controller:		Unit:
Partition	Size (Mb)	Mount Point	Notes	Partition	Size (Mb)	Mount Point	Notes
a				a			
b				b			
c			Entire Drive	c			Entire Drive
d				d			
e				e			
f				f			
g				g			
h				h			

Third-party or locally developed hardware:

Product	Supplier	System(s)	Driver Req'd	Date Available
_____	_____	_____	_____	_____
_____	_____	_____	_____	_____

Network Sketch

Primary Network IP Address:

Server   Client   Client

Subnetwork IP Address:

Sample hardware inventory form.

Hardware Requirements

Once you have identified the company's specific software and hardware requirements, you need to determine if your system(s) meet minimum Solaris requirements. To install and run Solaris, systems must have a SPARC, Power-PC (PPC), or Intel 486/Pentium processor with a minimum 150 Mb of disk space. Table 3-2, which follows, includes examples of systems capable of running Solaris.

Table 3-2: Solaris Architecture Information

Kernel Architecture	Application Architecture	System Model	64–bit Capable?	Model Name
Sun4c	Sun4	4/20, 4/25, 4/40, 4/50, 4/60, 4/65, 4/75	No*	SLC, ELC, IPC, IPX, SS1, SS1+, SS2
Sun4d	Sun4	SS1000, SS2000	No*	SparcCenter 1000 and 2000
Sun4m	Sun4	EC3, 4/15, 4/30, SS4, SS5, SS10, SS20, S240	No*	SparcEngine EC3, SparcClassic, LX, SS4, SS5, SS10, SS20, Voyager
Sun4u	Sun4	UltraSparc 1, UltraSparc 2,	Yes[†]	–
–	–	UltraSparc 5, 10, 30, 60, 80	Yes	–
–	–	Enterprise 220, 250, 420, 450, 3000, 4000, 5000, 6000, 10000	Yes	–
i86pc	i86pc	486, 586, 686	No	–
–	–	Pentium P5/P6	No	–
ppc	ppc	Power-PC 603	No[‡]	–
–	–	Power-PC 604	No[‡]	–

* The Sun4c architecture is supported under Solaris 7 in 32-bit mode only. The Sun4c architecture is not supported under Solaris 8.

[†] The UltraSparc 1 and UltraSparc 2 architectures may require a firmware upgrade in order to run 64-bit versions of Solaris.

[‡] The *ppc* architecture is supported under Solaris 2.5 and 2.5.1 only.

✓   *TIP: To determine if a particular x86 or PPC-based system is capable of running the Solaris operating system, consult the Sun web page at* http:/

/ access1.sun.com. *This site maintains hardware compatibility lists specifying which systems have been certified to be compatible with Solaris.*

Software Requirements

The Solaris operating system installation software defines four typical installation clusters: Core System Support, End User System Support, Developer System Support, and Entire Distribution. The Core cluster provides the minimum software required to boot and run Solaris. Each of the other clusters provides more utilities and functionality than the previous cluster in the list. The space requirements for each Solaris 8 cluster are summarized in table 3-3, which follows.

Table 3-3: Disk Space Requirements

Software Group	Installs	Space Required
Core System Support	Minimum software required to boot and run Solaris.	250 Mb
End User System Support	Core group plus recommended end user applications, including OpenWindows and DeskSet.	1.25 Gb
Developer System Support	End user software and libraries, including files, manual pages, and programming tools. (Compilers and debuggers are not included.)	1.7 Gb
Entire Distribution	Entire Solaris release. (Compilers and debuggers are not included.)	2.2 Gb

⟐   *NOTE: Some versions of Solaris distribution media include a fifth cluster named "Entire Distribution plus OEM support." This cluster should be used if installing Solaris on a SPARC clone system, and requires slightly more space than the Entire Distribution cluster. Consult the manuals distributed with the installation media for more information on this cluster.*

✓　　*TIP: It is possible to customize any of the predefined clusters at installation time. Customization allows the user to add or delete any optional packages and tailor the installation to individual requirements.*

The Solaris binaries are extracted from the distribution media and placed in predefined file systems on the disk drives by the installation software. These file systems, or disk partitions, must meet certain minimum size requirements. File system space requirements are summarized in table 3-4, which follows.

Table 3–4: Typical Partition Sizes Under Solaris

File System	Required On	Minimum Size (Mb)	Typical Size
/ (root)	All systems.	32	64 Mb.
swap	Optional on systems with minimum of 32 Mb memory. Recommended for reasonable performance.	0	2x size of memory.
/usr	All systems.	690	1024 Mb.
/opt	Optional software.	0	Site dependent.
/export/ exec	Required only on servers that support diskless clients.	15	32 Mb per supported client architecture.
/export/ root	Required only on servers that support diskless clients.	10	10 Mb base plus 32 Mb per diskless client.
/export/ swap	Required only on servers that support diskless clients.	0	32 Mb per diskless client.
/var	Optional. Often mounted as a separate partition to facilitate network log file archival, mail, and printer spooling.	0	128 Mb typical for small site, and several hundred Mb for larger site. Site dependent.

➥　　*NOTE: When installing a server system, providing a separate partition for the /var directory is often desirable. This directory stores log files and mailboxes and provides temporary storage for print jobs, as well as space for the NIS+ database and user accounting. At a typical large site, this partition may be 512 Mb (or more), and at smaller sites, 128 Mb.*

✓   *TIP 1: In order to reduce disk partitioning, the export file systems are sometimes combined into a single file system.*

✓   *TIP 2: In the case of a heterogeneous environment, it is usually desirable to have separate /opt and /usr/openwin file systems for each client architecture. This practice allows each client architecture to mount and access only the binaries that will execute on a particular client architecture.*

Other Important Hardware Considerations

Another requirement that must be met before running Solaris is the availability of hardware used as the software load media when performing the initial Solaris system installation. Unlike SunOS predecessors available on several distribution media, Solaris is distributed only on CD-ROM media.

➥   *NOTE 1: To install Solaris on x86 systems, a floppy disk drive is also required.*

➥   *NOTE 2: Solaris systems must contain at least 32 Mb of main memory, with 64 Mb or more recommended for better performance.*

Installation Type

Each of the following basic types of Solaris systems is designed for a specific function: standalone, diskless, and heterogeneous and homogeneous servers. Each type is discussed in the sections that follow.

➥   *NOTE: Solaris 2.5.1 was the last release that supported the dataless client architecture.*

Standalone

Under current versions of Solaris, a standalone system typically requires a minimum of 2 Gb of disk space. Standalone systems run entirely from a local disk, and do not rely on other network service providers. Some standalone systems are also servers.

Diskless

Under current versions of Solaris, these systems typically require at least 64 Mb of disk space on the server. Diskless systems are totally dependent on network services. The *root*, *swap*, and other file systems are provided by one of the servers on the network.

Heterogeneous and Homogeneous Servers

Under current versions of Solaris, servers usually have disk requirements in excess of 2 Gb. Servers are a special class of standalone system. The servers boot from local disks and provide other network hosts with the file systems they require. Because servers provide boot and application services to clients, they require more software on the local disks than non-server standalone configurations.

A homogeneous server is a system that serves only a single client kernel architecture. For instance, a Sun4m server with all Sun4m clients is considered a homogeneous installation.

In contrast, a heterogeneous server is a system that serves at least one client with a different kernel architecture than that of the server. An installation that includes a Sun4u server with Sun4m, PPC, and x86 clients would be considered a heterogeneous installation.

Example of Installation Evaluation

- -

Now that the terms used in the Solaris installation process are understood, the administrator needs to determine the specific requirements of each system that will be upgraded. For illustration purposes, consider an Enterprise 4000 server (hostname *mercury*), with one diskless Sun4c client (hostname *glenn*), and one x86 client (hostname *grissom*). The *mercury* system contains two 4.2-Gb disk drives and 256 Mb of main memory.

The diskless client will receive all services from the server's disks. The x86 client will receive user file space from the server, but will have its own copy of all essential system binaries on its internal disks. Assume that this site will be moving more clients to the *mercury* server as the management and users become more familiar with the new operating system. For illustration purposes, *mercury* will also be a master Network Information System (NIS/NIS+) name server. A configuration of the *mercury* system is shown in the illustration that follows.

➥   *NOTE: NIS+ is the replacement for the original NIS name server distributed under SunOS operating systems. The NIS+ package manages network-wide information such as passwords, file system availability, and names of hosts on the network. NIS+ is discussed in greater detail in Chapter 21, "Network Name Services."*

*Configuration
of the mercury
system.*

In the previous illustration, the *mercury* server will be installed as
an Entire Distribution cluster. For the purpose of simplicity,
there is no locally developed or third-party hardware. Of the two
applications to be installed on the system, one requires 50 Mb of
disk space and runs in binary compatibility mode, and the other
requires 75 Mb of disk space and runs in native mode. The inventory forms for the *mercury* system might resemble those shown in
the following illustrations.

Software Inventory Form

System Model: _____ENTERPRISE 4000_____

System Configuration: Server ☐   Standalone ☐   Diskless ☐

System Hostname: _____Mercury_____

System IP Address: __192.168.0.1__

Subnet Mask: _____255.255.0.0_____

Domain: _____astro.com_____

Name Service: NIS ☐   NIS+ ☐   NONE (includes DNS) ☐

Name Server Hostname: _____Mercury_____

Name Server IP Address: __192.168.0.250__

Time Zone: _____US/East Indiana_____

Disk Layout

Device	Partition Size		Mount Point	Device	Partition Size		Mount Point
Disk 1	a	64 Mb	/	Disk 2	a		
	b	128 Mb	swap		b		
	c				c	2048 Mb	/home
	d	256 Mb	/var		d		
	e	128 Mb	/usr/openwin		e		
	f	256 Mb	/export		f		
	g	256 Mb	/usr		g		
	h	768 Mb	/opt		h		

Locally Developed Software

Application	Language	Native/BCP	When Available	#Users	Size
analyze-cog	C++	Native	now	7	75 mb

Third-Party Software

Application	Language	Native/BCP	When Available	#Users	Size
design_a_cog	Binary only	BCP	Now	14	50 mb

SunSoft Products

Application	Language	Native/BCP	When Available	#Users	Size

Software inventory form for mercury system.

Hardware Inventory Form

System Name: _Mercury_   IP Address: _192.7.168.0.1_   System Memory: _64_ MB

Disk 1	Controller: 0	Unit: 0		**Disk 2**	Controller:	Unit: 1	
Partition	Size (Mb)	Mount Point	Notes	Partition	Size (Mb)	Mount Point	Notes
a	64	/		a			
b	256	swap		b			
c			Entire Drive	c	2048	/home	Entire Drive
d	256	/var		d			
e	128	/usr/open win		e			
f	256	/export		f			
g	256	/usr		g			
h	768	/opt		h			

Third-party or locally developed hardware:

Product	Supplier	System(s)	Driver Req'd	Date Available
N/A				
N/A				

Network Sketch

Primary Network IP Address: _192.168.0_

192.168.0.1   192.168.0.2   192.168.0.3

Server
mercury

Client
glenn

Client
grissom

Subnetwork IP Address: _N/A_

Hardware inventory form for mercury *system.*

✓   *TIP: Because the* mercury *system is to be the server for client systems, the client inventory information should be included with the* mercury *information to provide a complete picture of installation requirements.*

Determining Disk Partitions

Now that the type of installation and the disk space requirements have been determined, you must decide how to partition (or segment) the disks on the *mercury* system to provide for these needs. This procedure requires a rudimentary understanding of disk terminology.

➥   *NOTE: Disk geometry is discussed in more detail in Chapter 9, "Disk Subsystem Hardware."*

At a very simple level, the following illustration shows a disk drive divided into many data sectors. Each data sector is capable of storing 512 bytes of data. Disk drives may be divided into several (up to eight) UNIX file systems or partitions. Each of the UNIX file systems consists of numerous data sectors bound together by the *newfs* command. Recent versions of Solaris will try to install everything on a single partition. There are advantages and drawbacks to this type of installation.

A typical disk layout.

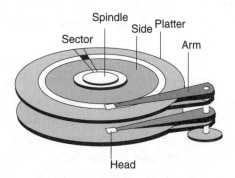

On a system with multiple partitions, the administrator has more control over disk space usage. A user should not be capable of running the system out of disk space with an errant job. Due to the buffering characteristics of the Solaris disk drivers, it is often desirable to make multiple-partition systems, and place the most active file systems in the middle of the disk drive. This allows for more optimal I/O performance, as the disk read-write heads are certain to pass over these partitions quite often. It is also usually desirable to have the *swap* space spread across several drives, and the easy way to do this is to create multiple partitions on the system's disks.

Another reason to create a multiple partition server is to enable control over exported file systems. When the disks are partitioned, the administrator has better control over which files are exported to other systems. This occurs because the Network File System (NFS) allows the administrator to export specific file systems. For example, on a server with separate */usr* and */opt* file systems, it would be possible to export the */opt* partition to all hosts without giving those hosts access to the */usr* file system on the server. On a single-partition system, the administrator would have to export the entire disk to other systems, thereby giving the client machines access to files that might be sensitive in nature.

One problem with a multiple-partition system is that it does not have the ability to make use of empty space in one partition when another partition is running in an overcommitted state. Multiple-partition installations also waste more disk space for partition information, and other file system overhead, than typical single-partition setups. For these reasons, some administrators prefer to create a large single-partition disk, and load everything into the */ (root)* partition. The Solaris installation process will let you create either type of scenario.

Following the previous argument, a 100-Mb file system would require a disk partition with at least 200,000 data sectors, or 200,000 X 512 = 102,400,000 bytes. In reality, the file system infrastructure will require extra sectors in order to provide a file system that will hold 100 Mb of data.

A typical rule of thumb when partitioning disks is to allow for 10 percent of the sectors in a file system to be consumed by overhead. It is also a good idea to allow room for future growth for each file system.

The *mercury* sample installation requires eight file systems: */, /usr, swap, /opt, /export, /var, /usr/openwin,* and */export/home/mercury.* By examining the cluster size information, the amount of memory in the system, the size of the local packages, and the number of clients supported by the *mercury* server, it is possible to develop the following reasonable disk layout for the server.

- 256 Mb for */*

- 1024 Mb for *swap*

- 1024 Mb for */usr*

- 256 Mb for */var*

- 256 Mb for */usr/openwin*

- 1024 Mb for */opt* (applications)

- 256 Mb for */export* (*root, swap,* and *exec* combined into a single file system)

- 4096 Mb for */export/home/mercury* (users' files)

- ➡ *NOTE: The space allocation previously listed is very generous. It is possible (and in many cases desirable) to create smaller file systems. Because the* mercury *system may support more users and clients in the near future, the previous layout was created to allow for extraordinary expansion. It is easier to make such decisions before installing Solaris than it is to repartition disks after the system has been in use.*

File System Sizes

For the example installation, one disk drive will be set aside for the user files (*/export/home/mercury*). This leaves one 4.2-Gb disk to contain the other seven file systems. Disk partition information and file system mount points appear in table 3-5, which follows.

Table 3-5: Example File System Sizes

File System Mount Point	Minimum Size (Mb)	mercury System Size (Mb)
/	32	256
swap	32	1024
/usr	190	1024
/usr/openwin	210	256
/opt	0	1024
/export	72	256
/var	0	256
/export/home	0	4096

At this point, pre-planning is complete for the *mercury* installation. If *mercury* is a new installation, the next task would be to load Solaris onto the system disks per the instructions shipped with the distribution media. If *mercury* were an upgrade installation, a few other tasks should be considered before beginning the Solaris installation.

Saving Critical Data

Before SunOS is shut down for the last time, there are several files you might want to save. For instance, because you may not have to change the partitioning to load Solaris, it would help to know how the disks on the system are currently partitioned. Saving the password file, NIS name server maps, and other critical information before you begin the SunOS shutdown procedure is recommended.

Saving the SunOS Disk Configuration

There are two methods of determining the current disk configuration information on a system. The *dmesg* command will display output from a recent reboot, and the *format* command will display all currently available disks. Because the *dmesg* information will not be available if the system has been up and running for several days, it is sometimes more convenient to use the *format* command, an example of which is shown in the following illustration.

Using format *command to identify available disks under SunOS.*

Once the device names of all disk drives on the system are known, obtain a hard copy of the current partitioning information for the disks by using the *dkinfo* command, as follows.

```
# /etc/dkinfo sd0 sd1 | lpr
```

Another bit of information that may prove useful is a copy of the current *fstab* file. This will help to prevent damage to the users' file systems during the conversion process. Obtain a hard copy of the current file system information by keying in the following command.

```
# lpr /etc/fstab
```

Saving Critical SunOS System Information

It is always wise to have a copy of all critical system information prior to installing a new version of an operating system. The following are a few of the more critical files to be saved.

- *./etc*
- *./.cshrc*
- *./.login*
- *./.logout*
- *./.profile*
- *./.rhosts*
- *./var/spool/mail*
- *./var/spool/calendar*
- *./var/spool/cron*
- *./var/spool/uucp*

- *./var/nis*

- User's files

✓ *TIP: It is best to perform a full file system backup before installing a new version of the operating system. The backup provides an extra level of security for the critical system information in case something should go wrong during the installation process.*

Creating a file containing the previous list of information can be automated with the following shell script.

```
# cd /
# tar cvf tape_drive `cat filename`
```

In this example, *tape_drive* refers to the device name of the drive to write the files on (e.g., */dev/rsto*). Next, *filename* is the name of the file containing the list of files and directories to be archived.

⇥ *Note: The list of critical files to be saved is only a suggested list. The actual list is highly dependent on your installation.*

Summary

. . . .

This chapter explored the prerequisites for installing Solaris. Determination of the type of installation, the system type(s) to be supported, and the planned use of the system(s) are factored into Solaris installation planning. An example installation was presented to illustrate typical record keeping and planning required before installing Solaris.

CHAPTER 4

Fundamental System Administration Tools

This chapter addresses the basic tools involved in Solaris system maintenance. The day-to-day management of a Solaris system requires the use of a variety of tools. Some of these are tools familiar to Solaris users, whereas others are special-purpose tools only used by administrators. Sometimes, common tools are used in a different manner by administrators as part of their work.

To maintain Solaris, the systems administrator must spend time studying, practicing, and planning basic tasks. Included are the installation of new systems and peripherals, preserving a set of system backups, installing user accounts, allocating disk space, monitoring system performance, and updating system and application software. Each of these tasks is covered elsewhere in this book in more detail. Discussion in this chapter is focused on selected fundamental tools and techniques applicable to a wide variety of tasks as part of maintaining any Solaris system.

Part of the work of maintaining a Solaris system includes reading books such as this one, familiarizing oneself with system documentation, keeping notes on what works and does not work, and recording how various tasks were accomplished. If the Solaris *AnswerBook* software is available, take a few moments now to read instructions on how to use it.

✓   *TIP: Get a laboratory style notebook and jot down the procedures to use when performing tasks such as installing new hardware or software. Your objective is to build a library of procedures specific to your situation.*

Study also means understanding a few basic UNIX tools and techniques all system administrators should know. This is like knowing how to change a flat tire. The knowledge may not be used daily, but it is handy when problems arise.

A system administrator must also maintain a discipline of carrying out daily, weekly, or monthly tasks to help to maintain the functioning and performance of the system and to prepare for unexpected problems. This might include a schedule of system backups, routine audits of user accounts and system security, checks to reveal capacity problems such as full disks or overloaded networks, and trimming system log files.

Discipline also refers to the way basic tasks are performed. A cautious, safe work procedure establishes techniques for easy recovery from problems or missteps. A few rules of thumb for performing system maintenance and administration tasks follow.

•   Avoid using *root* as much as possible. Use a less privileged account, where mistakes will be less drastic.

- Avoid using wildcard characters such as the asterisk (*) when using *root*. Accidents do happen, and nearly everyone has at least once inadvertently entered a command line such as *rm -r /usr/**.

- Create a backup copy of all edited files. Make it a habit to use something like the following command: *cp file file.orig ; vi file*. Using such a command makes starting over easy in the event of an editing mistake.

- Allow plenty of time for the first experience of installing a software package or any other new task. System administration tasks can be time consuming; schedule more down-time than necessary and make notes of how long tasks actually take. Users will appreciate it if the administrator has the system back on line ahead of schedule.

Moving Around the UNIX File Tree

The basis of file storage on Solaris and other UNIX systems is a hierarchical directory structure often depicted as a branching, downward-growing graph known as a "file tree," an example of which is shown in the following illustration. The top of the tree is the *root* (/). Files and directories can be referenced by listing names of the directories traversed from the *root* to the file to the desired directory. This is known as the full path name of a file or directory, such as */etc/inet/hosts* or */usr/local/bin*.

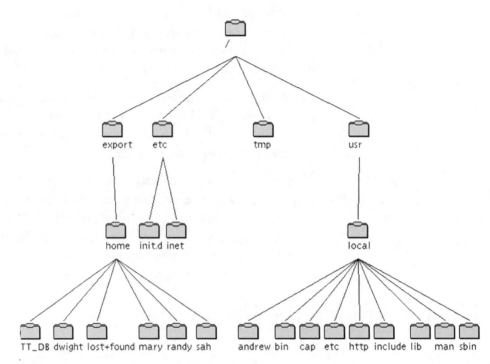

UNIX file tree with selected branches.

Several commands and graphical tools view, navigate, and manipulate files and directories in the UNIX file tree. These are discussed in the sections that follow.

File Manager

The Common Desktop Environment (CDE) provides a graphical tool, known as the File Manager, for browsing the UNIX file tree. The File Manager is similar to the point-and-click file management tools found in Microsoft Windows and the MacOS finder. Files and directories can be moved by dragging them with the mouse. At the top of the File Manager window, shown in the following illustration, is a graphical representation of the working directory's full path. This is the current location the File Manager is viewing in the UNIX file tree. The lower portion of the window shows the content of the working directory.

CDE File Manager window.

pwd, cd, and ls

Although graphical tools for viewing the file system are handy, they are often not available and a system administrator needs to be familiar with the traditional command line approach to navigating the file tree. The file tree is traversed by specifying the location to move to, either in relative terms from the current location or in absolute terms, listing the path from the *root* of the tree.

The UNIX command shells provide information about the working directory or current location in the file tree. The *pwd* command will print the location of the current working directory as a "full path name." This term describes the path through the file tree from the *root* to the current directory. If, for example, *pwd* returned */etc* as the working directory, the *hosts* file could be referred to by the relative path *inet/hosts* or *./inet/hosts*, where the period (.) signifies the name of the working directory. It could also be referred to by the absolute or full path, */etc/inet/hosts*.

All UNIX users should also know the *ls* and *cd* commands. In several formats, *ls* lists the content of the current directory or of a directory given as an argument. Although this command has numerous options, the *-l* option (to list the file ownership and protection mode bits) is the one most commonly used. The *cd* command allows the user to change directories. Examples of using the *cd* and *ls* commands appear in the following illustration.

Changing the working directory with cd *and viewing directory contents with* ls.

> ✓ *TIP: Every directory includes two special entries. First is the single period (.), which refers to the directory itself. Next is the double period entry (..), which refers to the directory one level up, or closer to the root, from the working directory. This feature simplifies navigation within a file system. For example, to move up one directory, type* cd .. *instead of the full path to the directory above the working directory.*

df and du

Two commands that may be unfamiliar to new Solaris system administrators are *df* and *du*. These two commands display the amount of disk space used in two different forms. The *df* command lists the file systems constituting the UNIX file tree. It shows the mount (start) point of the portion of the file tree on a particular disk or partition, as well as the space used and available.

The *du* command lists the amount of space used in a directory. Use *df* to search for nearly full disks and partitions that require action by the systems administrator. To view the space consumed by a given directory, use *du*. Examples of the two commands are shown in the following illustration.

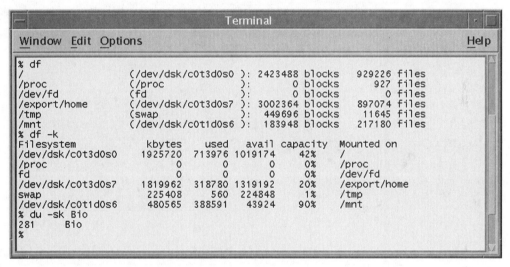

Examples of the df *and* du *commands in use.*

Editing Files

Numerous files control how Solaris functions. Examples include the password and group files that oversee user accounts, the hosts and network files for networking, and the configuration files for daemons such as *inetd* and *init*. Although special tools exist to manipulate files such as the password, most files are altered via an editor.

Using the Text Editor

On a workstation running CDE, text editing is made easy with the *text editor* found on the CDE "dashboard." The icon looks like a pad and pencil and is found on the sub-panel that pops up when the triangular button above the notecard icon is clicked. When accessed, an editor window appears on the screen, as shown in the following illustration.

Text Editor window.

To select the file to be edited, either drag the file from the File Manager onto the text editor icon on the dashboard or use the Open a File dialog box, shown in the following illustration, from the File menu.

To edit the file, position the cursor with the keys or the mouse, and then type. To select the text, hold down the right mouse button and sweep the cursor over the text. The Edit menu provides the familiar cut, copy, and paste functions found in other text processing programs. The File menu provides commands for saving the edited file and exiting the text editor.

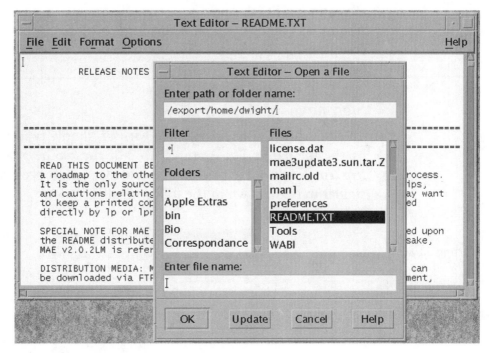

Open a File dialog box.

The vi Editor

Some administrators might prefer using the *textedit* command or the CDE text editor to modify system files. However, there are a surprising number of situations for which OpenWindows, CDE, and tools such as *textedit* are unavailable. Three examples of such situations are working on a server that does not have OpenWindows installed, working over a modem from a PC, or working on another vendor's UNIX system. At these times it helps to know how to use *vi*, a terminal-oriented editor common on most UNIX systems, including Solaris.

Although not as simple to use as *textedit*, *vi* is widely available and should be accessible on any UNIX system. The *vi* editor works easily on terminals with cursor positioning capability. Included are common PC terminal programs that emulate VT100 or ANSI terminals. The discussion in this segment is not intended to be a complete introduction to *vi*, but rather a description of selected basic functions.

Starting vi

To edit a previously existing file with *vi*, simply type *vi*, followed by the file name. For example, to edit the file */tmp/test*, type *vi/tmp/test*. If you want to create a new file, *vi* will present a screen similar to that shown in the following illustration.

Starting up vi.

> ◆   *NOTE: A very small window is used to make it easy to see the cursor position in the current and following examples. The* vi *editor adapts to the window size.*

The name of the file is shown at the bottom of the screen. The tilde characters along the left edge of the screen are *vi* markers indicating the empty region past the end of the file. The markers are not part of the file being created; *vi* displays them to help identify the boundaries of the screen and the file being worked on.

Entering Text in vi

The *vi* editor has two modes: command and input. When entering *vi*, begin in the command mode. To switch to the input mode, type either an *A* character (to append input to the line the cursor is on), or an *i* (to insert input at the current cursor position). To see how this works, follow along using *vi* on a workstation for the remainder of this section.

➥ *NOTE 1: In reality, there are several commands that will place vi in the input mode. The following discussion uses the A and I commands as examples.*

➥ *NOTE 2: The vi editor is modal. It has an input mode and a command mode. Keeping track of the current mode is the key to using vi.*

To edit */tmp/test*, type *A* to put *vi* into input mode. When inputting the text, shown in the following illustration, be sure to press <Return> at the end of each line. Unlike *text edit* and some other text editors, *vi* does not implement word wrap unless specifically set to do so. Typing in a paragraph without typing <Return> at the end of each line will result in a very long line. After finishing the last line, switch from input mode back to command mode by pressing the <Esc> key.

Entering text in vi.

✓   TIP: *When you forget the mode* vi *is in, press the <Esc> key. The* vi *editor will switch to command mode if it was in input mode, or cause the terminal bell to beep if it was already in command mode. Either way,* vi *will end up in command mode.*

In command mode, there are two basic functions: editing using the edit keys, and issuing commands that start with a colon (:).

Moving the Cursor

Moving the cursor is a basic function of the editing keys. The *vi* editor uses a set of keys to the right of the keyboard's home row. Table 4-1, which follows, describes the cursor movement keys for *vi*.

Table 4–1: Cursor Movement Keys for vi

Key	Cursor action
h	Move left one character
j	Move down one line
k	Move up one line
l	Move right one character

Try moving through the text you input. Finish the motion with the cursor on the *l* in *little* on the second line, as shown in the following illustration.

⟶   NOTE: *On many terminals, the arrow keys can also be used to move the cursor.*

✓   TIP: *Tired of moving around one character at a time? The* vi *editor has motional accelerator keys. The <b >key moves the cursor to the left by one word, and <w> moves the cursor to the right by one word.*

Moving the cursor.

Checking the Terminal Type

Improper cursor motion is a common problem for *vi* users because Solaris does not know what type of terminal or terminal emulator is being used. The *vi* editor depends on this information to emit the proper sequences of escape and control characters needed to move the cursor around the screen. These sequences vary according to terminal. The *vi* program determines the type of terminal in use by consulting an environment variable set in the command interpreter or shell.

To check the value of the *TERM* environment variable, exit *vi* by typing *<Esc> :wq* to save the changes and quit the editor. Next, type the following string.

```
% echo $TERM
```

If the terminal listed is the type being used, no changes are needed. If not, the *TERM* environment variable should be set to the correct terminal type. For example, if a VT100-type or VT100 terminal emulator is being used, and you use the Bourne shell (*/bin/sh*), you would input *TERM=vt100*. Export *TERM* to set the *TERM* environment variable.

Check the *TERM* environment variable if *vi* does not draw the screen correctly, refuses to work (i.e., asserts that the terminal in use does not have the required cursor functions), or moves the cursor incorrectly in response to the motion keys.

➥   *NOTE: It may also be necessary to use the* stty *command to set the terminal line disciplines. If the user resizes a telnet window, it may be necessary to use* stty *to set the number of rows and columns in the current window. For example,* stty *rows 24 columns 80 will set the number of lines of text to 24, and the number of characters per line to 80.*

After checking the *TERM* environment variable, restart *vi* with the same command as shown at the beginning of this section to continue with the following examples.

Changing Text

To change text, use a substitution command or the *c* edit key. The *c* key provides a choice of objects to be changed or the number to changed. Command options include *cw* to change the word, and *c* followed by the space bar to change a single character. The *cw* command will change the characters at the cursor point as far to the right as a space or punctuation mark.

Type *c$* to change the remainder of a line; that is, from the cursor position to the right end of the line. The *vi* program marks the characters to be changed by placing a dollar sign ($) at the right end of the group of characters to be changed and then enters input mode. Try typing *cw* with the cursor positioned on the *l* in *little* on the second line of the example file. An example of the *change* command used before entering new text is shown in the following illustration.

*Change command
before entering
new text.*

➡️   *NOTE: It is also possible to change the entire remainder of the line using
the* C *(uppercase C) command.*

Type the string *lot of.* The *vi* editor inserts the characters in the
space formerly occupied by *little.* You can insert as many characters
as necessary. Press the <Esc> key to exit input mode and return to
command mode. An example of the *change* command used after
new text is entered is shown in the following illustration.

*Change command
after entering
new text.*

Undoing an Action

Typographical errors can easily be remedied. The *vi* editor has a one-command *oops* buffer. Type *u* to undo the change. Type *u* again to *redo* the change. For the example presented here, the change will be kept.

Deleting Text

The *d* or delete edit key function is similar to the *c* key. The *d* key takes an argument to specify what is to be deleted. Try positioning the cursor on the *S* in *Susan*. Type *dw* (delete word). Type *u* to undo the change. The *x* key is an equivalent of *d* followed by a space bar (*d*). The *x* key deletes a single character. To delete the entire line, type *dd*. The following series of illustrations shows the screen changes after using each of these editing key combinations.

The *dw* key sequence deletes the first word and the space following the word. This neatly cleans up a misplaced word without an additional delete command, as shown in the following illustration.

Using dw *to delete a word.*

The undo command, *u*, pops the deleted text back into place, as shown in the following illustration.

Using u *to undo.*

 → *NOTE: Remember that* vi *is restricted to a one-item undo buffer. You*
    cannot recover a mistake made two or more commands ago with the u
    command.

The single character delete command, *x*, deletes the character at
the current cursor position. In the example shown in the following illustration, the deleted character is the *S* in *Susan*.

Using x *to delete a*
single character.

 ✓ *TIP: A handy way to fix transposed characters is to place the cursor on*
    the first letter, and then use xp. *This deletes the character under the cur-*

sor, and places the same character after the letter that ends up under the cursor.

The *dd* command deletes an entire line, as shown in the following illustration

Using dd *to delete an entire line.*

Review of vi Features

Table 4-2, which follows, summarizes the basic *vi* key sequence commands, including the commands used in the preceding examples.

Table 4-2: Basic vi Key Sequence Commands

Key(s)	vi Action
h	Move left one character
j	Move down one line
k	Move up one line
l	Move right one character
c or *r*	Change character
cw	Change word
c$ or *c*	Change remainder of line
x	Delete character

Key(s)	vi Action
dw	Delete word
d$ or d	Delete remainder of line
dd	Delete entire line

✓ TIP: *The dot or period (.) editing key repeats the last edit function. If a word was just inserted or the remainder of a line deleted, or a word is moving to a new cursor location, type a period (.) to repeat the insertion or deletion.*

vi *Line Commands*

Moving the cursor and making changes with the editing keys works well for small changes. To make changes throughout a large file, *vi* line commands are entered by typing a colon (:). An example of this is shown in the following illustration.

The vi *editor is ready to accept a line command.*

The colon drops the cursor down to the bottom of the screen to allow input of a line command. Once a line command is entered, press the <Return> key to execute it and *vi* will automatically return to command mode at the location in the file where the line command finished working. In many instances, line command changes leave the cursor positioned on the last line in the file.

Substituting Text Using a Regular Expression

The most common editing command is a string substitution that takes the form of a *regular expression*. This specialized string substitution and pattern-matching language is found in several UNIX commands, including the *grep* family and the stream editor, *sed*. Regular expressions are a rich language with many features. A commonly used feature is their ability to match arbitrary strings when carrying out substitutions. For example, the following command would change the text in the example file from past to present tense.

```
1,$ s/had/has/
```

This command could be translated as: "Starting with the first line and ending with the last line, substitute the string *had* with the string *has* once per line." An example of this function is shown in the following illustration.

Using a regular expression to substitute has *for* had.

There are two parts to a regular expression. The first, *(1,$)*, is an address range. The second, *s/has/had*, is the substitution expression. The address range specifies the line numbers over which the expression will be applied. In the example, the entire file is specified by stating the *1,$* range. The dollar sign represents the last line of the file when used in the address range portion of a regular expression.

To restrict a regular expression to the line the cursor is on, type a colon and omit the address range section altogether (e.g., *:s/had/ has/*). To specify the range as an offset from the position of the cursor before the colon was typed, use a plus (+) and minus (-) sign as well as line numbers. An example follows.

```
-1,+2 s/foo/bar/
```

This expression instructs *vi* to substitute *bar* for the first occurrence of *foo* beginning with the line before the cursor position (*-1*) through to two lines below the cursor position (*+2*).

➠ *NOTE 1: To substitute for a period (.) or an asterisk (*), the user needs to prevent those characters from being interpreted as special matching characters. This is done by placing a backslash (\ ) before each character. For example, s/./,/ would transform the first character encountered into a comma (,) in the current line, whereas s/\./,/ would change the first period (.) to a comma (,).*

➠ *NOTE 2: Remember that* vi *has a one-command oops buffer. If a substitution is executed that changes more than is intended, type* u *and try again.*

Exiting vi

There are several ways to save a file and exit from *vi*. Changes can be saved into the original file or into a different file, or disregarded altogether to leave the original file untouched. A variety of *vi* exit commands are described in table 4-3, which follows. The most common exit is to type *ZZ* in command mode, which tells *vi* to save the changes and exit.

Table 4-3: Exiting vi

Characters Typed	vi Action
:w	Write the current file out.
:w!	Write the current file out even if it is marked as *read only*.
:w foo	Write the current editor content to the file *foo*.
:q	Quit the editor.
:q!	Quit even if unsaved changes are present.
ZZ	Save the current file and exit the editor.

Final Comments on vi

The *vi* program is a difficult editor to learn. However, its wide use and general availability on many types of UNIX platforms makes it worth the effort to learn. Use *vi* to edit practice files, focusing on the following key points.

- Remember to check which mode *vi* is in. To return *vi* to command mode, press <Esc>.

- Check the context of the character being typed. A dollar sign ($) can mean either "move to the end of the line" or "move to the last line of a file" in the address part of a regular expression in command mode, or simply the dollar sign character in input mode. Practice using overloaded characters such as the dollar sign in every possible context.

- Remember what *vi* is doing with the characters being typed. Are the characters being inserted into the file or interpreted as commands?

In general, caution is required when editing system files. It is easy to make mistakes in *vi*, especially for beginners. A good approach to editing a system file is to make a backup copy of any file *before* you edit it. To change the */etc/inet/hosts* file, which contains system host names and their IP addresses, execute the following commands.

```
# cp /etc/inet/hosts /etc/inet/hosts.000405
# vi /etc/inet/hosts
```

Preceding *vi* with the *cp* command creates a backup copy, which is invaluable as a quick solution if changes to the file do not work as expected or should problems arise while editing. This is better than retrieving the file from a recent system backup to recover from a typo. The numbers added to the file name are a code for the date. Two digits each represent the year, month, and day, in that order.

Old files are easy to clean up by using a naming convention such as the numerical form of the date. For example, backup files from 1998 might be removed by typing the following.

```
# rm -i /etc/*.98????
```

This string will match any file name that contains a *98* date code on the end and will prompt the user for verification before removing any files. The question marks are wildcard characters and will match any single character. In the example, four question marks match the four digits of the month and date. This command should be implemented *after* a system backup so that the files will be on a backup tape in the event they are needed.

✓   *TIP: The -i option on* rm *is a good one to use with wildcard characters such as the asterisk (*) or question mark (?) while working as root. The root account has the privilege to remove anything, including things that should not be removed such as critical system files, device entries, and even the system kernel. The -i option asks permission before each dele-*

tion. This gives the user a chance to avoid undesirable results from a wildcard character.

UNIX Power Tools

With file editing in hand, the next collection of basic tools a systems manager needs can best be described as UNIX power tools. The following section demonstrates the basic concepts and uses of these tools.

Modern workstations provide window-based terminals and graphical monitoring tools. The command shell, pipes, and filters in Solaris and other UNIX variants form a flexible and powerful tool set that can be readily applied to common system management problems.

Juggling Tasks with Multiple Windows

Systems administrators will often need to refer to several tasks at once. The CDE terminal window and multiple screen features (see the following illustration) help by providing a method for organizing work and easily switching between tasks. CDE also provides advanced e-mail and performance monitoring tools.

This illustration shows two CDE terminal windows. The output of the *df* command is in one window and a directory listing is in the other. The terminal window provides a useful scroll bar for reviewing commands or output that have scrolled off the window.

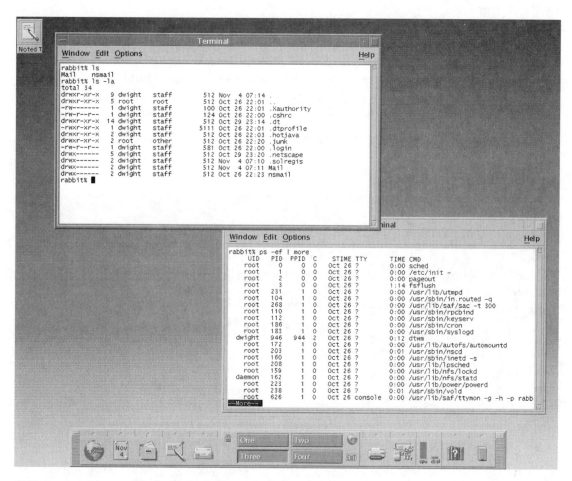

CDE screen with two terminal windows.

Although the user can open more windows, CDE offers another option. The buttons in the middle of the CDE dashboard allow for switching between several virtual screens, each containing its own set of windows. The following illustration shows another virtual screen containing only a web browser. Because the windows can be moved between screens and the screens can be given names, work is organized in logical groupings (e.g., web browsing, e-mail, and so forth).

Electronic mail is another basic tool of systems administration. Many beginner systems administrators are familiar with the shell command, *mailx*, which is used to read e-mail. Systems administrators who receive large volumes of e-mail should consider more powerful mail handling tools such as the CDE mail program, shown in the following illustration. The mail program icon is located next to the screen buttons at the left of the CDE dashboard.

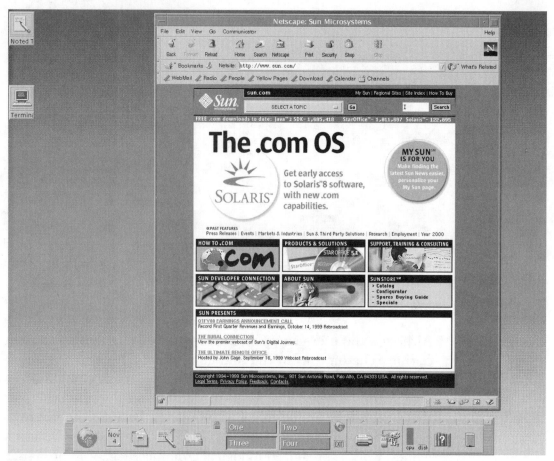

CDE screen showing Netscape Navigator.

CDE mail program and options dialog box.

The CDE mail program displays a list of the subject lines of e-mail messages, which can be easily scanned to pick out important messages quickly. It has a built-in editor similar to the CDE text editor, and option dialog boxes for controlling message and header display, mailboxes, and e-mail address aliases.

Another important window-based tool, the "performance meter," appears in its basic form on the CDE dashboard. By clicking on the bar graph display on the CDE dashboard, the full version of the performance meter can be displayed. The display style, parameters monitored, and monitoring rate can be changed via the performance meter's Properties dialog box, shown in the following illustration. This is one way an administrator can keep an eye on the health of a system while performing other tasks.

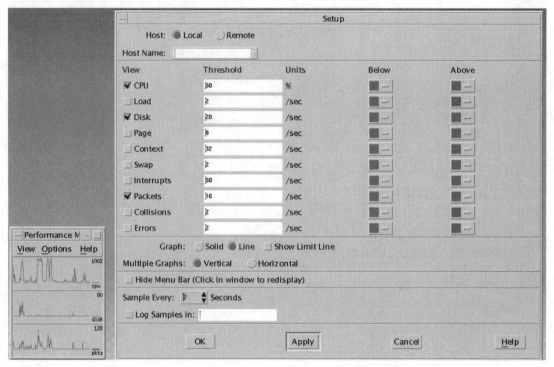

Performance meter and properties dialog box.

Pipes

The real power of Solaris is found in command combinations. The principal method for combining tools is called a "pipe." A pipe connects commands by directing the flow of data from one command into another. With UNIX command shells, the flow is from left to right along the command line; the pipe connects the output of the command on the left side of the pipe to the input of the command on the right side of the pipe. The examples that follow clarify pipe use.

A common problem on a Solaris system is the identification of a running process by its process ID number, which can be used to stop the process or send it a signal. Suppose the web server process needs to be identified and stopped. The process ID can be found using the *ps-ef* command. The *-ef* options to *ps* list all processes along with the process IDs, as well as additional information associated with each process.

Filtering with more

Unfortunately, the list of processes, even on a workstation, is too long to fit in a 24-line window, an example of which is shown in the following illustration. Although you can scroll backward in a window, you simply lose the output if you are working on a terminal. Another approach is to filter the output of *ps* to show only items of interest. A good first choice is to use the *more* program, which displays output a page at a time. The *more* program will read from its standard input if no file name is provided. The output of *ps* is connected to the *more* input by using a pipe (|).

```
┌─────────────────────────────────────────────────────────────────────────────┐
│─│                                Terminal                                │ ▫ □│
├─────────────────────────────────────────────────────────────────────────────┤
│ Window  Edit  Options                                                    Help │
├─────────────────────────────────────────────────────────────────────────────┤
│al/http                                                                        │
│     http  5624   197   0    Mar 10 ?          0:00 /usr/local/http/httpd -d /usr/loc│
│al/http                                                                        │
│     http  5121   197   0    Mar 09 ?          0:00 /usr/local/http/httpd -d /usr/loc│
│al/http                                                                        │
│     http  5109   197   0    Mar 09 ?          0:00 /usr/local/http/httpd -d /usr/loc│
│al/http                                                                        │
│     root  1078   106   0    Mar 02 ?          0:00 in.rlogind                  │
│   dwight  3756   323   0    Mar 07 ?          0:00 /bin/sh -c dtfile -noview    │
│     http  5110   197   0    Mar 09 ?          0:00 /usr/local/http/httpd -d /usr/loc│
│al/http                                                                        │
│     http  5108   197   0    Mar 09 ?          0:00 /usr/local/http/httpd -d /usr/loc│
│al/http                                                                        │
│   dwight  3757  3756   0    Mar 07 ?          0:05 dtfile -noview              │
│   dwight  5830   334   0    Mar 10 pts/6      0:00 /bin/csh                    │
│     root  7225   106   0 06:41:25 ?           0:00 rpc.rstatd                  │
│     root  5101     1   0    Mar 09 ?          0:00 /usr/sbin/aspppd -d 1        │
│   dwight  5836  5835   0    Mar 10 ?          0:18 /usr/openwin/bin/snapshot    │
│   dwight  5842  5830   0    Mar 10 pts/6      0:12 imagetool                   │
│   dwight  5814  5771   0    Mar 10 ?          3:10 /opt/SUNWwabi/bin/wabiprog -c -s│
│d:\msoffice\winword\winword.exe /export/home/d                                 │
│   dwight  5771  5770   0    Mar 10 ?          0:00 /bin/sh /opt/SUNWwabi/bin/wabi -s│
│d:\msoffice\winword\winword.exe /export/home/                                  │
│% ▊                                                                            │
└─────────────────────────────────────────────────────────────────────────────┘
```

The list of processes is so long it scrolls off the screen.

This process makes column headings at the top of the *ps* listing and the first few processes easy to read. By pressing the space bar, the *more* program will display the next page of output. However, to locate a particular process, you may have to view several pages of output. At this point, you may want to further customize the list of processes. The tool used to create such a list is *grep*. An example of a *more* filtering is shown in the following illustration.

```
─                              Terminal                                    ·  □

Window  Edit  Options                                              Help

% ps -ef | more
     UID    PID   PPID  C     STIME TTY      TIME CMD
    root      0      0  0     Mar 01 ?        0:01 sched
    root      1      0  0     Mar 01 ?        0:10 /etc/init -
    root      2      0  0     Mar 01 ?        0:06 pageout
    root      3      0  0     Mar 01 ?       11:49 fsflush
    root    259      1  0     Mar 01 console  0:00 /usr/lib/saf/ttymon -g -h -p tigg
er.net-kitchen.com console login:    -T sun -d /
    root    160      1  0     Mar 01 ?        0:01 /usr/lib/lpsched
    root    258      1  0     Mar 01 ?        0:00 /usr/lib/saf/sac -t 300
    root    100      1  0     Mar 01 ?        0:17 /usr/sbin/in.named
    root     80      1  0     Mar 01 ?        0:01 /usr/sbin/rpcbind
    root     82      1  0     Mar 01 ?        0:00 /usr/sbin/keyserv
    root    197      1  0     Mar 01 ?        0:01 /usr/local/http/httpd -d /usr/loc
al/http
    root     88      1  0     Mar 01 ?        0:00 /usr/sbin/kerbd
    root    106      1  0     Mar 01 ?        0:03 /usr/sbin/inetd -s
    root    109      1  0     Mar 01 ?        0:00 /usr/lib/nfs/statd
    root    111      1  0     Mar 01 ?        0:00 /usr/lib/nfs/lockd
    root    130      1  0     Mar 01 ?        0:00 /usr/lib/autofs/automountd
    root    186      1  0     Mar 01 ?        0:01 /usr/sbin/vold
    root    134      1  0     Mar 01 ?        0:09 /usr/sbin/syslogd
    root    150      1  0     Mar 01 ?        0:33 /usr/sbin/nscd
--More--
```

Output of ps -ef *filtered by* more.

Filtering with grep

The *grep* family of commands (*grep*, *egrep*, *fgrep*) is typically used by beginning UNIX users to find text strings in files. An example of a commonly used command sequence follows.

```
# grep mercury /etc/inet/hosts
```

This command would look for the string *mercury* in the host table found in *etc/inet/hosts*. This command might be used when seeking a host's IP address. Similar to *more*, *grep* can be used to search for any lines with matching text strings in data sent to it via a pipe, an example of which is shown in the following illustration.

```
─                                    Terminal                                    ⌐ □

 Window  Edit  Options                                                       Help

% ps -ef | grep httpd
      root    197      1   0    Mar 01 ?          0:01 /usr/local/http/httpd -d /usr/loc
al/http
      http   5115    197   0    Mar 09 ?          0:00 /usr/local/http/httpd -d /usr/loc
al/http
    dwight   7237   5830   0 06:56:40 pts/6        0:00 grep httpd
      http   5111    197   0    Mar 09 ?          0:00 /usr/local/http/httpd -d /usr/loc
al/http
      http   5112    197   0    Mar 09 ?          0:00 /usr/local/http/httpd -d /usr/loc
al/http
      http   5623    197   0    Mar 10 ?          0:00 /usr/local/http/httpd -d /usr/loc
al/http
      http   5624    197   0    Mar 10 ?          0:00 /usr/local/http/httpd -d /usr/loc
al/http
      http   5121    197   0    Mar 09 ?          0:00 /usr/local/http/httpd -d /usr/loc
al/http
      http   5109    197   0    Mar 09 ?          0:00 /usr/local/http/httpd -d /usr/loc
al/http
      http   5110    197   0    Mar 09 ?          0:00 /usr/local/http/httpd -d /usr/loc
al/http
      http   5108    197   0    Mar 09 ?          0:00 /usr/local/http/httpd -d /usr/loc
al/http
% ▮
```

Output of ps -ef *filtered by* grep *to capture a specific process by name.*

The output lines of *ps -ef*, which contained the name of the process, *httpd*, have been filtered out of the longer listing. The only item missing is the column header that *ps* prints, identifying each column of output. To filter the output of *ps* to include both the header and the process of interest requires another member of the *grep* family.

The *egrep* variant of *grep* accepts regular expressions in its search string, similar to those used in *vi*. A regular expression that consists of a string, a vertical bar, and another string will cause *egrep* to match either the first or the second string. An example of this is shown in the following illustration.

```
Terminal

Window  Edit  Options                                              Help

% ps -ef | egrep "UID|httpd"
    UID    PID  PPID  C     STIME TTY        TIME CMD
   root    197     1  0   Mar 01 ?          0:01 /usr/local/http/httpd -d /usr/loc
al/http
   http   5115   197  0   Mar 09 ?          0:00 /usr/local/http/httpd -d /usr/loc
al/http
   http   5111   197  0   Mar 09 ?          0:00 /usr/local/http/httpd -d /usr/loc
al/http
   http   5112   197  0   Mar 09 ?          0:00 /usr/local/http/httpd -d /usr/loc
al/http
   http   5623   197  0   Mar 10 ?          0:00 /usr/local/http/httpd -d /usr/loc
al/http
   http   5624   197  0   Mar 10 ?          0:00 /usr/local/http/httpd -d /usr/loc
al/http
   http   5121   197  0   Mar 09 ?          0:00 /usr/local/http/httpd -d /usr/loc
al/http
   http   5109   197  0   Mar 09 ?          0:00 /usr/local/http/httpd -d /usr/loc
al/http
   http   5110   197  0   Mar 09 ?          0:00 /usr/local/http/httpd -d /usr/loc
al/http
   http   5108   197  0   Mar 09 ?          0:00 /usr/local/http/httpd -d /usr/loc
al/http
 dwight   7241  5830  0 06:57:48 pts/6      0:00 egrep UID|httpd
%
```

Output of ps -ef by grep filtered to capture the head line and a specific process.

Take time to note how the pipe character is being used. Without quotation marks, the shell would attempt to interpret the vertical bar as another pipe. Like the dollar sign character ($) in *vi*, the vertical bar means different things in different contexts.

In the previous illustration, the vertical bar between the *ps* command and the *egep* command indicates a connection of the output of *ps* to the input of *egrep*. However, the vertical bar found inside the quoted search string after the *egrep* command is part of a regular expression that directs the *egrep* command to "match this or that." The meaning of the vertical bar is controlled by the quotation marks placed around the search string. The quotation marks tell the shell to pass the string contained within them to *egrep*.

tar Tricks

Another common system management problem is moving directories while preserving the ownership and permissions assigned to the files and subdirectories. This problem often arises when there is a need to move a user's home directory from one disk to another, or even to another machine. The usual solution to this problem is to use the *tar* command to copy the directory onto an archive file on tape, and then extract the copy of the directory from the archive file on tape to the new location.

A faster method, however, is to combine the two *tar* commands with a pipe. With this procedure, the administrator avoids storing the archive file created by *tar*, and instead sends it directly from the *tar* that creates the archive to the *tar* that will extract the archive. The following illustration shows a simple example of moving a directory between file systems on a single system.

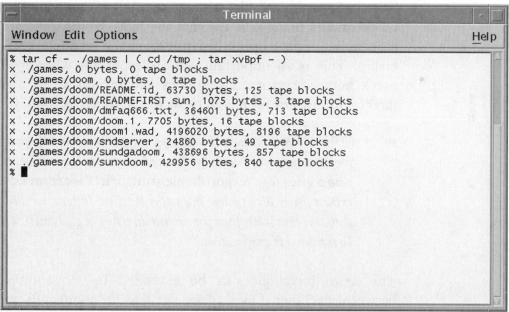

Moving a directory between disks using two tar *commands and a pipe.*

Reading the command from left to right, *tar* shows the *cf* option flags, which mean "Create an archive on the specified device or file." The hyphen (-) indicates that the file is standard output. The *./games* element indicates the directory to be archived. The output of this *tar* command flows through the pipe to the second *tar* command. The commands on the right side of the pipe within the parentheses constitute a single command to the pipe.

The parentheses create a new shell, which interprets the commands within. The first command is the familiar *cd*, which indicates where to place the moved directory. Next comes the second *tar* command, which reads the archive fed to it by the pipe from the first *tar* command. The second *tar* uses the *xvBpf* option flags. This is interpreted as "Extract the archive, print verbose messages, keep reading even if a full block is not sent, preserve the file modes and ownership, and read from the specified device or file." In this case, the hyphen (-) indicates standard input.

The *B* option flag avoids potential problems from a less-than-full block of data being sent through the pipe. If *root* is running the *tar* command, the *p* option flag preserves the file modes and ownership. This is an important option, especially when moving a user's home directory. The modes and ownership of a moved directory must exactly match those of the source directory.

> ►❖   NOTE: *When creating archives using* tar, *be aware that the path given for the directory or files to be archived is saved in the archive. If a full path is given (e.g.,* /export/home/stuff), *it will not be possible to extract the archive content in another location. Instead, use a relative path, one that starts from the current directory (e.g.,* ./stuff). *This can be extracted in any location.*

The same technique can be extended to move directories between machines. Once again, use two *tar* commands and add the *rsh* command. The *rsh* command runs shell commands on a remote machine with access to the command's standard input and output. This procedure, an example of which is shown in the following illustration, is similar to the single-machine example. The first *tar* command creates an archive on its standard output using the *cf* option flags, the parentheses, and the relative path of the directory to be moved.

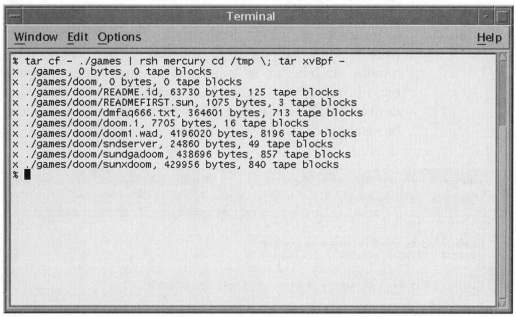

Moving a directory between machines with rsh, *two* tar *commands, and a pipe.*

On the receiving end of the pipe is an *rsh* command. This command connects with the indicated remote system, and then runs the shell commands listed after the machine's name on the remote system. As before, *cd* positions the shell in the proper location for the directory being moved.

The backslash in \; prevents the shell from applying a special interpretation to the semicolon character (;). In brief, the semicolon is not processed by the shell on the sending machine, but is passed along via *rsh* to the remote machine so that two commands, *cd /tmp* and *tar xvBpf -*, can be run there. As before, the second *tar* command uses the *xvBpf* option flags and a hyphen (-).

rsh, login, telnet, rcp, and ftp

The *rsh* command is one of several used to access machines over a network. As shown in the preceding section, *rsh* can be used to run a command on a remote machine. The *rlogin* and *telnet* commands are used to begin a terminal session on a remote machine, an example of which is shown in the following illustration. Because they use different protocols when connecting to a remote machine, it is important to note which protocol the remote machine or networked device supports. Some network devices (such as network hubs, terminal servers, and routers) can be managed over a network, but only accept *telnet* connections.

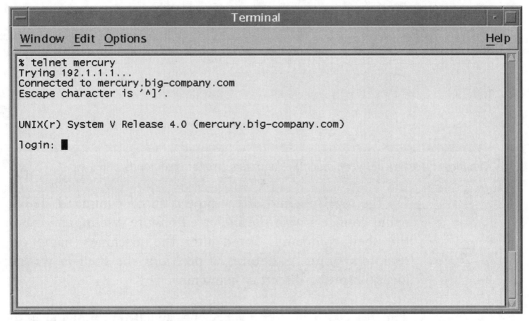

A telnet *connection to a remote machine.*

Files can be moved between machines on a network by using the *rcp* and *ftp* commands. The *rcp* command acts like a *cp* command but adds syntax to the file name description to include a machine name. The machine name is added prior to the file's path name and separated from it by a colon. An example of copying a file named *motd* is shown in the following illustration.

Copying /etc/motd *between machines using* rcp.

➥   *NOTE: The* rcp, rsh, *and* rlogin *commands make use of the* .rhosts *file to determine access. Access is granted if the machine and user being contacted have a* .rhosts *file, and if that file contains a line listing both the machine the connection is coming from and the user name being used. Exercise great care before allowing access to the root account via entries in the* /.rhosts *file. See Chapter 18, "Network Security," for more information on root account security.*

The *ftp* command provides more extensive file manipulation between machines. This command has an exclusive command interpreter allowing mode setting, directory listing, and file copying. The *ftp* command is a streaming protocol, and is thus efficient for moving files over wide-area networks and the Internet.

FTP access is accomplished via a user name and password dialog similar to that of logging in for a terminal session. An example of copying a file using *ftp* is shown in the following illustration.

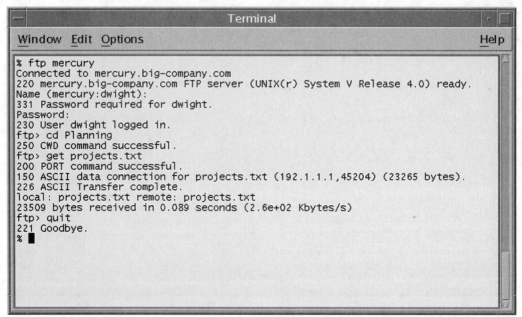

```
% ftp mercury
Connected to mercury.big-company.com
220 mercury.big-company.com FTP server (UNIX(r) System V Release 4.0) ready.
Name (mercury:dwight):
331 Password required for dwight.
Password:
230 User dwight logged in.
ftp> cd Planning
250 CWD command successful.
ftp> get projects.txt
200 PORT command successful.
150 ASCII data connection for projects.txt (192.1.1.1,45204) (23265 bytes).
226 ASCII Transfer complete.
local: projects.txt remote: projects.txt
23509 bytes received in 0.089 seconds (2.6e+02 Kbytes/s)
ftp> quit
221 Goodbye.
%
```

Copying a file via ftp.

✓   *TIP: By default,* ftp *assumes the files being copied are text files containing seven-bit ASCII characters. To copy a binary file, such as a program, be sure to set* ftp *to binary mode by typing* bin.

Anonymous ftp

Some machines on the Internet offer anonymous FTP access, often for purposes of distributing software. In this case, input *anonymous* for the user name, and provide the user's e-mail address as the password, as shown in the following illustration.

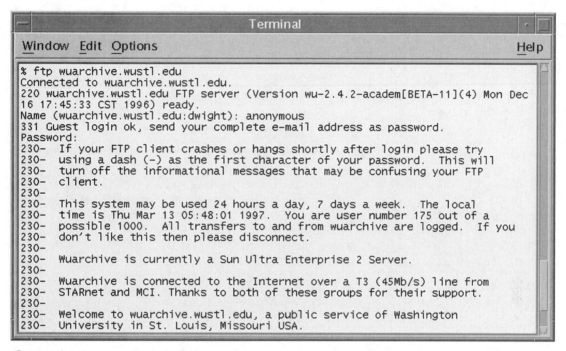

Connecting to an anonymous ftp *server.*

Summary

This chapter covered a number of basic UNIX tools found on Solaris systems that provide general-purpose aids to system management. Because the need to edit files is common in system administration, this chapter provided an introduction to editing with *vi*. Task management using multiple windows was also discussed, and a selection of useful commands and command combinations presented.

Consider taking the time to practice using these basic tools, and to study the options not covered here and the basic commands in the on-line manual pages. System management involves a commitment to continually learning about the capabilities and features of available tools.

PART 11

System Management

Refining System Boot and Shutdown Procedures

Unlike starting up a personal computer, the boot sequence of a Solaris system involves many steps. The Solaris boot sequence can be modified in several places to suit various needs. This chapter explores the terminology used in the Solaris boot sequence, and changes that can be made in the sequence. Several methods of altering the system run level and respective impacts on boot and shutdown sequences are also explored.

Boot Sequence

The term *boot* is derived from the expression "pull yourself up by the bootstraps." To boot a computer is to start it running from a stopped state, usually immediately after turning the power on. The steps required in the Solaris boot sequence are shown in the following illustration.

Boot sequence.

Step 1: PROM monitor executes
power-on self-test (POST).

Step 2: PROM monitor loads boot
block from boot device.

Step 3: Boot block loads */ufsboot*
from boot device.

Step 4: */ufsboot* loads SunOS kernel
from boot device.

Step 5: SunOS kernel configures devices,
initializes memory, and starts *init*.

Step 6: *init* reads */etc/inittab,* moves to
default run level, executes *rc* files.

Step 7: *rc* files execute scripts from
/etc/rc{run_level}.d directories.

The first step of the boot process relies on the physical hardware of the workstation to initialize itself and load a small program. This program is usually stored in a ROM (read-only memory) or PROM (programmable read-only memory) chip. The program loaded at power-on is called a *PROM monitor.*

➡   *NOTE: Personal computer users may be more familiar with the term basic input/output system (BIOS). The BIOS of an x86 system is functionally equivalent to the SparcStation PROM monitor.*

Step 1: PROM Monitor

The PROM monitor has several functions. It can be used to modify basic hardware parameters such as serial port configurations or the amount of memory that should be tested upon system power-up. Another PROM configurable parameter is the system *boot device* specification. This parameter tells the PROM monitor where it should look for the next stage of the boot process. Most important, the PROM monitor has routines to load the next stage into memory and start it running.

The next stage of the boot process may be loaded from a disk attached to the system, a floppy diskette, a CD-ROM drive, a tape drive, or over the network from a boot server (in the case of a diskless workstation). The load source depends on the setting of the PROM monitor's boot device parameter.

On a SparcStation, the PROM monitor initially starts in a restricted state. This state is sometimes referred to as the "old" command mode, because this mode was the only one available on early model SPARC workstations. The old command mode restricts the operations that can be taken to accomplish the following tasks.

- Start the boot process.

- Continue the running program if the system has accidentally entered the PROM monitor.

- Switch to the new command mode.

The boot process is invoked with the *b* command. The new command mode is invoked with the *n* command. If the operation of the system has been interrupted, it may be returned to the running state by invoking the *c* (continue) command. You can tell you are in the old command mode by the prompt character ">".

b (boot) *Option*

The *b* command accepts an optional device argument, which indicates the device to use when booting, and an optional file name, which dictates the program to be loaded and started. The *b* command also allows an option flag that controls the boot process. A common use of the option flag is to specify the *-s* option to cause the system to enter run level 1 (single-user mode) so that system maintenance activities may be performed. The format for the *b* command is *b [device] [file] [options]*. A summary of the *b* command appears in table 5-1, which follows.

Table 5-1: Boot Options

Parameter	Allowed Values	Default Value
[device]	Cdrom, disk, floppy, net, tape	Determined by the PROM monitor boot device parameter. Typically set to SCSI disk 3.
[file]	Any bootable file name	Determined by the PROM monitor boot file and boot mode parameters. Typically set to *vmunix* if the system is in normal mode, and to */stand/diag* if the system is booted in diagnostic mode.
[option]	-a (Prompt for further information) -r (Rebuild the kernel device tables) -s (Enter run state 1; single user)	Not applicable.

For example, booting a system from the restricted mode of the PROM monitor is done by typing *b* at the command prompt. The following command line contains the *-r* option, which forces the system to perform a reconfiguration boot.

```
> b -r
```

The flag indicates that the *boot* process needs to check for changes in the available system devices and take actions such as loading device drivers and creating or deleting entries in the */dev* directory to match any changes that have occurred since the system was last booted.

➡ *NOTE: See Chapter 8, "Working with Solaris Device Names," for a detailed discussion on device reconfiguration.*

n (new) *Option*

The *new* command mode, also known as the *forth monitor*, allows an extra level of control and customization. The new command mode is identified by the *ok* prompt instead of the > prompt used in the old command mode. From *new* command mode, two parameters control the system boot state. One of the parameters is a security mode setting that prevents a user from specifying an alternate boot. The other parameter is called *auto-boot*, and if it is set the workstation will bypass the restricted mode and continue on directly with the boot process. Table 5-2, which follows, presents a few of the most common commands available in *new* command mode.

Table 5-2: Forth Monitor Options

Command	Description
boot	Similar to the *b* command in restricted mode.
probe-scsi probe-ide	Probes the internal SCSI (IDE) bus and lists the devices found. The *probe-scsi-all* variant will probe all SCSI bus adapters and report on all devices connected to any SCSI bus on the system.
reset	Causes the PROM monitor to initialize the system as if power had been cycled.
printenv	Lists the current value of PROM monitor parameters.
setenv	Sets the value of a PROM monitor parameter.

➡ *NOTE: Use caution when setting PROM monitor parameters, especially* boot-device *and* boot-file. *The PROM monitor does not check to determine whether parameters are valid; a mistake could pre-vent the system from automatically booting.*

The following is an example of starting the boot process from the *new* command mode.

```
ok boot
```

If the *auto-boot* parameter is set to *true*, or if the *b* or *boot* commands are issued, the PROM monitor proceeds to the next stage of the boot process. The next stage consists of loading and starting a program known as the *boot block*.

Step 2: Boot Block

Similar to the PROM monitor, the boot block gets its name from the location in which it is stored. Typically, the boot block is stored in the first few blocks (sectors 1 to 15) on the hard disk attached to the workstation. The boot block's job is to initialize some of the system's peripherals and memory, and to load the program, which will in turn load the SunOS kernel. A boot block is placed on the disk as part of the Solaris installation process, or in some circumstances, by the system administrator using the *installboot* program.

Although a diskless machine does not read its boot block from a disk, the term is still used to describe the small program the PROM monitor loads into the workstation's memory from a boot server over the network. A boot server is another system on the network that provides network boot services via the trivial file transfer protocol (*tftp*) daemon. When the diskless client's PROM monitor issues a network request for its boot block, the server replies by sending the boot block through the network to the client workstation. The PROM monitor on the client workstation loads the program into memory and starts its execution.

Step 3: Boot Program

Depending on the location of the boot block, its next action is to load a boot program such as */ufsboot* into memory and execute it. The boot program includes a device driver as required for the device (such as a disk drive or network adapter) that contains the SunOS kernel. Once started, the boot program loads the SunOS kernel into memory, and then starts it running. On a diskless workstation, the boot program handles setting up NFS access over the network to the workstation's root (/) file system and *swap* area.

Step 4: SunOS Kernel

The *kernel* is the foundation of both SunOS and Solaris. Once loaded into memory by the boot program, the kernel has several tasks to perform before the final stages of the boot process can proceed. First, the kernel initializes memory and the hardware associated with memory management. Next, the kernel performs a series of device probes. These routines check for the presence of various devices such as graphics displays, Ethernet controllers, disk controllers, disk drives, tape devices, and so on. This search for memory and devices is sometimes referred to as *auto-configuration*.

✓   *TIP 1: On certain systems the status lights may blink on devices such as disk drives while the kernel probes for and locates the devices.*

✓   *TIP 2: Solaris 7 and Solaris 8 will, by default, boot a 64-bit kernel on recent UltraSparc systems. In order to force these systems to run a 32-bit kernel, change the* boot-file *in the* eeprom *such that the system boots the file* /platform/sun4u/kernel/unix *instead of the default file* /platform/sun4u/kernel/sparcv9/unix.

With memory and devices identified and configured, the kernel finishes its start-up routine by creating *init*, the first system process. The *init* process is given process ID number 1 and is the parent of all processes on the system. Process 1 (*init*) is also responsible for the remainder of the boot process.

Step 5: init and /etc/inittab

The *init* process and the files it reads and the shell scripts it executes are the most configurable part of the boot process. Management of the processes that offer the *login* prompt to terminals, the start-up for daemons, network configuration, disk checking, and more occur during this stage of the boot sequence.

Upon invocation, the *init* process reads a configuration file named */etc/inittab*. The */etc/inittab* file contains a series of *init* directives. The format for these directives is *id:run-level:action:process*. The *init* directives are described in the following.

- *id*: A one- to four-character label that uniquely identifies the entry.

- *run-level*: The system run level at which the current process will be invoked.

- *action*: This parameter indicates one of a variety of actions to be taken by *init*.

- *process*: The name of the program that will be invoked depending on the values of the previous two fields (*run level* and *action*).

Solaris Run Levels

The run level is a number that specifies a particular system state. Actions to be taken by *init* are keyed to various run levels. The run levels used in Solaris are described in table 5-3, which follows.

Table 5-3: Solaris Run Levels

Run Level	Description
0	Power-down or shutdown state: *init* stops all system activity and causes the kernel to exit, returning control to the PROM monitor.
1	Single user: *init* executes actions that unmount all file systems except *root* (/) and / *usr*, and allows *root* access only on the system console.
2	Multi-user: resources are not exported to other systems.
3	Alternate multi-user. Not currently used.
4	Multi-user: resources are exported to other systems.
5	"Green" or power saving mode for Sun4m and Sun4u systems.
6	Reboot: shut the system down and reboot.
s,S	Single-user mode with all file systems available: terminal logins are disabled, but network logins are enabled.

More than one run level can be listed for */etc/inittab* entries, in which case the process is invoked when *init* enters any of the run levels listed. An entry with no associated run level is assumed to be valid at all run levels.

init *Actions*

When *init* moves to a given run level, each line in the */etc/inittab* file with the current run level is processed. The processes invoked by some *inittab* entries are run once, whereas other processes must be restarted whenever they stop running. The actions listed in table 5-4, which follows, describe the things *init* can do with a process.

Table 5-4: init Actions

Action Code	Description
respawn	Start the process upon entering this run level. Do not wait for process to finish before moving on to next entry in *inittab*. Restart process if it should die.
wait	Process is started and *init* will wait for process to complete before moving on to next entry in *inittab*.
once	Process is started but *init* does not wait for process to complete. No *init* action is taken if process dies.
boot	Process is executed only during the first reading of the *inittab* file.
bootwait	Process is executed when *init* moves from single user to multi-user, and *init* waits for process to complete before moving on to next entry in *inittab*. Process is not restarted if it dies.
powerfail	Process is executed when *init* receives a power fail signal.
powerwait	Similar to *powerfail* except that *init* waits for process to finish before proceeding to next entry in *inittab*.
off	If process is running, *init* will stop it when entering this run level.
initdefault	Special entry to denote default run level. The run level argument of the line with the *initdefault* action is interpreted as the default run level.
sysinit	Process listed in this entry is to be run before *init* starts processes associated with allowing log-ins.

A Closer Look at /etc/inittab

Action codes and run levels are easier to understand when the *inittab* file is examined line by line. Study the following sample *inittab*.

```
ap::sysinit:/sbin/autopush -f /etc/iu.ap
fs::sysinit:/sbin/rcS          >/dev/console 2>&1 </dev/console
is:3:initdefault:
p3:s1234:powerfail:/sbin/shutdown -y -i5 -g0 >/dev/console 2>&1
s0:0:wait:/sbin/rc0            >/dev/console 2>&1 </dev/console
s1:1:wait:/sbin/shutdown -y -iS -g0   >/dev/console 2>&1 </dev/console
```

```
s2:23:wait:/sbin/rc2         >/dev/console 2>&1 </dev/console
s3:3:wait:/sbin/rc3          >/dev/console 2>&1 </dev/console
s5:5:wait:/sbin/rc5          >/dev/console 2>&1 </dev/console
s6:6:wait:/sbin/rc6          >/dev/console 2>&1 </dev/console
of:0:wait:/sbin/uadmin 2 0 >/dev/console 2>&1 </dev/console
fw:5:wait:/sbin/uadmin 2 6 >/dev/console 2>&1 </dev/console
RB:6:wait:/sbin/sh -c 'echo "\nThe system is being restarted." ' >/dev/
   console 2>&1
rb:6:wait:/sbin/uadmin 2 1>/dev/console 2>&1 </dev/console
sc:234:respawn:/usr/lib/saf/sac -t 300
co:234:respawn:/usr/lib/saf/ttymon -g -h -p " `uname -n` console login: "
   - T T "uT sun -d /dev/console -l console -m ldterm,ttcompat
```

In the following discussion, the file is dissected line by line in much the same way the *init* process reads the file. A line's two-character *id* field is used to identify which line is being discussed.

```
ap::sysinit:/sbin/autopush -f /etc/iu.ap
fs::sysinit:/sbin/rcS >/dev/console 2>&1 </dev/console
```

Because the first two lines, *ap* and *fs*, have no associated run levels, and both lines use the action code *sysinit*, *init* executes each process, and then waits for the process to finish before moving on to process more lines. The *ap* line initializes the keyboard and mouse drivers for use with the console and other terminal lines. The process on the *fs* line is an *rc* (run command) shell script, which is discussed in detail later in the chapter. The *rcS* script checks the / (*root*) and */usr* file systems for consistency.

```
is:3:initdefault:
```

The *is* line sets the default system run level to 3 (multi-user) via the *initdefault* action code. The *is* line does not change the run level, but rather specifies which run level *init* will move to by default when it is started. By specifying run level 3, *init* will move from level 0 to 1, 1 to 2, and 2 to 3 as the system progresses through the boot process.

```
p3:s1234:powerfail:/sbin/shutdown -y -i0 -g5 >/dev/console 2>&1
```

Upon booting, *init* moves from run level 0 to the default run level 3 in the example *inittab* file. Because the *p3* line lists several run levels, including run level 3, *init* performs the *powerfail* action. The *powerfail* action causes *init* to "remember" the process listed on the *p3* line as the process to be run if a power failure signal (*sig-pwr*) is sent to *init*. This action is used at sites where a backup power supply and special monitoring software are installed to automatically shut the system down in the event of a power failure.

```
s0:0:wait:/sbin/rc0 off         >/dev/console 2>&1 </dev/console
s1:1:wait:/sbin/shutdown -y -iS -g0   >/dev/console 2>&1 </dev/console
```

The *s0* line is skipped because it is not valid for run level 3. The *s1* line is skipped for the same reason.

```
s2:23:wait:/sbin/rc2         >/dev/console 2>&1 </dev/console
```

The *s2* line invokes an *rc* (run control) shell script. The *rc2* script in turn invokes a set of scripts to set the system time zone, and starts several system daemons.

```
s3:3:wait:/sbin/rc3          >/dev/console 2>&1 </dev/console
```

The *s3* line invokes the */sbin/rc3* script, which in turn invokes a series of other scripts to configure the network, and starts additional service daemons.

```
s5:5:wait:/sbin/rc5 ask         >/dev/console 2>&1 </dev/console
s6:6:wait:/sbin/rc6 reboot      >/dev/console 2>&1 </dev/console
of:0:wait:/sbin/uadmin 2 0      >/dev/console 2>&1 </dev/console
fw:5:wait:/sbin/uadmin 2 2      >/dev/console 2>&1 </dev/console
RB:6:wait:/sbin/sh -c 'echo "\nThe system is being restarted." ' >/dev/
    console 2>&1
rb:6:wait:/sbin/uadmin 2 1      >/dev/console 2>&1 </dev/console
```

The *s5*, *s6*, *of*, *fw*, and *RB* lines are skipped because they are not valid for run level 3.

```
sc:234: respawn:/usr/lib/saf/sac -t 300
co:234:>respawn<D>:/usr/lib/saf/ttymon -g -h -p "`uname -n` console login:
   " -T sun -d /dev/console -l console -m ldterm,ttcompat
```

The last two lines, *sc* and *co*, are scanned, and the processes listed are invoked by *init*. These processes are automatically restarted should they die as a result of the *respawn* action specified in the *inittab* entries. The *sc* line starts the process that monitors the terminal lines and provides a log-in prompt for users. The *co* line starts the process that monitors the system console and provides a log-in prompt on that device. When *init* finishes processing the entries in */etc/inittab*, the system boot process is complete.

Modifying /etc/inittab

A common boot process modification is to add entries to the */etc/init.d* script files such that new processes may be started at specific run levels. Verify that these new entries are placed in the correct sequence with respect to other */etc/inittab* entries.

For example, to add a line for a network license manager daemon, it would be wise to ensure that the daemon is started after things it depends on (such as network configuration or the NIS+ daemons). Most third-party software packages that require modification provide explicit directions on where respective entries must be placed in */etc/inittab*, and which run level to use. Although modifications to */etc/inittab* can be used to change the final stages of the boot sequence, there are significant drawbacks to this method. These follow.

- Difficulty in testing *inittab* entries

- Single-line limit for entries

- Difficulty in making entries conditional upon factors other than the run level

For these reasons, the use of run control (*rc*) files are the preferred method of implementing boot process modifications.

Run Control Files

The run control (*rc*) files and associated scripts are the most commonly modified component of the boot sequence. The local administrator can add or modify many system configuration processes in much the same way as for an *inittab* entry, but these files are easily tested and allow the programming features of the Bourne shell. Run control files allow greater flexibility in configuring the system during the boot process.

Each system run level has an associated *rc* file. Most *rc* files contain a directory bearing the name of the *rc* file with a *.d* extension. The *rc* file is a short Bourne shell that scans its associated directory and invokes the scripts found there in a certain sequence. The *rc* files are located in the */sbin* directory, and respective script directories are located in the */etc* directory. The *s3* line in the sample */etc/inittab* file invokes the */sbin/rc3* script when the system enters run level 3.

```
s3:3:wait:/sbin/rc3        >/dev/console 2>&1 </dev/console
```

When *init* encounters the line previously shown upon entering run level 3, it will invoke the */sbin/rc3* script and wait until the script finishes. The *rc3* script will in turn scan the */etc/rc3.d* directory in search of scripts to be run. Each script in the */etc/rc3.d* directory handles a specific system function, such as the management of the system print service or NFS services.

For each run level, there is an *rc* file and (usually) a directory of scripts to handle the start-up or shutdown of the services associated with that run level. Table 5-5, which follows, presents the *rc* file, script directory, and a brief description of the functions the *rc* file and scripts perform for each run level.

Table 5-5: Run Control Script File Relationships

Run Level	RC Script	RC Directory	Common Functions
0	*/sbin/rc0*	*/etc/rc0.d*	Kills all running processes except *init* and unmounts all file systems.
1	*/sbin/rc1*	*/etc/rc1.d*	Kills all nonessential system daemons.
2	*/sbin/rc2*	*/etc/rc2.d*	Kills the NFS, volume management, and *lp* services; mounts file systems; and starts most system services except NFS server upon start-up.
3	*/sbin/rc3*	*/etc/rc3.d*	Starts the NFS server process.
4	*None*	None	Not currently used.
5	*/sbin/rc5*	None	Same as *rc0*.
6	*/sbin/rc6*	None	Same as *rc0*.
S	*/sbin/rcS*	*/etc/rcS.d*	Starts basic network services, mounts */usr*, and executes reconfiguration commands required for a reconfiguration boot.

The scripts in these directories are run in their ASCII sorting order within two divisions. Files that start with *K* contain commands to kill off processes and are run first when the corresponding run level is entered. The *K* scripts are invoked with the *stop* argument (e.g., *K2olp stop*). Files that start with *S* contain commands to start processes and are invoked after all *K* scripts for a particular run level have been invoked. The *S* scripts are invoked with the *start* argument (e.g., *S2olp start*). The flow of control from *init* through an *rc* file onto a script in an *rc* directory is shown in the following illustration.

Control flow for rc files.

init reads /etc/inittab, executes rc file.

rc script scans associated directory, executes script files.

Scripts are run in alphabetic, then numeric, order.

Although it is sometimes necessary to directly modify an *rc* file, the more common approach is to add, delete, or modify a script located in the appropriate */etc/rc?.d* directories. By convention, the same script is used to start and stop a given service. The script contains a simple *case* statement and takes different actions depending on the argument it is passed at invocation (*start* or *stop*). The same script may be used at several run levels. To avoid multiple copies of a script and to ease maintenance, symbolic links are used.

All scripts used by the *rc* files are located in the */etc/init.d* directory. A symbolic link is made from the script in */etc/init.d* to a file in the appropriate */etc/rc?.d* directory.

Symbolic Links

A *symbolic link*, or simply *link*, is a term used in UNIX for another directory entry that points to the same file. A link may be considered an alias because it is another name by which the same information is known. Links are created using the *ln* command, shown in the following illustration.

Creating and examining a file link.

Note that once the link is made, using either the file name or the link name, the content of the same file is shown. This property allows links to be placed in locations where files would normally be found. Because a link is just another name for the file, all changes made to the file are seen through all respective links. In the case of scripts found in *rc* directories, this makes maintenance easier because only the original script needs to be modified in order to resolve problems.

✓   *TIP: Although Sun ships Solaris with hard links between the files in the* init.d *and* rc.d *directories, many system administrators prefer to use symbolic links instead of hard links when linking* rc *scripts to* /etc/init.d *entries.*

By convention, files that contain all scripts used by the *rc* files are located in the */etc/init.d* directory. Their names are provided by the function they perform or the system service they start and stop (e.g., */etc/init.d/nfs.server*).

Each *rc* script file reads and processes links whose names begin with a *K* to stop processes that should not be running at this run level. Next, the *rc* file reads and processes the links that begin with an *S*, to start processes that should be running at this run level.

For example, the script that starts the *cron* program, which runs other programs at scheduled times, is found in */etc/init.d/cron*. Because the *cron* program should not be running at run levels 0 or 1, links are made from */etc/init.d/cron* to *K* files in the */etc/rc0.d* and */etc/rc1.d* directories (i.e., */etc/rc0.d/K70cron* and */etc/rc1.d/K70cron*). A cron script and links are shown in the illustration that follows.

✓   *TIP: It is always good practice to verify that any editor used to modify files in the /etc/rc?.d directories does not break the symbolic link and create a new file as part of the editing process. Text Edit and vi preserve links.*

The cron script and links in the rc directories.

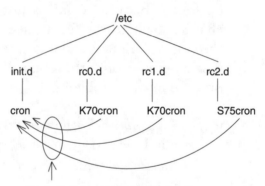

Symbolic links pointing to */etc/init.d/cron*

The *70* in the name of the *K70cron* link controls the order in which the links are processed by the *rc* scripts. For example, when the *rc1* script is run by *init*, it runs the scripts in the */etc/rc1.d* directory in the following order.

1.   *K00ANNOUNCE*

2.   *K42audit*

3.   *K47asppp*

4.   *K55syslog*

5.   *K57sendmail*

6. *K65nfs.server*

7. *K67rpc*

8. *K68autofs*

9. *K70cron*

10. *K80nfs.client*

11. *S01MOUNTFSYS*

The *K* links are read and processed first in numerical order. The *rc* script runs each *K* script with the single argument *stop*. For example, the *cron* link in */etc/rc1.d* is run as if the */etc/rc1.d/K70cron stop* command had been invoked. Because of the link to */etc/init.d/cron*, the command is equivalent to typing */etc/init.d/cron stop*.

The *cron* process should be started at run level 2; consequently, a link is made in the */etc/rc2.d* directory for */etc/rc2.d/S75cron*. The *S* scripts are read and processed after the *K* scripts in numerical order based on the two-digit number following the *S*. Like the *K* scripts, the *S* scripts are run as if the */etc/rc2.d/S75cron start* command had been invoked, which in turn is equivalent to */etc/init.d/cron start*.

Modifying Run Control Files

Modifications to the run control files are generally made by adding, changing, or deleting one of the scripts in */etc/init.d*, and adding or deleting the corresponding links in the *rc* directories as needed. To add a new system daemon, such as a license manager, a new script is created in */etc/init.d*. At the very least, the script must be able to handle the following arguments.

- *start*: This argument is given to the script when it is called as an *S* script from one of the *rc* files. When the script is invoked with the *start* argument, it should take whatever start-up action is required, such as starting processes, clearing out directories, testing for the availability of configuration files, and so on.

- *stop*: This argument is given to the script when it is called as a *K* script from one of the *rc* files. When the script is invoked with the *stop* argument, it should take whatever shutdown and clean-up action is required, such as killing off processes, clearing out temporary files, saving state information, and so on.

As an example, the script that follows is used to start up a license manager daemon (portions of the script are numbered for the purpose of further explanation). The daemon controls how many copies of certain programs are running. The programs contact this daemon over the network for permission to run. This script is placed in */etc/init.d/lmf*. It is written using Bourne shell syntax because the *rc* files will run these scripts using the Bourne shell.

```
# Start and stop the license manager daemon.
# Save the first argument in the variable state.
state=$1
# Handle each possible value for state.
                (1)


case $state in
                (2)


'start')
    # Check for the lmf configuration file and if present, start lmfd.
                (4)


    [ -f /usr/lib/lmf.conf ] && /usr/lib/lmfd
                (2)


    ;;
```

                        (3)

```
'stop')
   # Find lmfd in the output of ps.
```
                        (5)

```
   id=`ps -e | grep lmfd | awk '{print $1}'`
   # If a process id was found, then kill it.
   if test -n "$id"
   then
      kill $id
   fi
```
                        (3)

   ;;

                        (1)


```
esac
```

The heart of the script is the block of lines starting at the *case* and ending at *esac* (see 1). This block is further divided into two sections: the first runs from *'start')* to the double semicolon (see 2), and the second runs from *'stop')* to the double semicolon (see 3).

The *start* block is executed if the first argument given to the script and saved in the *state* variable is the word *start*. The *[-f /usr/ lib/lmf.conf] &&* phrase (see 4) is Bourne shell syntax for "Check to see if the file /usr/lib/ lmf.conf exists, and if so, do the following." The phrase that follows, */usr/lib/lmfd*, invokes the license manager daemon.

Simply stated, the entire block checks for the *lmf* configuration file. If the file exists, the block invokes the *lmf* daemon program.

The *stop* block is a bit more involved. The line that begins with *id=* (see 5) executes three commands joined by pipes, and places the result in the *id* variable. The first command is *ps -e*, which lists all processes currently running. The output of this command is passed to *grep lmfd*, which filters out the line listing the license manager daemon. That line is then passed along to *awk '{print $1}'*.

The *awk* term is a programmable filter. The program is the *{print $1}* phrase. Using this program, *awk* prints the first word, or group of characters separated by spaces in the line passed to it. In this case, the output of *awk* is the process ID number. Thus, the entire series of commands finds the process ID number for the license manager daemon and puts that number in the *id* variable.

However, what if the license manager daemon was not running? In this case, nothing would be found, and the value of *id* would be empty. The *if* line tests for this condition to ensure that the *kill* command is invoked only if *id* contains a value. The *kill* command sends the *SIGKILL* signal to the license manager daemon, and the daemon stops running.

Benefits of Using rc Directory Scripts

One of the benefits of using *rc* directory scripts is that they are easily tested. The scripts may be manually invoked with the *stop* and *start* arguments as a check to determine whether they function correctly before creating the symbolic links and trying them under actual system boot conditions.

✓   TIP: *This procedure is recommended because it can help you catch mistakes that might interrupt the boot process and leave the system unusable.*

The final step in adding a script is to add links in the appropriate places in the *rc* directories. Because the daemon in this example needs to communicate over the network, it should be started *after* the network utilities are started. One possible choice is to create the following links.

```
ln -s /etc/init.d/lmf /etc/rc0.d/K21lmf
ln -s /etc/init.d/lmf /etc/rc1.d/K21lmf
ln -s /etc/init.d/lmf /etc/rc3.d/S95lmf
```

This set of links ensures that the license manager is not running when the system is in run level 1 (single-user mode). Next, when you use these links, the license manager is started after the network services have been started when the system is in run level 3 (multi-user mode).

Changing Run Levels

Up to this point, system run levels have been discussed only with respect to the normal boot sequence. In addition to the boot process, several other methods may be employed to change the system run level.

➽   *NOTE: The commands discussed in the following sections are typically reserved for system maintenance activities.*

shutdown Command

The *shutdown* command is the preferred method for shutting the system down. This command will bring the system down to a particular run level in an orderly manner. The *shutdown* command sends messages to all users warning them of the impending system shutdown. The command will also notify the system *syslog* daemon of the shutdown.

When the specified time arrives, *shutdown* will cause the system to sync the memory buffers to disk, and then change the system run level to the desired state. Three of the most common uses of shutdown are to (1) bring the system to a halt, (2) bring the system down to single-user mode, and (3) reboot the system. The use of the *shutdown* command with no command line options is shown in the following illustration.

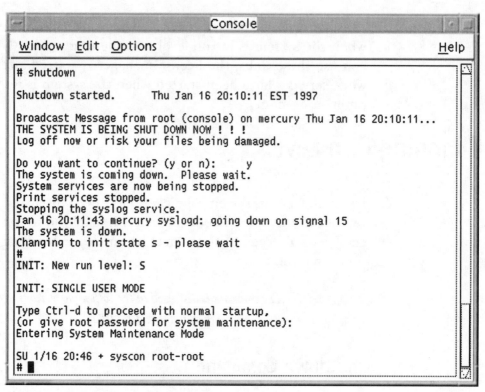

```
┌──────────────────────── Console ────────────────────────┐
│ Window  Edit  Options                              Help  │
├──────────────────────────────────────────────────────────┤
│ # shutdown                                               │
│                                                          │
│ Shutdown started.       Thu Jan 16 20:10:11 EST 1997     │
│                                                          │
│ Broadcast Message from root (console) on mercury Thu Jan 16 20:10:11... │
│ THE SYSTEM IS BEING SHUT DOWN NOW ! ! !                  │
│ Log off now or risk your files being damaged.            │
│                                                          │
│ Do you want to continue? (y or n):      y                │
│ The system is coming down.  Please wait.                 │
│ System services are now being stopped.                   │
│ Print services stopped.                                  │
│ Stopping the syslog service.                             │
│ Jan 16 20:11:43 mercury syslogd: going down on signal 15 │
│ The system is down.                                      │
│ Changing to init state s - please wait                   │
│ #                                                        │
│ INIT: New run level: S                                   │
│                                                          │
│ INIT: SINGLE USER MODE                                   │
│                                                          │
│ Type Ctrl-d to proceed with normal startup,              │
│ (or give root password for system maintenance):          │
│ Entering System Maintenance Mode                         │
│                                                          │
│ SU 1/16 20:46 + syscon root-root                         │
│ # ▊                                                      │
└──────────────────────────────────────────────────────────┘
```

Using shutdown *with no command line options.*

➠   *NOTE 1: The* /usr/sbin/shutdown *command can only be executed with root access.*

➠   *NOTE 2: In order to ensure proper operation of the* shutdown *command, the operator should first exit any windowing environment and change directory to the / (root) directory.*

The *shutdown* command recognizes several flags. The use of the *shutdown* command with command line modifiers is shown in the illustration that follows.

- *-y* tells the system to pre-answer all confirmations with a yes.

- *-g N* tells the system to execute the shutdown in *N* seconds. The default is 60 seconds.

- *-i N* tells the system to change to the *init* state specified by *N*.

✓ *TIP: It is always a good idea to warn users in advance of a shutdown. This allows them time to save files and log off, avoiding the risk of file damage or loss. If a grace period of 900 seconds is used, the users will be notified with a 15-minute warning of the impending shutdown.*

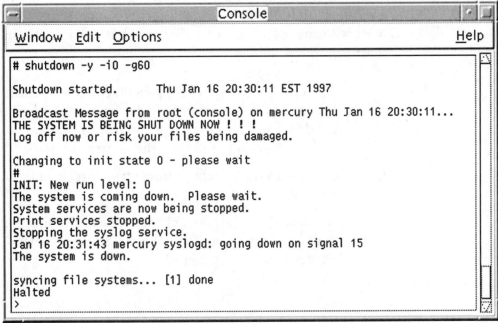

```
# shutdown -y -i0 -g60

Shutdown started.       Thu Jan 16 20:30:11 EST 1997

Broadcast Message from root (console) on mercury Thu Jan 16 20:30:11...
THE SYSTEM IS BEING SHUT DOWN NOW ! ! !
Log off now or risk your files being damaged.

Changing to init state 0 - please wait
#
INIT: New run level: 0
The system is coming down.  Please wait.
System services are now being stopped.
Print services stopped.
Stopping the syslog service.
Jan 16 20:31:43 mercury syslogd: going down on signal 15
The system is down.

syncing file systems... [1] done
Halted
>
```

Using shutdown *with command line modifiers.*

init 0 Command

Another method for bringing the system to a different run level is through the *init* command. As with *shutdown*, *root* access is required to execute this command. The *init* command does not send warning messages to users. Using the *init* command to change the system run level is best reserved for instances in which nothing else has worked.

✓   *TIP: Some system administrators prefer* init 0 *over the* halt *command because* init 0 *will execute the scripts in* /etc/rc?.d. *These scripts cleanly shut down databases and other systems before the system is halted.*

The *telinit* program is a link to the */sbin/init* program, and allows a single-character argument. This argument is the alphanumeric identifier of one of the *init* states. In addition to the *init* state identifiers understood by *shutdown*, *telinit* understands the following arguments. The use of *telinit* for shutdown is shown in the illustration that follows.

- *a,b,c*: The *a,b,c init* states are pseudo run levels. Using one of these states in the */etc/inittab* file or with the *telinit* command tells the system to run commands that are tagged as level *a,b,c* but do not change the system run state.

- *q,Q:* Causes the system to immediately reexamine the */etc/inittab* file.

- *s,S:* Causes the system to change *state* to the single-user mode. The terminal that issues the *telinit* command becomes the system console.

Using telinit _to bring the system down._

halt Command

Sometimes it is not feasible or necessary to warn users of an impending shutdown. The _/usr/sbin/halt_ command initiates an orderly shutdown of the operating system, and does not send warning messages to users. The _halt_ command will not execute scripts in _/etc/rc?.d_, nor does it leave the system in _init_ state _s_ (single user). The use of _halt_ without flags is shown in the following illustration.

Using halt _without flags._

As with _shutdown_ and _init_, root access is required to execute the _halt_ command. The _halt_ command recognizes the following four flags, which control how the system is shut down. The illustration that follows shows the use of _halt_ with command line flags.

- *-l*: Forces *halt* to *not* inform *syslog* of the system halt.

- *-n*: Forces *halt* to *not* write (sync) the memory buffers to disk before halting.

- *-q*: Forces *halt* to perform a rapid halt. This flag will also cause the system to *not* sync the memory buffers to disk, and may cause some processes to exit less than gracefully.

- *-y*: Forces *halt* to halt the system, even if the terminal issuing the *halt* command is a dial-up line.

Using halt *with command line flags.*

✗   WARNING: *Halting a system without writing the memory buffers to disk can be dangerous to the file system! The* halt -n *method is generally used only if memory content is known to be corrupt. Consequently, executing a* sync *before halting is not desirable.*

reboot Command

The *reboot* command is another method for bringing the system down. This command causes the system to shut down and reboot to the default state specified in the */etc/inittab* file. The *reboot* command does not execute the *shutdown* scripts in */etc/rc?.d*. The following four flags are recognized by the *reboot* command.

- *-d*: Causes the reboot process to generate a core dump. This option is provided for compatibility, but is not supported.

- *-l*: Causes the suppression of notification to the *syslog* daemon.

- *-n*: Causes the system to halt and then reboot without syncing the memory buffers to disk.

- *-q*: Causes the system to ungracefully halt and reboot without syncing the memory buffers to disk.

✗   WARNING: *Halting a system without writing the memory buffers to disk can be dangerous to the file system! The* reboot -n *method is generally used only if memory content is known to be corrupt. Consequently, executing a* sync *before halting is not desirable.*

Results of using the *reboot* command to bring the system down and back up follow.

```
Booting from: sd(0,0,0)vmunix
Sun0S Release 5.8 Version Beta [UNIX(R) System V Release 4.0]
Copyright (c) 1983-1999, Sun Microsystems, Inc.
configuring network interfaces: hme0.
Hostname: redstone
The system is coming up. Please wait.
checking ufs filesystems
/dev/rdsk/c0t2d0s86: is clean.
/dev/rdsk/c0t3d0s1: is clean.
/dev/rdsk/c0t2d0s3: is clean.
starting routing daemon.
starting rpc services: rpcbind keyserv done.
Setting netmask of hme0 to 255.255.0.0
Setting default interface for multicast: add net 224.0.0.0: gateway red-
    stone syslog service starting.
Print services started.
volume management starting.
Tue Jan 28 14:20:49 2000
The system is ready.
redstone console login:
```

Stop-A Command

Sometimes a system will lock up, making the use of the commands previously described impossible. When all else fails, the system can usually be stopped by holding down the <STOP> key (upper left corner of the keyboard) while simultaneously pressing the <A> key. This sequence is known as *Stop-A* (or Lı-A). The key sequence stops the system processor, and puts the system under control of the PROM monitor.

```
# STOP-A pressed
Abort at PC OxF0031870.
>
```

Once the system has been stopped in this manner, it is best to type the *sync* command to the PROM monitor. This will cause the monitor to sync the memory buffers to disk, and then crash the system. This action will force a system crash dump to be saved for post-mortem inspection. It may be possible to determine the reason the system was locked up by examining the crash dump information. Consult the manual page on *crash(1M)* for more information on crash dumps. Using the PROM monitor to cause a crash dump is demonstrated in the following.

```
>sync
panic: zero
syncing file systems ...2 2 2 2 2 2 2 done.
 586 static and sysmap kernel pages
  34 dynamic kernel data pages
 179 kernel-pageable pages
   0 segkmap kernel pages
   0 segvn kernel pages
 799 total pages (799 chunks)
dumping to vp ff1b9004, offset 181288
117 pages left    109 pages left  101 pages left
5 pages left
```

```
799 total pages, dump succeeded
rebooting...
```

Summary

. . . .

Modifying the system boot process is an important means of tailoring Solaris to a unique environment. Certain software products and procedures may be started or performed at system boot by adding them to the boot sequence. Through the run control files and their associated directories, new boot procedures can easily be added.

By using Bourne shell scripts, a new boot procedure can conditionally handle a variety of situations, and is easily developed and tested without the need to repeatedly boot a system. The preferred method of changing the system run level is to use the *shutdown* command. When this is not possible, there are other options available that allow the operator to change the run level of the system.

# CHAPTER 6

Creating, Deleting, and Managing User Accounts

Solaris is a multi-user operating environment. Not only can multiple windows and processes be running on a Solaris system, but multiple people can use such a system simultaneously. Unlike a single-user system, such as a PC or Macintosh, Solaris needs to keep track of each user's preferences and the location of each user's files and electronic mail. Each user must also have some way of identifying and authenticating herself to Solaris.

A user account is the name given to the collection of information that defines a user on a Solaris system. This information is employed when a user logs into the system, sends and receives electronic mail, and starts up OpenWindows. The maintenance of user account information is an ongoing task for the system manager. Solaris provides numerous tools for making user account maintenance easier.

User Account

A user account is a collection of information that defines a user on a Solaris system. The definition starts with the user's log-in name, or user name, which is used when logging in and sending or receiving electronic mail. This definition contains various pieces of information used to control access to files, as well as a password used for authenticating the user. The following are the components of a user account.

- A name by which the user is known to the system.

- A set of credentials consisting of a user identification number, group identification(s), and access lists that control ownership and access to directories, files, and devices (such as printers and tape drives).

- The location of a home directory, where basic user files such as *.login* and *.cshrc* are stored.

- The name of a command interpreter, called a shell in UNIX terminology. This is the program run when the user logs in. Different users may have different shells.

- An optional disk space usage limit or quota.

- Optional information such as a password "age" to force the account's owner to periodically change his/ her password.

An account can be considered analogous to the collections of items in a wallet or purse, the concept of which is shown in the following illustration. The driver's license number is like a user name. The picture on the driver's license allows other people to verify who the license holder is in much the same way the password allows Solaris to verify the user's log-in name. A club membership card grants the card holder certain access to shared resources in the same way a UNIX group allows a user shared access. Spending limits on charge cards are similar to the usage quotas some systems place on disk usage.

License and credit card analogy.

admintool

- -

Managing the collection of user account information on earlier UNIX systems required editing several files and using a couple of different commands. Under Solaris, account management is brought together under a single tool, *admintool*.

In large networks, consider installing and using Solstice Admin-Suite by Sun Microsystems. This product has an interface similar to that provided by *admintool*, but adds the ability to manage accounts across multiple systems from a single location.

➥ *NOTE: Solaris 2.5, 2.5.1, 2.6, and 7 have a serious security vulnerability in the service daemon used by the Solstice AdminSuite, sadmind. Be sure that if you use Solstice AdminSuite that you are using the current*

set of patches that addresses this vulnerability. See http://
www.cert.org/advisories/CA-99-16-sadmind.html *or* http://
www.ciac.org/ciac/bulletins/k-013.shtml *for details.*

Overview of admintool

The *admintool* is a jumping-off station for accessing a collection of
administrative functions. These functions include user account
management; editing and updating various system databases; setting up printers, modems, terminals, and other devices; and managing host names and software packages. The *admintool* is
discussed throughout this book as each of these functions is
addressed. This section focuses on editing the group database file
and using the user account management functions.

Using admintool to Manage User Accounts

To begin working with the *admintool*, log in as *root* and start
OpenWindows, or use the *su* command to gain *root* privileges
from a terminal window if OpenWindows is already running.
Next, type *admintool &* at the *root* prompt (#) to start the *admintool* and put the process in the background. The base *admintool*
window is shown in the following illustration. This window presents a list of user accounts that are ready to be added to, deleted
from, or edited. Other *admintool* functions are accessed via the
Browse menu.

Before diving into setting up an account, there are a number of
housekeeping chores that should be taken care of and policy decisions to be made. Accomplishing these tasks now will speed up
account creation and lay the foundation for easier management
of accounts in the future.

```
┌─────────────────────────────────────────────────────────────┐
│ ─              Admintool: Users               · □│
│  File  Edit  Browse                                    Help  │
│ ┌───────────────────────────────────────────────────────────┐│
│  User Name         User ID    Comment                        │
│ ┌──────────────────────────────────────────────────────────┐ │
│ │adm                    4     Admin                        │ │
│ │bin                    2                                   │ │
│ │daemon                 1                                   │ │
│ │listen                37     Network Admin                 │ │
│ │lp                    71     Line Printer Admin            │ │
│ │noaccess           60002     No Access User                │ │
│ │nobody             60001     Nobody                        │ │
│ │nobody4            65534     SunOS 4.x Nobody              │ │
│ │nuucp                  9     uucp Admin                    │ │
│ │root                   0     Super-User                    │ │
│ │smtp                   0     Mail Daemon User              │ │
│ │sys                    3                                   │ │
│ │uucp                   5     uucp Admin                    │ │
│ └──────────────────────────────────────────────────────────┘ │
│                              Host: glenn.net-kitchen.com     │
└─────────────────────────────────────────────────────────────┘
```

The admintool *opening screen.*

Names, UIDs, and GIDs

Each user account must be assigned a name. This name is used to identify the account for purposes such as electronic mail and logging into a machine. Although just about any string of two to eight characters can be used for an account name, many sites prefer to use a systematic naming convention, such as the initials of the person who owns the account (e.g., *ddm*), the owner's last name (e.g., *mckay*) or first name (e.g., *dwight*), or a combination of initials and names (e.g., *ddmckay*, or *dwightm*).

A number can be added if there is more than one user with the same name or initials (e.g., *ddm2*). This sort of naming scheme makes it easy for people to recognize a user's account name when sending electronic mail, transferring files, and so on. For example, for a person whose name is John Doe, the system administrator can construct account names such as *doej, johnd, jd, jdoe, doe, john,* and *jd1*.

User IDs

Along with a name, each account is assigned a user identification number, or UID. Solaris, like other versions of UNIX, uses the UID number to tag files owned by the account and to control access permissions to files and devices. The first 100 UID numbers (1 through 100) are reserved by the Solaris system to be assigned to system accounts such as *root* and *uucp*. Likewise, UIDs above 60000 are assigned to special purposes. However, assigning a UID is a bit more complicated than simply picking a number between 100 and 60000.

- UID numbers *must* be unique for each user account. File ownership is recorded and controlled by the UID number. Accounts with the same UID number are effectively the same in regard to file ownership and other owned items such as processes and accounting records.

- In order for NFS file sharing to work correctly, UID numbers *must* be consistent between machines on a network. NFS uses the UID number when determining if a given user has access to a given file.

- ➥   *NOTE: Avoid reusing UID numbers whenever possible. The UID number is kept in several data files, all of which must be carefully checked and cleaned up before a UID can be reused. Reuse occurs when a UID number from a deleted account is assigned to a new account. One example of this is the disk quota database that stores a user's disk usage limits by UID number. Reusing a UID can result in an incorrect disk quota being assigned to a user account.*

Group IDs

A second number, called a *group identification number* (GID), is also assigned to each account. Unlike UID numbers, multiple users can share GID numbers. Groups and respective GID numbers are used to provide shared file access. Like UID numbers, Solaris also tags files with a GID number to allow access to other users who share the same GID number. In addition, the first 100 GID numbers (1 through 100) are reserved by the Solaris system to be assigned to special uses. GIDs must also be consistent between machines that share file systems via NFS. GIDs are an important and powerful mechanism in Solaris for controlling file sharing.

Upon examining a long file listing produced by *ls -l*, you will see that each file listed has both an owner and a group associated with it. There are separate permission bits for reading, writing, and execution associated with the file's owner, group, and everyone else. These bits can be set to make files readable or writable by members of a group, while remaining inaccessible to other user accounts. The GID given to a user's account specifies the *primary* group a user is in. A user can also be made a member of up to 15 secondary groups (to make a total of 16 groups) by using the *admintool* to add that user's account name to a group listed in the group file.

Devising a scheme for grouping accounts before creating new accounts is highly recommended. Consider the ways in which the various user accounts need to be able to work together. Do groups exist for each project? What about functional groupings such as accounting, engineering, or clerical? Creating groups first and then assigning accounts to groups makes file sharing and resource management easier later on. Consider the following guidelines, which are similar to those for UID numbers.

- GID numbers *must* be unique for each group.

- GID numbers *must* be consistent between machines on a network in order for NFS file sharing to work correctly.

- *NOTE: Avoid reuse of GID numbers as well. Although the GID number is kept in fewer data files than the UID number, the system admin-*

istrator should nevertheless carefully check for potential problems before reusing a GID number.

✓   *TIP: Consistency in UID and GID numbering can be easily maintained by using NIS+ to manage the various account database files. See Chapter 21, "Network Name Services," for details.*

Creating a Group

To create a group, select Groups from the Browse menu in the Admintool window. This brings up the Group Database window, shown in the following illustration.

Group Name	Group ID	Members
adm	4	root,adm,daemon
bin	2	root,bin,daemon
daemon	12	root,daemon
lp	8	root,lp,adm
mail	6	root
noaccess	60002	
nobody	60001	
nogroup	65534	
nuucp	9	root,nuucp
other	1	
root	0	root
staff	10	
sys	3	root,bin,sys,adm

Host: glenn.net-kitchen.com

Group Database window.

As an example, assume that a group of users will be using an accounting software package. They need permission to access the package's data, but that data should be protected from access by other users. To meet this need, create a group named *account* by selecting Add from the Edit menu. A small window containing a form appears. Fill in the Add Group form, shown in the following illustration, to specify the group name and identification number.

Add Group form.

If additional groups are required, the process can be repeated. The *admintool* has the very helpful ability to add groups and user accounts in any order at any time, making it unnecessary to define the entire grouping and account naming and numbering scheme at the outset. Accounts and groups can be added as needed.

Preconfiguring with Skeleton Files

When a user logs into her account, a command interpreter or shell is started. The setup of the shell, how it behaves, where it looks for programs, and so forth can be modified to fit a particular environment, and automatically configured for each user account when it is created. As with account naming and UID and GID numbering, planning and preparation prior to setting up new accounts will save time later on.

Solaris provides two levels of shell customization: a system-wide set of files read by every shell started by every user, and a set of files that can be copied into the user's home directory when an account is created. The global files are */etc/.login* for the C shell, and */etc/profile* for the Bourne and Korn shells. The global files should contain only the settings needed for *all* shells to access when they start. These global initialization files have the advantage of not being modifiable by individual users. However, any initialization files the user creates will override the shell settings given by the global files.

In */etc/skel*, the system administrator can place the initialization files, which will be copied into a user account's home directory when the user account is created. *Admintool* copies the files */etc/skel/local.cshrc* and */etc/skel/local.login* into a user's home directory as *.login* and *.cshrc*, respectively, if the user is assigned */bin/csh* as his shell. If the user is assigned */bin/sh*, */etc/skel/local.profile* is copied to his home directory as *.profile*.

The advantage of skeleton files over global initialization files is that skeleton files are given to the user and can be modified by the user if and when necessary. The advantage to the system manager is that each user is given a predefined set of shell initialization files to ensure that items such as user search paths are set correctly, without requiring users to create initialization files from scratch.

Sun provides sample initialization files for Bourne, Korn, and C shells in */etc/skel*, which serve as a basis for customized files for a particular site. In version 2.3 and later of Solaris, these initialization files include useful start-up procedures for OpenWindows, as well as search path settings for commonly used Solaris programs and tools.

Creating an Account

With the preparatory work in place, select Add User from the Edit menu in *admintool* to begin adding a new user account. The Add User form, shown in the following illustration, is divided into three sections, each dealing with a different aspect of the account to be created. Each section is discussed in subsequent sections as the form is filled in to create an account.

Add User form.

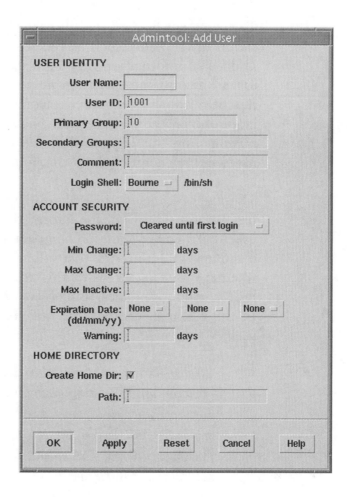

User Identity

The User Identity section contains the basic name and number information for the account. The User Name field is limited to eight characters. The User ID is the number discussed earlier that identifies this user. The primary group is the group owner-ship given by default to files created by this user. In the example, the *account* group—created earlier using *admintool*—will be used. The Secondary Groups are other groups in which this account is a member. Under Solaris, a user can be a member of as many as 16 groups.

The comment field is typically used to contain additional identification information, such as the real name of the account user. Additional information may be placed in this field, such as the user's office phone extension number. Be aware that all data in this field will show up when anyone uses the *finger* command to inquire about the user account. The administrator can hide information from this command by inserting a comma between the name and any additional information.

The *finger* command will only display the information up to the first comma. However, the information will still be visible in the user's entry in */etc/passwd*. The Login Shell menu selects the command interpreter or shell that will be started by the log-in process when the user logs in using this account. Every shell has advantages and disadvantages; some users prefer one shell over another. In the example, the account will be given the default Bourne shell.

✗   *WARNING: Be very careful when choosing whether to change the shell of the root user. The setting provided with Solaris is chosen to allow for system operation even if problems such as disk failures occur, which may make other shells inaccessible. It is possible to render a system unusable by changing the root shell without first ensuring that the shell is available under all possible conditions.*

✓   *TIP: It is often useful to have a root window available while experimenting with changes to the root shell and/or system security. If something goes wrong, root access to the system is still available to carry out repairs.*

Account Security

The Account Security section of the Add User form deals with the password used by the account user to authenticate herself to the system. The password itself can be set in the following four different states via the Add User form.

- *Cleared until first log-in*: This is the default setting. An account created in this manner will prompt the user to input a password the first time she/he logs in to the account. Although this is handy, consider the implications of setting an account's password to this state, especially in an environment where the account may not be used and is likely to be discovered by someone other than its owner. See the discussions of system security in Chapter 7, "Managing System Security," and Chapter 18, "Network Security."

- *Account is locked*: In this state, the account exists but cannot be logged into until and unless a password is set for the account. An account can be put in this state automatically, as described in material that follows.

- *No passwd—set uid only*: This setting creates an account that cannot be logged into. It is used when an account is needed for a daemon or system program to use. An example of this is the *uucp* account, which allows the *uucp* program to own files and directories required for that program to function.

- *Normal Password*: This selection brings up a small form in which a password can be assigned to the account.

The three date fields that follow the Password menu control a function called "password aging." The aging process allows the system manager to enforce the practice of changing account passwords on a regular basis. Reusable passwords, such as those found in Solaris, are vulnerable to being discovered if they remain unchanged for long periods. Changing the password periodically helps reduce this risk.

The downside to password aging is the psychological factor. Some users dislike changing passwords. Being asked to change with no warning may contribute to a user choosing a simpler, easily guessed password, or the user may simply enter a new password and then change back to the old password immediately afterward. Password aging is most effective when the account user understands the reasons for periodically changing a password, the elements of a good password, and is given a chance to choose a good password.

The Solaris password aging system gives the system administrator control over three aspects of the process. The Min Change field specifies the minimum time in days between password changes. Setting this to a value other than zero prevents a user from setting a new password and then immediately switching back to the old password until the time specified has passed. The Max Change field specifies the age the password must reach before the password aging system requests that the user change the password. The Warning field, at the bottom of this section, sets the number of days a user will be notified prior to being required to change the password. This period gives users time to think of a new password before they are required to make the change.

For further protection, Solaris adds an inactivity timer and an expiration date. These features help to prevent inactive accounts or improper use of accounts owned by people who may no longer use them. The Max Inactive field sets the maximum time in days an account can be left idle before it is locked. The expiration date sets a date at which the account is considered expired and is locked. These mechanisms are especially handy for accounts transient in nature, such as student accounts in an academic environment and accounts created for temporary employees.

Home Directory

The Home Directory section allows the option of having the User Account Manager automatically create and populate the user's home directory as part of account creation. This option provides considerable time savings and is worth examining in detail.

The Path field contains the UNIX directory path to the user's home directory. Typically the path is something like */export/home/ smithb*, as it would be in the case of the user account being set up in this example. The Create Home Dir check box tells the *admin-tool* to create the user's home directory. Uncheck the box if the user's home directory is being provided by another machine via NFS or has otherwise already been created. A completed Add User form is shown in the following illustration.

Completed Add User form.

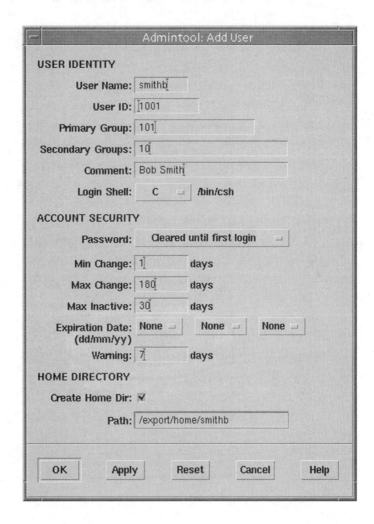

Modifying and Deleting Accounts

With the foundation work required to set up accounts completed, using the *admintool* to modify and delete accounts becomes very easy. Modifying an account is simply a matter of opening the *admintool* as before and double clicking on a user name in the list or selecting Modify from the Edit menu. The selection of an account is shown in the following illustration.

Selecting an account to be modified.

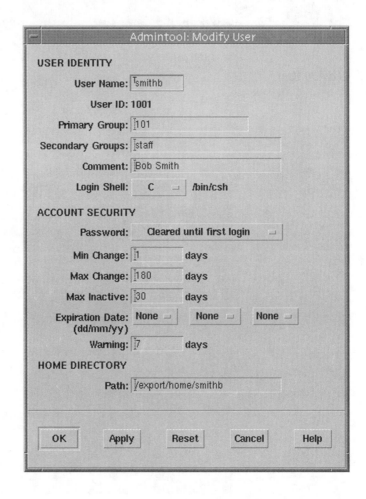

All fields filled in during account creation can now be edited and changed. Common activities include locking and unlocking accounts, and changing an account's password using the Password menu. Selecting Normal Password will access a small form in which a new password can be entered.

However, the Modify User form does not completely handle all changes the administrator might want to make. Watch out for changes to the fields listed in table 6-1, which follows.

Table 6-1: Password Fields to Watch Out For

Field Name	Description
User ID	Although the Modify User form will change the UID number for the user in the /etc/passwd file or in the NIS+ database, it will not change the ownership of the files in the account's home directory or elsewhere (e.g., / tmp, /var/ mail). This must be done manually, as described in material that follows.
Primary Group	The same rule applies for the primary group. The /etc/passwd or NIS+ database files will be changed, but not the group ownership assigned to files in the account's home directory or elsewhere (e.g., /tmp, /var/mail).
Path	The user's home directory path will be changed only in the /etc/passwd and NIS+ database files. The directory itself must be moved manually *before* changing the account form, or the account form will report an error because it will not be able to find the new path.

Changing File Ownership

If the UID or GID numbers assigned to an account are modified, the ownership of the files and directories in the account's home directory must be modified manually so that the account will continue to have access to those files and directories. This is easy to accomplish with the *chown* and *chgrp* commands. These two commands work in a similar manner: both require the name or number of the UID or GID to change the name of a file. Common usage when changing the UID or GID number for an account is to use the *-R* option to *chown* and *chgrp* to recursively change the ownership of all files in a sub-tree.

- The *chown* command sets the file ownership by changing the UID number the file is tagged with. The UID can be specified as a number or as a user name. If the user name is specified, *chown* will search for the number in the password file. The *-R* option will recursively descend through any directories it finds in the list of files given it, and change the ownership of the directory and the files under the directory. Examples of typical *chown* commands follow. No output is produced unless there is an error.

```
chown dwight file1 file2 file3
chown -R dwight big-directory
```

- The *chgrp* command sets the file group ownership by changing the GID number the file is tagged with. As with *chown*, the GID can be specified as a number or as a group name. If the group name is used, *chgrp* will search for the number. This command also uses the *-R* option to recursively descend through directories and change the group ownership of all files and directories found. Typical *chgrp* commands also produce no output unless there is an error. Examples follow.

```
chgrp 105 file
chgrp system file4 file9
chgrp -R accounting secret-directory
```

✓   *TIP: A common mistake is to specify .* as the list of files to be changed when using* chown -R *or* chgrp -R, *as in* chown -R dwight .*. *Unfortunately, the .* expression will match not only the files beginning with a period (.) in the current directory, such as .cshrc or .login, but also the directory above the current directory as referenced by its nickname, two periods (..). To avoid this, use a more restrictive expression such as .??*, which will match any file that begins with a period (.) and* *has at least two characters after the period. As always, be very careful using wildcard expressions such as the asterisk (*) while working as the root user.*

File Ownership Change Example

For example, if the UID and GID numbers were changed to 1002 and 1000 for the *smithb* account, you might use the sequence of commands shown in the following illustration to change the user and group ownerships of the files in the *smithb* account home directory.

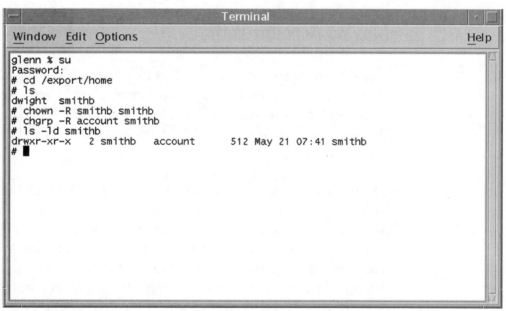

Changing user and group ownerships.

The sequence of commands used in the previous illustration follows: (1) The *su* command is used to become *root*. The *root* account has permission to change the ownership of any file. (2) The current directory is changed to the directory *above* the account's home directory. This positions the shell to simplify the next two commands. (3) The *chown* and *chgrp* commands are used to change the UID and then the GID numbers assigned to the files in the *smithb* directory. Directories contained within the *smithb* directory can be changed by using the -R option flag with the same command.

➥   *NOTE: If the* smithb *account's UID and GID have already been changed using the User Account Manager, you can use the account name,* smithb, *in place of the UID number, and the group name,* account, *in place of the GID number. Both the* chown *and* chgrp *commands will search for the numbers from the group and password databases.*

Moving Account Paths

Moving an account's home directory from one location to another on the same server or between servers requires a bit more work than setting UID and GID numbers. With the use of the *tar* command, and some shell magic, directories can be moved without resorting to tapes or temporary files.

The basic sequence is to create a *tar* archive of the files and directories in the current location, and by means of a pipe (|) or connection between processes, send that archive to a second invocation of *tar* that extracts the archive in the new location. The procedure appeared in Chapter 4, "Fundamental System Administration Tools," but bears repeating and expanding upon here. In the following illustration, the *smithb* directory is moved to another file system on the same server.

✓   *TIP: Although you could use the* cp *command to move user accounts, using* tar *provides a major advantage. The* tar *command preserves symbolic links and file ownerships, whereas* cp *does not.*

```
┌─────────────────────────────────────────────────────────────────────┐
│ ─                             Terminal                          · □ │
├─────────────────────────────────────────────────────────────────────┤
│ Window  Edit  Options                                           Help │
├─────────────────────────────────────────────────────────────────────┤
│ # cd /export/home                                                    │
│ # ls                                                                 │
│ dwight  smithb                                                       │
│ # tar cf - ./smithb | ( cd /export/home2 ; tar xvBpf - )            │
│ tar: blocksize = 8                                                   │
│ x ./smithb, 0 bytes, 0 tape blocks                                   │
│ x ./smithb/.login, 575 bytes, 2 tape blocks                          │
│ x ./smithb/.cshrc, 124 bytes, 1 tape blocks                          │
│ # cd /export/home2                                                   │
│ # ls                                                                 │
│ smithb                                                               │
│ # ▮                                                                  │
│                                                                      │
│                                                                      │
│                                                                      │
│                                                                      │
│                                                                      │
└─────────────────────────────────────────────────────────────────────┘
```

Moving an account's home directory.

The Process Continued

After moving to one of the home file systems on this machine, a long command is issued consisting of several parts. The first part has the *tar* command creating an archive file of the *smithb* directory and putting it on the standard output. The *./smithb* path is used because it is a relative path, which will allow the archive created by *tar* to be extracted in any location. Along the line to the right is a pipe (|), or the shell pipe character. The pipe takes the standard output of the *tar* command and routes it into the next command to the right in the command line.

The two commands inside the parentheses are executed in a new shell created for them. They act as a single command and will read from the pipe. The two commands move to the new location for this path (*cd*) and then carefully extract the archive file coming in from the pipe (*tar*). The options used with the second *tar* command cause it to extract the archive (*x*), maintain the same file ownership and permission bits (*p*), continue to read even if the archive file is slow in coming through the pipe (*B*), and produce a verbose list of the files as they are extracted from the archive (*v*).

Once this operation is executed, modify the account's path using the User Account Manager and then delete the old directory. However, prior to removing the original path, verifying that the account can be logged into and that no mistakes occurred is good practice. A few minutes of checking can save hours of work recovering files and directories from backup tapes.

Moving a Home Directory between Servers

Moving an account's home directory between servers uses the same general principle of creating a *tar* archive file and using a pipe to send it to another *tar* command to extract the content. Appearing in the following illustration is an example of moving the *smitha* account from the server named *glenn* to the server named *grissom*.

The change of directory and the first *tar* command used are the same as in the previous example. On the right side of the pipe, the *rsh* command is used to execute commands on the *grissom* server instead of creating a second shell on the local server. The \; is used to prevent the semicolon from being processed on the local machine. This allows two commands to be passed to the remote machine via *rsh*: one to change to the appropriate directory, and the other to extract the archive file coming through the pipe, using *tar*.

```
┌─────────────────────────────────────────────────────────────────┐
│ ─                          Terminal                        · □   │
├─────────────────────────────────────────────────────────────────┤
│ Window  Edit  Options                                     Help   │
├─────────────────────────────────────────────────────────────────┤
│ # cd /export/home                                                │
│ # ls                                                             │
│ dwight   smithb                                                  │
│ # tar cf - ./smithb | rsh grissom cd /export/home2 \; tar xvBpf -│
│ tar: blocksize = 8                                               │
│ x ./smithb, 0 bytes, 0 tape blocks                               │
│ x ./smithb/.login, 575 bytes, 2 tape blocks                     │
│ x ./smithb/.cshrc, 124 bytes, 1 tape blocks                     │
│ # rsh grissom ls /export/home2                                   │
│ smithb                                                           │
│ # ▮                                                              │
│                                                                  │
└─────────────────────────────────────────────────────────────────┘
```

Moving a home directory between servers.

Compared to creating or modifying an account, deleting an account is simple. Just click the account to be deleted from the User Account Manager window and select Delete User from the Edit menu. The Delete User form that appears presents the option of having the account manager delete the account's home directory, or it can be manually deleted by the system administrator. Depending on the particular environment, it is often desirable to make a backup copy of the content of any account before deleting it, just in case.

Managing Accounts without admintool

On systems lacking access to OpenWindows, tools such as *admintool* are not available. In these cases, such as file servers without any graphics displays, you must resort to the *useradd*, *usermod*, and *userdel* commands for account management; *groupadd*, *groupmod*, and *groupdel* for group management; and *passwd* or the *vipw* command provided in the Berkeley UNIX compatibility package to control passwords.

The basic strategy for account management is the same as that previously presented for use with the *admintool*. The system administrator should carry out preparatory work to decide on an account naming scheme, UID numbers, required groups, and GID numbers. Then, local modifications to the shell initialization files must be made. With these foundations in place, groups and accounts can be created.

groupadd, groupmod, and groupdel

This trio of commands works on the group database. Unlike the *admintool*, the command line group management commands *do not* update the NIS+ databases. These databases must be updated manually, as described later in this book.

Adding a group is accomplished by specifying the group name and GID number as options to the *groupadd* command. The following command line creates the *account* group.

```
/usr/sbin/groupadd -g 1000 account
```

Group modification is executed by using *groupmod*. Changing the *account* group GID to 1234 could be accomplished with the following.

```
/usr/sbin/groupmod -g 1234 account
```

→   *NOTE: The* groupmod *command changes the GID number in the /etc/ group file only. File ownership must be changed manually using the* chgrp *command, as shown in the examples in the preceding section.*

Similarly, groups are deleted using the *groupdel* command. Deleting the *account* group would be accomplished as follows.

```
/usr/sbin/groupdel account
```

useradd, usermod, and userdel

This trio of commands performs the same functions as the User Account Manager via a series of command line options similar to the *groupadd* family. In the following string, the *smithb* account is created with the *useradd* command in the same way an account was created with the User Account Manager.

```
/usr/sbin/useradd -u 1000 -g account -c "Bob W. Smith" \
-s /bin/sh -m -d /export/home/smithb smithb
```

This command is lengthy, but easy to follow. As described in table 6-2, which follows, the individual option flags parallel the four sections of the Add User form.

Table 6-2: Option Flags for the useradd Command

Option	Description
-u 1000	Sets the UID number for *account*.
-g account	Sets the primary group.
-c Bob W. Smith	Sets the comment field.
-s /bin/sh	Sets the account shell to the Bourne shell.
-m	Creates the home directory.
-d /export/home/smithb	Sets the path to the account's home directory.
Smithb	Account user name.

➥   *NOTE: The* useradd *command copies all files from the skeleton direc-*
      tory /etc/skel, *but it does not rename the files to their proper names*
      (e.g., local.cshrc *to* .cshrc), *as the* admintool *does. One workaround*
      option is to manually rename each file. Another workaround would be
      to set up a separate directory for each shell containing the dot file (e.g.,
      / etc/skel/csh/.cshrc) and use the -k *option of* useradd *to tell it to*
      copy the dot files from the specified directory.

The *useradd* command also has options to set the account's expiration date (-*e*) and the maximum number of days the account can
be idle before being locked (-*f*). This command does not have
options for setting the password aging time period. In addition,
the *useradd* command does not handle setting the user's password.

The *usermod* command uses the same set of option flags as *useradd*. An example of changing the comment field for *smitha* via
usermod follows.

```
/usr/sbin/usermod -c "William W. Smith" smitha
```

Finally, the *userdel* command simply takes the name of the
account to be deleted, as in the following.

```
/usr/sbin/userdel smitha
```

passwd

To set a user's password from the shell, use the *passwd* command
from the *root* account and give the user account's name as the
first argument, as in the following.

```
passwd smitha
```

The *passwd* command is also used to control the password aging
time limits. The -*n* option sets the minimum number of days

between password changes. The *-x* option sets the maximum number of days between password changes. The *-w* sets the number of days before the expiration date that the user will be warned about password expiration. The command line for setting these values for the example user account follows.

```
passwd -n 2 -x 180 -w 7 smitha
```

As demonstrated in the following, the password aging feature can be turned off by using *-1* for the maximum age.

```
passwd -x -1 smitha
```

The *passwd* command also has the ability to give the *root* user password status information. The *-s* option prints a single line of information on the account listed. The line contains the account name, status (*NP* = no password, *LK* = locked, *PS* = normal), the date of the last password change, and the minimum, maximum, and warning password aging values, as in the following.

```
# passwd —s smitha
smitha PS 7/1/94 2 180 7
```

You can obtain a list of the password status for all accounts on a system by using the *-a* option along with the *-s* option, as in the following.

```
# passwd -s -a
root PS
daemon LK
dwight PS
smitha PS 7/1/94 180 7
 (...)
```

/etc/shadow

The password for an account is not stored in the */etc/passwd* file because the file is typically *world* readable. A simple method of breaking system security is to obtain a copy of the encrypted passwords, and then run a program to crack the encryption by guessing passwords, encrypting them, and comparing the results.

To provide more secure passwords, Solaris stores the encrypted passwords in the */etc/shadow* file. This file is readable by the *root* account only. All log-in functions use password retrieval routines that know to check the */etc/shadow* file for the password. By keeping the passwords in a locked file, system crackers cannot easily obtain a copy of encrypted passwords, and system security is maintained.

For Experts Only: vipw

If the Berkeley UNIX compatibility package is installed, the *vipw* command can be used to add, modify, and delete account records from the */etc/passwd* file. This command simply allows the administrator to directly edit this data file using an editor, typically the *vi* editor, from which the command gets its name.

Direct editing of the password file is prone to errors. One must meticulously adhere to the file format and be careful not to leave incomplete or malformed entries in the file. The *vipw* command performs no checking other than to verify that the *root* account is sufficiently correct to allow *root* to log in. Incorrect entries in the password file can prevent users from logging in to their accounts. Unless you are accustomed to this type of file editing from working on other UNIX systems, stay away from *vipw* and use the commands previously described. These commands are designed to prevent errors.

Summary

An account is a collection of identification, authentication, and permission information stored in data files or an NIS+ database. By using OpenWindows, the *admintool* provides an easy-to-use graphical series of forms for adding, modifying, and deleting user accounts. However, attention must be devoted to preparing for new accounts and handling situations not covered by the *admintool*. If OpenWindows is not installed on the system being managed, you can use a series of command line tools that cover nearly all functions found in the *admintool*.

Summary

Managing System Security

System security is probably the most talked about topic in the computer industry today. Workstations and servers running Solaris present a variety of security challenges for the system manager. It is essential to understand the level of security required by a particular site and what must be done to establish that security. Maintaining the appropriate level of security for a particular environment requires a balance between technical considerations and the needs of the organization.

Information Sources

In this chapter, the discussion is limited to basic security concepts and tools for a single system. This includes file and directory security, user account security, the *aset* and *tripwire* security tools, and the tools used to install patches or fixes for bugs found in Solaris. Chapter 17 builds on these concepts and tools and covers issues of system security in a networked environment. Taken together, these two chapters serve as a primer on Solaris system security.

System security is a rapidly developing field, with new threats, tools, and techniques being developed continuously. Readers whose Solaris systems are connected to the Internet will want to go beyond the basics presented here. References that specifically address UNIX system security should be considered required reading for managers of systems connected to the Internet. A short list of suggested readings appears at the end of Chapter 18, "Network Security."

How Much Security?

When considering system security on a single system, the content of users' files constitutes the most basic item being protected. The system administrator needs to ask several questions. Who must be able to read or modify the file content? Would knowledge of the content of a certain file allow someone to gain access to other files? Would knowledge of the content of the file release information that should not be disseminated?

Think about an ordinary paper document (file) for a moment. If the file is very sensitive, one might lock it in a file cabinet or even a safe. The same holds true for the computer equivalent of such files. If the content of a file on the system is highly sensitive, should the file be available on-line at all? Consider the hazards. Workstations, disk drives, and especially notebook computers are easy to carry off. Networked computers allow files to be copied quickly and invisibly.

How can system administrators protect the most sensitive data? Files that demand the highest security yet still need to be manipulated on a computer should be stored on removable media such as a floppy, removable disk, or tape that can be physically locked up in the same way as a paper file of similar importance.

Encrypting Sensitive Data

Another approach to securing sensitive files is to encrypt them using the Solaris *crypt* command. The command uses a mathematical function to scramble the file content in such a way as to make it unreadable until the file is decrypted.

Although encryption might seem to be a less bothersome method of protecting a sensitive file than removing the file from the system, be warned that encryption only protects a file for a limited period. Given time, the key or password used to encrypt the file will be discovered. Automated methods of deciphering such passwords are available.

The length of time required to identify an encryption key is a function of the key length, the encryption method used, other information known about the content of the encrypted file, and the computer resources available to the person committed to discovering the key. Strong encryption techniques using long keys may require years to circumvent. When something is known about the key or file content, or when weak methods with short keys are used, encryption can be circumvented in hours or less.

Owners, Groups, and Permissions

The content of most files typically does not require high-level protection. It is sufficient to ensure that the files are readable or modifiable by certain users only. Access to the content of a file on a Solaris system is controlled by a combination of the file's user and group ownership and the protection bits associated with the file. The output of *ls -l* in a typical directory appears in the following illustration.

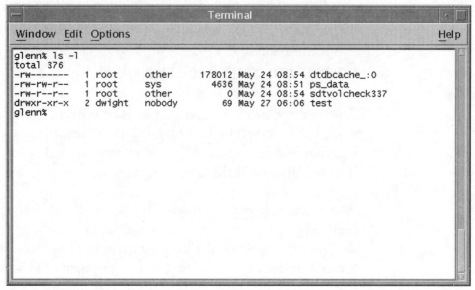

Output of ls -l.

The long listing option (*-l*) of the *ls* command produces several columns that contain various pieces of information about the files in the current directory. The columns from left to right are as follows.

1. Symbolic representation of the permission bits, sometimes called the permission mode or simply the mode

2. Number of file links made to the file

3. File owner

4. File group ownership

5. File size in bytes

6. Date and time the file was last modified

7. Name of the file

The first, third, and fourth columns from the left contain the security information.

File Protection Bits

Every file has three levels of security associated with it, which match the three types of users that may access that file. File protection is assigned according to the following three categories of users. File protection bits displayed by *ls-l* are shown in the illustration that follows.

- *Owner:* The assigned owner of the file.

- *Group:* A logical group the file and user belong to (such as a department within a company).

- *Other:* All other users who either do not own the file or are not a member of the file's group.

A closer look at the file protection bits displayed by ls-l.

The previous illustration shows the grouping of protection bits as represented by *ls -l*. Each character in the security field represents a significant feature of the file. The first character indicates type information. Type codes are listed in table 7-1, which follows.

Table 7-1: Protection Type Codes

Type Code	Meaning
-	File.
d	Directory.
b	Block special device entry. A file system entry SunOS maps to a specific device. Block specials are usually buffered devices such as disks.
c	Character special device entry, or a file system entry SunOS maps to a specific device. Character specials are "raw" or unbuffered devices such as terminal ports.
p	Named pipe. A special entry used for interprocess communication.
l	Symbolic link. Unlike a file link, a symbolic link contains the path to another file instead of a direct pointer to it. Symbolic links can span file systems and even point to files available via NFS from other machines.

➡️ *NOTE: In the case of a symbolic link, the permissions of the referenced file apply. Permissions pertaining to the link are not used to determine file access.*

The nine remaining characters are treated as three groups of three. As shown in the diagram, the first group indicates permissions for the file owner, the second for group ownership, and the third for other users. Each character triplet denotes various permissions for each entity. The presence of a character such as *r* indicates that permission is granted, whereas a dash (-) in place of the character indicates that permission is denied.

Each triplet is independent; all three permission bits can be set individually for the owner, group members, and others. However, the letter spaces in the triplet are "overloaded"; that is, several less commonly used bits show up as letters other than the common *rwx* to signify special permissions. The permissions and their meanings for files are listed in table 7-2, which follows.

Table 7-2: File Permissions

Permission	Meaning
r	Read. Programs can open and read the content of a file.
w	Write. Programs can store new content in the file.
x	Execute. Depending on the "magic number" stored in the first two bytes of the file, one of two things happens. If the number indicates that this is a program binary, it is loaded into memory and starts running. If the magic number is not present or indicates that the file is a shell script, a shell is started and the file is read by the shell.
s	*Setuid* or *setgid*.
t	The "sticky" bit found in the third position of the third triple. This bit indicates that the file is to be left on the *swap* device after it finishes execution. It was used to improve performance for frequently executed commands on older versions of UNIX, but has since become obsolete as a result of improved forms of management.
l	Mandatory file locking. This letter indicates that the *setgid* is on and the group execution bit is off. If a program locks this file, the kernel will prevent other programs from accessing it until the lock is cleared.

The "magic number" is a special value placed at the beginning of a file. This number tells the system how to load and run the content of the file. It is consulted if the file's execution bit(s) are set and the user enters the name of the file as a command. The two most common magic numbers are used for program binaries and shell scripts.

By default, */bin/sh* is used to run files that do not appear to have a magic number. To use another shell, place the string #! and the path to the shell as the first line of the file. The string #! is itself a magic number and causes the loader to use the remainder of the first line as the path to the command to be run. For example, #!/ *bin/csh* as the first line of a file would cause */bin/csh* to be started and the remainder of the file to be read by */bin/csh*.

✗   WARNING: *The* setuid *and* setgid *bits are very powerful and easily misused, and can result in security problems. The* setuid *bit is most often used to allow unprivileged users to execute a command as if they had root privileges. Do not use this facility lightly. Read the detailed description of SetUID problems later in this chapter.*

As shown in table 7-3, which follows, protection bits have similar meanings for directories.

Table 7-3: Directory Permissions

Function	Meaning
r	Read permission. Files in the directory can be listed.
w	Write permission. Files or links can be added or removed from the directory.
x	Execute permission. Files can be opened in the directory. The shell can change to this directory (cd).
s	Setgid. In a directory, setting the setgid bit causes all files created in the directory to be in the same group as the directory itself.
t	The sticky bit. Setting the sticky bit on a directory means that only the owner of a file in the directory can remove it. The sticky bit is most often used for directories such as /tmp.

Note the difference between *read* and *execute* permissions for a directory. In the following illustration, note how a directory with only the *read* permission allows files to be listed but not examined, even though the files have the required permissions.

➥   NOTE: *The accessibility of a given file is a combination of the permissions and ownerships of all directories along the path to the file, as well as the permissions and ownerships of the file itself. When planning file protection schemes, consider which files need to be read and which files should be hidden. Grouping files in directories for security is an approach worth considering.*

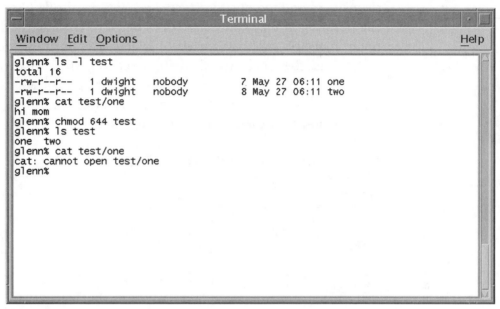

Directory with read permission.

chown *and* chgrp *Commands*

Changing file ownerships and groups is easily executed with the use of the *chown* and *chgrp* commands, as follows.

```
# chown username item
# chgrp group item
```

The *item* in the previous input is a file or a directory. The *username* and *group* are the user account name and group name, respectively, as found in */etc/passwd* and */etc/group*, or in the NIS or NIS+ databases or numerical UID or GID numbers. To change an entire directory and all of its subdirectories, both *chown* and *chgrp* accept the -*R* option.

Setting Protection Using Symbolic Names

Specifying changes to protection bits is more involved. Make the changes with the *chmod* command, which has an -*R* option such as *chown* for handling entire directories. Two methods of specifying the permission bit changes are possible. The first method is symbolic. A set of symbols is used to specify the entity (user, group, other) and the mode (read, write, execute) to be changed. The symbols are listed in tables 7-4 and 7-5, which follow. Table 7-4 presents symbols for *chmod* command entities, and table 7-5 presents symbols for *chmod* command modes.

Table 7–4: Symbols for chmod Command Entities

Entity	Symbol
User	u
Group	g
Everyone (other)	o
All entities	a

Table 7–5: Symbols for chmod Command Modes

Mode	Symbol
Read	r
Write	w
Execute	x
Mandatory locking	l
SetUID or *SetGID*	s
Sticky bit	t

The symbols are combined with the plus sign (+) operator to add the permission, the dash (-) operator to remove the permission, and the equals sign (=) operator to assign the permission. For example, you could remove the *read* permission for *other* from the file named *bob- stuff* with the following command line.

```
chmod o-r bob-stuff
```

Multiple permissions can be changed at the same time by specifying lists of symbols and operators. An example of setting the execute bits for all and restricting the read access on a file named *test99* follows.

```
chmod a+x,g-r,o-r test99
```

Developing a habit of checking modes and owners on files after setting them is recommended. Use *ls -l* and check the symbolic permission bit column to see who can read and execute the file. Check the owner and group. It is very easy to make a typographical error and set the modes to unintended values.

Using Octal Numbers

The second form of specifying permission bits is to provide them as a "base eight" (or octal) number, as shown in the previous directory example. This looks confusing at first, but it is not difficult to learn and has the advantage of compactly representing all protection bits in a single easy-to-type form. A simple means of deriving this octal number is to add up the numbers assigned to each permission bit from a table, such as table 7-6, which follows.

Table 7-6: Octal Number Permissions

Number	Bit
4000	SetUID
2000	SetGID
400	Read permission for owner
200	Write permission for owner
100	Execute permission for owner
40	Read permission for group
20	Write permission for group
10	Execute permission for group
4	Read permission for other

Number	Bit
2	Write permission for other
1	Execute permission for other

Using this table, it is possible to specify permissions for a file that has read, write, and execute permissions for the owner, and read and execute permissions for *group* and *other*, as follows: 400 + 200 + 100 + 40 + 10 + 4 + 1 = 755.

Use the total number with *chmod* as follows.

```
chmod 755 demofile
```

Setting the Default File Mode

Created files are given a default set of permission bits that is easily customized. The default is set in one of the shell initialization files with the *umask* command. This command takes a single argument, which is the complement of the mode the newly created files will receive. The argument can easily be calculated by subtracting the desired file mode in octal from the number *777* for directories and *666* for files. The numbering is such that by default directories will have the execute bit set to "on," and files will not.

For example, to create a very secure *600* file mode that would give read and write permission to the file owner only, calculate the *umask* as follows: 666 - 600 = 066.

A more friendly or less secure mode of *644* would be set with a value of *022*. To set the value, simply use it as the first argument to *umask*, as follows.

```
umask 022
```

Many sites set their local default value for permission bits by placing a line that includes the *umask* command in the system-side shell initialization files. On a Solaris system, the administrator can also set *umask* in the */etc/default/ login* file. This sets it for all users who log into a system.

➥ *NOTE: See Chapter 6, "Creating, Deleting, and Managing User Accounts," for additional information on these files and other initialization files where* umask *might be placed.*

In general, you should consider the content of a file and who should have access to it when specifying the owner, group, and permission bits. System files that do not need to be read by user programs should be made unreadable.

Implementing File Security Using Groups and File Protection Modes

Effective file security using UNIX groups and file protection modes requires forethought and organization. First, consider the structure of information within the organization. Next, consider which users need to share information and which users should not share nor participate in the exchange of information (i.e., guest accounts, remote log-ins, and so on). Users who need to share files should be placed in a UNIX group. For example, you might put the members of the accounting department into a group named *account*. See Chapter 6, "Creating, Deleting, and Managing User Accounts" for instructions on how to accomplish this.

Continuing with the example, files such as an accounting database or other records can be given the group ownership *account*. File modes can be set to allow members of the *account* group access to the accounting database. By selecting a mode that gives read and write permissions to the owner and group, but not to others, files can be effectively shared within a selected group but protected from other users.

✓   *TIP: Using the combination of a group and the group file protection bits is an often overlooked method of obtaining secure but shared files.*

Like the *rm* command, *chmod* does not check the option flags provided to determine whether they make sense; it simply makes the change. Sometimes the changes to the protection bits lead to unexpected results. Be careful not to fall into the following common traps.

•   You can inadvertently remove permissions for the file or directory owner. If you suddenly learn of a file that someone owns but cannot read or write, check that the owner still has permission to read or write the file.

•   Directories require execute permission to access content. Removing the execute permission on a directory leaves it in the confusing state of being readable but inaccessible.

Directories require read permission for their content to be seen by *ls* and other commands. If you want to create a directory that others can read and execute items from, but do not want them to see the content of the directory, remove the read permission while leaving the execute permission. Be warned that this does *not* prevent others from guessing at the names of other files in the directory, nor does it protect them from being read. This procedure only hides them from being listed by commands such as *ls*.

Dangers of SetUID and SetGID

The following is a scenario that could easily endanger system security. A system administrator logs in to a workstation as *root*. The administrator needs to leave the workstation for a while and forgets to log off or lock the screen. Has the administrator left the system in peril? Could someone use the unsecured session to gain *root* access later on without the administrator's knowledge? Sadly, the answer is yes. One approach available to this "someone" is to use the *setuid* bit. For example, assume that an unscrupulous person typed the following into an unsecured *root* session.

```
# cp /bin/sh /tmp/gotcha
# chmod 4755 /tmp/gotcha
```

These commands create a copy of the Bourne shell that when run will execute with full *root* privileges. The person who typed in these two statements will not require the *root* password, nor will she or he be required to log in as *root*. The person need only run the copy of the shell binary that contains the *setuid* bit set.

There are many other methods a determined person could use to exploit the *setuid* bit. Great care should be exercised when considering which programs are given this ability. For example, suppose a *setuid root* copy of the *vi* editor is made. Be aware that a user can start up a shell from within *vi*. What privileges would that shell have?

Taking Precautions

The following are general guidelines for avoiding *setuid/setgid* problems.

- Avoid using *setuid/setgid* whenever possible. If there is another way to implement the function that does not require *setuid/setgid*, use it.

- Make the program *setuid/setgid* to the least privileged user account possible. Avoid *setuid root* programs unless the program *must* have that privilege to function.

- *Never* make a shell script *setuid* or *setgid*. Shell scripts are too easily subverted and used to create *setuid* shells.

- When developing new *setuid/setgid* programs, consider releasing the privilege as soon as the privileged functions are completed. There are specific system library functions for setting and removing privileges in this manner.

- Consider having all programs that are *setuid* or *setgid* write to a log file when they are run. This provides an audit trail allowing the system administrator to monitor privileged usage.

Rooting Out Potential Problems

Another precaution worth taking is to scan the files on all systems for *setuid* and *setgid* programs. One method of accomplishing this is with the *ncheck* command. The *ncheck* command scans on a file-system by file-system basis. An example of *ncheck* use follows.

```
# ncheck -F ufs -s /dev/dsk/c0t3d0s6
```

The last argument is the device entry for the file system to be checked. To identify the device entry for a given file system, type *df*. The list produced by *ncheck* shows the *inode* (or index node) number for each file it found, and the file name relative to the mount point of the file system. In other words, if the device entry of the */usr* file system were entered, the list of file names produced would start at */usr*. A name listed as */lib/acct/accton* would have a full path of */usr/lib/acct/accton*.

Once a list of files has been generated, use *ls* to examine each file to identify its owner, group, and permission bits. To simultaneously accomplish these tasks, process the output of *ncheck* and feed it directly to *ls*. An example of this processing as it would be used to check the */usr* file system on a typical system follows.

```
# ls -l `ncheck -F ufs -s /dev/dsk/c0t3d0s6 | cut -f2 | sed 's:^:/usr:'`
```

Seemingly complex, this command line elegantly handles the task. The *ls -l* command will produce listings of the owner, group, and permissions for each file name given to it as the output of the phrase between the back quotes. That phrase runs *ncheck* to produce the list of *setuid* or *setgid* files. The second column of that list, the file names, is clipped out by using the *cut* command. Then the list of files is passed through *sed*, which uses a regular expression to prepend */usr* to each file name to make them full path names. The result is a list of the *setuid* and *setgid* files on the /*usr* file system, complete with respective owners, groups, and permissions.

Account Security

Chapter 6 discussed how user accounts were created and managed. One of the features Solaris offers to aid account management is a variety of controls over the account password. The password is the mechanism by which the system authenticates the user of a given account. The file security mechanisms described previously depend on this authentication to ensure that the person who has logged on to a particular account is really the file's owner.

Solaris stores account passwords and most other account information in separate places. On a system not using NIS or NIS+, these places are the */etc/passwd* and */etc/shadow* files. Examine the owners and permissions for these files. The */etc/passwd* file is readable by all because many programs make use of the information in this file. The */etc/shadow* file is protected, and contains the encrypted version of every user account password, along with password aging information. Because the encrypted passwords, like all encrypted text, are susceptible to being discovered, the content of the */etc/shadow* file must be kept unreadable (except by the *root* account).

Educate Users About Safe Passwords

It is also important to educate users with regard to password security. Writing the password on paper should be discouraged, as should sharing of passwords among users. How important this is depends on the environment and policies of the system owner organization. The more secure the environment must be, the more thorough user education and password protection should be.

The next line of defense is to ensure that the password chosen by each user is not easily guessed. Personal information should not be used as a password. Items such as the user's real name, phone number, license plate number, address, spouse's or children's names, office number, and so on should be avoided because these items are too easy for a determined individual to guess.

The use of common words or names is not a good approach either. A determined individual can make use of high-speed password-guessing software in combination with large dictionaries of words. Although this approach is time-consuming, it is getting less so as password guessing tools improve and faster computer hardware becomes readily available. Be especially careful to avoid famous names such as Superman because these will most certainly be in even the shortest dictionary used to guess passwords.

Better choices for passwords include the following properties: punctuation characters, numbers, and modified words. Some suggestions are word combinations such as *best1boy*; the first letter or number of words in a phrase (e.g., *towamfn* from "There once was a man from Nantucket"); or weird combinations of numbers and letter phrases (e.g., *10%grade*, or *any14tos* from "Anyone for tennis?"). Do not use anything suggested here or in other books, as you can be sure a determined individual trying to guess at an account password will know of these examples.

Using the Solaris Password Aging Feature

As discussed previously, encrypted text only remains protected for the period required to discover the encryption key or password. For this reason, user account passwords should be changed on a regular basis. This is the purpose of the Solaris password aging mechanism. The frequency of password changing depends on the security requirements of the environment. More frequent changes are better for security, but more annoying for the account user.

Frequent changes may even prompt an account user to select a poor password just to get back to work. Be sure to spend the time to educate and explain the reasons behind good passwords and changing passwords to all users. The security of the system as a whole depends in part on each user's choice of a password.

For the system administrator, no password is more important to protect than the password for the *root* account. The *root* password gives access to *all* data stored on the system and may provide access to other systems on the network. The importance of preceding comments about user account passwords is doubled for the *root* account. Carefully choose a *root* password and change it frequently. Make a habit of frequently checking the log files in */var/adm*, especially */var/adm/sulog*, which logs valid and failed uses of the *su* command. The *su* command is often used to gain *root* privileges from an ordinary user account.

Tracking Security Using aset and tripwire

Checking for security problems such as *setuid* files and monitoring file ownerships and permission bits can be quite a chore. Fortunately, there are tools to automate these tasks and provide assurance that your systems are secure. Solaris provides *aset*, a security tool that will help with these chores and act on problems automatically. Tripwire Security's *tripwire* tool is a popular and important tool for verifying file owners and permissions, as well as the integrity of system files.

aset

The *aset* tool consists of a set of tasks or separate small programs that check specific areas of system security. The available tasks are *tune, cklist, usrgrp, sysconf, env, eeprom,* and *firewall*. The *aset* tool is run at a specific security level: low, medium, or high. The tasks report and take actions to aid in securing the system based on the selected security level. Table 7-7, which follows, describes the tasks and respective actions at each security level.

Table 7-7: aset Tasks by Security Level

Task	Function	Low	Medium	High
tune	Sets system file permissions.	Sets permissions to release values.	Tightens permissions to improve security.	Sets permissions to be highly restrictive.
cklist	Checks system files against master list.	Compares some files against previously saved permissions settings and reports changes.	Increases the number of areas checked.	Checks the most files.
usrgrp	User/group checks.	Checks for consistency and accuracy of password and group files. Also checks NIS+ password if the *ypcheck* environment variable is set.	Ibid.	Ibid.

Task	Function	Low	Medium	High
sysconf	System configuration file checks.	Checks numerous files in /etc.	Makes changes to restrict access.	Makes the most restrictive set of changes.
env	Root environment check.	Checks "dot" files for *path* and *umask* settings; checks *root* and other user accounts as listed.	Ibid.	Ibid.
eeprom	EEPROM security check.	Checks the *eeprom* security setting and password. Does not make changes, but reports recommendations.	Ibid.	Ibid.
firewall	Sets system to act as network firewall.	Reports changes required to convert system to a network firewall.	Ibid.	Takes actions required to convert system to a network firewall.

Taking a walk through a typical *aset* run will further explain the items in the previous table, and will also show how *aset* is used and can be modified to fit a given environment. Starting *aset* is a matter of becoming *root* and typing the command, as follows.

```
# /usr/aset/aset
```

To start *aset* at the medium or high security levels, use the *-l* option. An example follows.

```
# /usr/aset/aset -l medium
```

For the */bin/csh* shell, use the following to set ASETSECLEVEL.

```
# setenv ASETSECLEVEL medium
```

For the */bin/sh* shell, use the following to set ASETSECLEVEL.

```
# ASETSECLEVEL=medium ; export ASETSECLEVEL
```

While *aset* is running, it is possible to check on its progress by using the *taskstat* utility, shown in the following illustration.

Checking aset *progress with* taskstat.

```
# /usr/aset/util/taskstat

Checking ASET tasks status ...
Task firewall is done.
Task env is done.
Task sysconf is done.
Task usrgrp is done.

The following tasks are done:
        firewall
        env
        sysconf
        usrgrp

The following tasks are not done:
        tune
        cklist
        eeprom
#
```

As *aset* runs it will generate a large amount of disk activity. Consequently, running *aset* when system use is at a low level is recommended. It is also a good idea to run *aset* periodically, especially in a networked environment, to check for any changes that may have occurred. As shown in the following, both of these conditions are met by using the *-p* option to *aset*.

```
# /usr/aset/aset -p
```

The previous statement will put an entry for *aset* in the *root crontab* file. The *cron* daemon reads each user's *crontab* files and runs programs at the times listed in the *crontab* file. The *aset -p* command inserts a scheduled run of *aset* to be performed nightly at midnight. This can be checked by typing *crontab -l* and looking for the line that contains *aset*. To remove this regularly scheduled *aset* run from the *root crontab*, type *crontab -e* as *root*, and delete the line containing *aset*.

Each *aset* run creates a set of log files that detail any problems found by *aset* and all corrective actions taken. The log files are found in directories in */usr/aset/reports*. The directory names correspond to the date and time of the *aset* run that created the reports. For instance, the directory *0716_07:47* contains the reports for the 7:47 a.m. *aset* run on July 16. The symbolic link */usr/aset/reports/latest* is set to point to the most recent report directory.

The report directory contains files with the *.rpt* extension. These are the reports generated by various *aset* tasks. At the low security level used in the previous example, very little is written into most reports. However, even at this level, the system has a few minor problems worth correcting. The content of *env.rpt*, the *root* environment report, for the *aset* in the previous example follows.

```
# more env.rpt
*** Begin Environment Check ***
Warning! umask set to umask 022 in /etc/profile - not recommended.
*** End Environment Check ***
#
```

This report indicates a questionable setting for *umask* in */etc/profile*. This setting would create files that are globally readable. If global read access is not desirable in a given environment, a change to the *umask* line in */etc/profile* is in order. Another error was spotted by the *sysconf* (system file configuration task), and reported in the *sysconf.rpt*, as seen in the following.

```
# more sysconf.rpt
*** Begin System Scripts Check ***
Warning! The use of /.rhosts file is not recommended for system security.
*** End System Scripts Check ***
#
```

Although a */.rhosts* file is a convenience, allowing remote access to *root* across a network requires that the remote machine be every bit as secure as the local machine. This warning should be carefully considered and the */.rhosts* file removed if it is not needed in a given environment.

Reading through all reports the first time you run them is worthwhile to get a sense of what *aset* checks for and the problems that may need correction. For periodic monitoring, however, reading reports is time consuming. Another approach is to use the *diff* command to compare reports and see if any significant changes have occurred since the last *aset* run. This command helps to highlight problems that have recently emerged and reduces the repetitive nature of reading the entire report series. A quick shell script that compares the reports in a named directory against the latest reports follows.

```
#!/bin/sh -f
#
# Compare files in $1 with latest
cd /usr/aset/reports/$1
for i
in *.rpt
do
        echo == $i
        diff $i ../latest/$i
done
```

Running this script provides the following results on the example system.

```
# /tmp/sample 0716_07:47
== cklist.rpt
4,6c4,7
<< No checklist master - comparison not performed.
<< ... Checklist master is being created now. Wait ...
```

```
<< ... Checklist master created.
---
>> ... Checklist snapshot is being created. Wait ...
>> ... Checklist snapshot created.
>>
>> No differences in the checklist.
== eeprom.rpt
== env.rpt
== firewall.rpt
== sysconf.rpt
== tune.rpt
== usrgrp.rpt
#
```

Because the July 16 run was the first time *aset* was run on this system, the *cklist* task had not yet created a master checklist. The *diff* command caught the change and reported the changed lines. No changes were found in the other reports. A short display of differences such as this can be a real time saver, especially if you are managing numerous systems. With a little work, this simple shell script can be extended to mail any changes to a specific user account. Mail could be sent only if changes were found, further reducing the effort required for security monitoring of large collections of workstations.

aset *Customization*

Of course, the items checked by *aset* may not be appropriate for a given site. There are two ways to control what *aset* does and which files and directories it checks.

The first method is to use the *aset* environment variables. ASETSECLEVEL was already mentioned; setting it controls the security level at which *aset* runs. In addition, the ASETDIR variable specifies an alternative directory for *aset* to use when looking for files and writing reports.

⊷   *NOTE: The environment variables are typically used for interactive runs of* aset. *When* aset *is run by the* cron *daemon, it uses command line options to select the security level (e.g., -l medium) and alternative directory (e.g., -d /home/security).*

The second method of customization is to modify the various configuration files for *aset*. The first of these is *asetenv*. This file is usually found in */usr/aset*. It contains a set of variables near the top, which may be customized to suit the local environment. The variables are listed in table 7-8, which follows.

Table 7-8: asetenv Configuration Variables

Variable	Description
TASKS	Sets the list of tasks to be performed by *aset*.
PERIODIC_SCHEDULE	Sets the schedule to be entered into the *root crontab* file when *aset -p* is run.
UID_ALIASES	Sets the name of the file to consult for a list of UIDs that can be shared by different account names.
YPCHECK	Specifies whether a check of the INS or NIS+ database will be made. This check will only report problems; it does not take action to fix problems.
CKLISTPATH_LOW	Sets the list of paths used at the low security level by the *cklist* task.
CKLISTPATH_MED	Sets the list of paths used at the medium security level by the *cklist* task.
CKLISTPATH_HIGH	Sets the list of paths used at the high security level by the *cklist* task.

The CKLISTPATH variables contain a set of directories separated by colons. For example, to set CHKLISTPATH_LOW to check */usr/bin*, */sbin*, and */etc*, enter the following command line.

```
setenv CHKLISTPATH_LOW /usr/bin:/sbin:/etc
```

The second set of customization files is found in */usr/aset/master* and consists of the *tune* family of files. These files contain the lists of files the *tune* task checks for integrity. The format is described in the top few lines of the file and consists of the path name to the file, its permission bits in octal notation, owner, group, and type. You can add path names of files to be checked by following the format shown. Add files in the *tune* file for the security level at which *aset* will run: *tune.low*, *tune.med*, or *tune.high*.

Restoring Permissions to Pre-aset Values

To conclude the discussion of *aset*, one more handy *aset* tool exists: */usr/aset/aset.restore*. This program restores system file permissions to the state they were in before *aset* was first run. You might want to use this tool to restore settings before removing *aset*, or if file permission settings were found to be too restrictive after an initial *aset* run.

tripwire

Another highly effective security tool is a program called *tripwire*. The current version is available in commercial form from Tripwire Security, but older versions are freely available. *Tripwire* tries to deal with the tough problem of detecting system file modification or outright replacement.

Suppose a cracker had gained access to a system; how might this person use the system and remain hidden? One technique is to replace the *ps* command with one that does not list processes owned by the cracker. Especially devious crackers will not only replace the program but set the mode, owner, and even time stamps on the file to hide the change. How can this rogue version of *ps* be detected by the system administrator?

One solution to this problem might be to perform a bit-level comparison between each file on the system disk and the copy on the Solaris installation CD-ROMs. This is very time consuming and will fail to cope with any changes that occur as a result of patches installed on the system.

Tripwire offers a more effective solution to the problem. *Tripwire* records the file size, permissions date information, and a "finger-print" of the file using cryptographic checksums. The crypto-graphic checksum is similar in concept to an ordinary checksum; it is a small code (computed from the content of the file) that allows changes to a file to be detected without the need to save a complete copy of the file. The use of a cryptographic algorithm improves on the basic checksum concept by greatly reducing the possibility that two files will produce the same "fingerprint."

➥   *NOTE:* Tripwire *is a file integrity checker. Unlike* aset, *it performs no check for the correctness of file modes or ownership, and makes no changes to file or directory modes or ownership.*

Obtaining tripwire

Tripwire is available from Tripwire Security Systems (*www.trip-wiresecurity.com*). There is a free version (typically an older release), and a commercial version (which adds additional fea-tures, as well as wider support for different platforms) available.

tripwire *Initialization*

The initial installation of *tripwire* and the creation of the file fin-gerprints is a critical operation. The assurance that *tripwire* pro-vides is completely dependent on the state of the system files at the point in time when the *tripwire* database is created. *Tripwire* only identifies changes, it does not check for correctness. For this reason it is recommended that *tripwire* be installed immediately after Solaris is installed, and before the system is placed on the network. If *tripwire* is installed on a system that has already been compromised, it will not be able to detect changes made during the compromise of the system.

With that limitation in mind, a critical evaluation of the current state of a system's security is needed before installing *tripwire*.

Installation on a newly installed, not yet networked, system is the best approach, but *tripwire* can be installed on systems that are already in use with the understanding that *tripwire* will only detect changes that occurred *after* its installation. The detection of future changes may well be worth the lower level of assurance that results from a late installation of *tripwire*.

✗   *WARNING:* Tripwire *is not a "correctness" checker like ASET; it only detects changes to a system. For maximum assurance,* tripwire *must be installed directly after installing Solaris and before a system is networked.*

The actual initialization of the *tripwire* database is straightforward. First, a configuration file is created for the system. Several sample configuration files are provided with both versions of *tripwire*. *Tripwire* is then installed and executed. The commercial and free versions of *tripwire* include good documentation for this phase of the installation process. Once the execution completes, *tripwire* will create a database of file fingerprints.

This fingerprint database needs to be stored on a secure file system. The safest approach is to place the database, configuration file, and *tripwire* binary on a removable medium such as a floppy and remove it from the system. The commercial version of *tripwire* provides a method of encrypting the database. However, an off-line copy provides higher assurance that the database has not been tampered with.

The following is the output of a typical *tripwire* database initialization. The *-init* option is used to generate a new database, whereas the *-c* option specifies the configuration file to be used.

```
# tripwire —init —c tw.config_glenn
### Phase 1:   Reading configuration file
### Phase 2:   Generating file list
```

```
### Phase 3:   Creating file information database
###
### Warning:   Database file placed in ./databases/tw.db_glenn.
###
###        Make sure to move this file and the configuration
###        to secure media!
###
###        (Tripwire expects to find it in '/tmp/databases'.)
```

Checking for Changes

Once installed, *tripwire* can be used to verify that none of the files in its database have been changed since the last database update (or creation). This is a two-part process. First, the CD-ROM or floppy diskette containing the *tripwire* database, configuration file, and program is loaded on the system. *Tripwire* is executed to create a current database of file fingerprints, and this database is compared to the version stored on the CD-ROM/diskette. The result will be a report that details the changes that have occurred since the time the CD-ROM/diskette containing the *tripwire* database was created.

In reading the report, note that many of the changes are to files that will be changing frequently regardless of the security of the system. Files such as */etc/passwd* and */etc/shadow* are likely to change as a part of the normal use and administration of the system. Changes to system programs such as *ps*, or the addition of files or directories in system areas, are signs of potential problems and are easily spotted by *tripwire*.

The following sample output shows excerpts from a *tripwire* report. These excerpts illustrate the checks performed, and the problems reported by *tripwire*. During the first three phases of execution, *tripwire* gathers information on the files previously "fingerprinted." In phases 4 and 5, *tripwire* compares the file date stamps, owners, permission bits, and fingerprints. Note that *tripwire* also discovers newly added files in the system directories and provides a two-column listing of all changes found in phase 5 of the report.

```
# tripwire —c tw.config_glenn
Tripwire(tm) ASR (Academic Source Release) 1.3.1
File Integrity Assessment Software
(c) 1992, Purdue Research Foundation, (c) 1997, 1999 Tripwire
Security Systems, Inc. All Rights Reserved. Use Restricted to
Authorized Licensees.
### Phase 1:  Reading configuration file
### Phase 2:  Generating file list
### Phase 3:  Creating file information database
### Phase 4:  Searching for inconsistencies
###
###                      Total files scanned:7508
###                         Files added:         5
###                         Files deleted:       5
###                         Files changed:      56
###
###                      Total file violations:      66
###
added:    -rw-r--r-- root     2449 Oct 13 10:39:54 1999 /etc/amd.conf
changed: -rw-r--r-- root      108 Jan 4 16:08:04 2000 /.rhosts
changed: -rw-r--r-- root     4737 Jul 26 13:39:33 1999 /etc/inetd.conf
changed: -rw-r--r-- root      706 Feb 17 16:29:02 2000 /etc/mtab
      … part of the report skipped …
### Phase 5:  Generating observed/expected pairs for changed files
###
### Attr     Observed (what it is)     Expected (what it should be)
### ======== =============================================================
/.rhosts
    st_size: 108                        138
   st_mtime: Tue Jan 4 16:08:04 2000   Wed Jan 27 14:08:50 1999
   st_ctime: Tue Jan 4 16:08:04 2000   Wed Jan 27 14:08:50 1999
```

```
    md5 (sig1):    0DqX61RNgN9ZCPJBWWroek      2mzLgHacQUf1RhE:CIjYGs
    snefru (sig2): 3R9FchJm.X3:mHUvc7:AdM      1kDUkH0FMuVlI3wt0n6OIo


/etc/inetd.conf
        st_ino: 4637557              4600429
       st_size: 4737                 4733
      st_mtime: Mon Jul 26 13:39:33 1999    Mon Jul 26 13:39:07 1999
      st_ctime: Fri Oct 29 01:01:34 1999    Mon Jul 26 13:39:07 1999
    md5 (sig1): 0kVmCaNPw0.3FksM18120K        2V4eZn:X5C2vkK3CW2uYHc
    snefru (sig2): 3pgcYm833.6bAkaodu42yY     3MTGF6mte2IbHsUKO9KjmY


/etc/mtab
        st_ino: 4386763              4576784
```

Updating the tripwire *Database*

The continuing administration of a system will result in a growing number of files and directories that have changed. This will result in a lengthening list of false positive changes appearing in *tripwire* reports. Updating the *tripwire* database deals with this problem. *Tripwire* can be run in an interactive update mode that allows the administrator to okay changes to files. In update mode, *tripwire* will fingerprint the files and update their entries in the database. As with initialization, the updated database will need to be removed from the system to ensure it has not been tampered with. It is also a good idea to keep earlier copies of the database to guard against mistakes in the update process.

Installing Patches

One of the most critical steps toward securing a system is to keep the system binaries up to date. Software bugs play a critical role in system security. No operating system or suite of programs is completely bug free, and attackers make use of flaws in programs to obtain access to systems. Common programming mistakes such as buffer overflows are frequently used as the entry point for attackers to gain access to a system remotely, or to gain privileges once they have access to a system.

To fix these inevitable problems, Sun makes patches available. The patch distributions contain corrected programs, configuration files, library routines, and other files. Solaris ships with several tools to ease the installation of these patches, and several noteworthy patch installation and management tools are also available from third parties.

Basic Patch Tools

Three basic tools for Solaris patch management are *showrev*, *patchadd*, and *patchrm*. These commands are summarized, including some of the typically used options, in table 7-9, which follows.

Table 7-9: Patch Commands

Command	Description
showrev -p	Shows all patches installed on a system.
patchadd /path_to_patch_file and *patchadd -B / path_to_backout_directory / path_to_patch_file*	The first form of the *patchadd* command installs the patch in the file given as an argument. The second form uses the *-B* option to save the patch *backout* information in an alternate directory.
patchrm nnnnnn-nn and *patchrm -B / path_to_backout_directory nnnnnn-nn*	The first form of the *patchrm* command removes the patch specified by the patch ID number. The second form uses the *-B* option to specify an alternative directory to be used for the patch *backout* information.

The *showrev* command may be used to determine if a given patch has been installed on a system. For example, to determine what patches have been installed on the SUNWkvm package, you could issue the following command.

```
# showrev -p | grep SUNWkvm
Patch: 104796-01 Obsoletes:   Packages: SUNWkvm
```

Each line of output from *showrev* has three sections. The first section, Patch, lists the identification number of any installed patches. The second section, Obsoletes, lists any older patches this patch replaces. The final section, Packages, lists the packages this patch modified.

The *patchadd* and *patchrm* commands are used to install and remove patches. To allow for patch removal, the old versions of patched programs and files must be saved. *Patchadd* also produces a log that details the changes resulting from the patch installation. The *logfile*, and copies of old system files, are stored in the */var/sadm* directory. For this reason (among others), the */var* file system needs to be much larger then the Sun-recommended default. How large you make the */var* file system depends on several factors, including other uses of the */var*, such as mail storage, spool space for printers, and so on.

Recommended Patch Clusters

Although the *patchadd* command can be used to install individual patches, there are often several patches that need to be applied in a group. Sun provides a pre-built group of updated patches called a Recommended Patch Cluster.

➠   *NOTE: The Recommended Patch Clusters can be found on the WWW at* http://sunsolve.Sun.COM/pub-cgi/show.pl?target=patches/ patch-access.

The Recommended Patch Cluster comes with its own tool, *install_cluster*, to simplify cluster installation. This shell script installs the suite of patches in the order specified by the *patch_order* file supplied with the recommended patch cluster. Like the *patchadd* and *patchrm* commands, you must have sufficient space in */var* to accommodate the logs and patch *backout* information.

➥ *NOTE: The patch files are distributed as a compressed archive file. The user will need to uncompress this file to access individual patch files and the installation tools.*

Additional Patch Tools

The patch installation process is quite slow, in part because the *patchadd* process is overly cautious. A faster tool is *fastpatch*. *Fastpatch* is a perl program that emulates most of what *patchadd* does, but runs as much as five to ten times faster than *patchadd*. Be sure to read the documentation supplied with *fastpatch* to see what features it does not implement.

➥ *NOTE:* Fastpatch *is part of an auto-install suite and can be found at* ftp://ftp.wins.uva.nl/pub/solaris/auto-install/.

Along with installing patches comes the problem of keeping up to date with which patches need to be installed. Three tools that help with this problem are *patchdiag*, *patchreport*, and GASP. *Patchdiag* is supplied to support contract customers by Sun. It compares the information supplied by *showrev* to that available in Sun's support web site. *Patchdiag* generates a report detailing which patches need to be installed to bring the system up to the Sun-recommended patch level.

Patchreport, a perl program, takes the process a bit further. Not only will *patchreport* make a comparison of the patches available to those installed, it can also download and install the missing patches.

➟   *NOTE:* Patchreport *is described further and is available for download at* ftp://ftp.cs.duke.edu/pub/PatchReport/ index.html.

The Georgetown Automated Solaris Patchtool (GASP) is a graphical program with capabilities similar to *patchreport*.

➟   *NOTE: GASP was written at Georgetown University and is available at* http://www.georgetown.edu/reillyb/gasp/.

✗   *WARNING: Although tools such as* patchreport *and GASP can be very convenient, they can also be very dangerous. The system administrator needs to ensure that the patches installed by these tools do not cause other problems on the system. Automated patch tools should be monitored closely in order to minimize potential problems.*

Summary

This chapter discussed the basics of the UNIX file permission scheme, and how to view and modify file permissions. The dangers of the *setuid* and *setgid* bits were discussed; these should not be overlooked when creating programs that have these attributes. By using the password aging feature, *aset*, and *tripwire*, many aspects of Solaris system security can be controlled. These tools and techniques are sufficient for many environments. However, system administrators whose sites are connected to the Internet will need to go further to secure such systems.

CHAPTER 8

Working with Solaris Devices

One of the major functions performed as a system administrator is the installation of devices on the system. This chapter presents an overview of Solaris devices, followed by discussion of device types and methods used to access these devices. Finally, the chapter covers naming conventions used for system devices under Solaris, how to determine which devices are connected to systems, and how to inform Solaris that those devices are available.

Device Terminology

Understanding device terminology is the first step toward allowing an administrator to troubleshoot problems with devices on a system. Most devices are named with an acronym derived from the device type. For instance, an SCSI disk drive is an *sd* device. An SCSI disk is known as a *dsk* device. An SCSI tape drive is an *st* device, but this same tape drive also has an *rmt* device name associated with it.

In brief, under Solaris a device can have several names. The name of the device is also known as the device identifier. Table 8-1, which follows, lists a few of the common devices on SPARC-based systems.

Table 8-1: Device Names

Device Name	Description	Device Name	Description
cn	System console	st	SCSI tape
esp	System SCSI adapter	zs	On-board serial ports
/dev/fd0a /dev/ fd0b /dev/fd0	Solaris 1 compatibility links to floppy device	/fd0, /fd1, /fd2	Standard input, standard output, and standard error
ecpp0	Parallel port	ebus	On-board bus bridge
hme	100 Mbit Ethernet	dad	IDE disk drive
/dev/diskette / dev/rdiskette	Solaris floppy diskette device	rmt	Magnetic tape
pci	System PCI adapter	sd	SCSI disk drive
le (obsolete)	Lance Ethernet devices	/dev/ttya or /dev/term/a	Serial port A
mbus	System mbus adapter	qe	Quad Ethernet devices
sbus	System sbus adapter	cua	Calling unit (modem)

Hardware-level Device Access

SPARC-based systems allow the administrator to access system devices from two levels: hardware-level device access, and operating system device access. Hardware-level access involves the use of OpenBoot monitor commands. These commands allow the administrator to view and manipulate devices without booting the operating system, and can be very useful when new hardware is being added to the system.

The operating system device access scheme employs utilities available once the operating system is executing. These commands allow the operator to add, query, and troubleshoot device problems while the operating system is executing. Both access schemes are grounded on an understanding of how these devices are connected to the system, and on the concept of a "device information tree."

Device Information Tree

On SPARC-based systems, the input/output devices are attached to a set of interconnected buses. The OpenBoot monitor recognizes these interconnected buses and respective devices as a tree of nodes. This representation of the device interconnection is called the *device information tree*, an example of which follows.

Typical device information tree as illustrated by the OpenBoot monitor show- devs command.

```
ok show-devs
/options
/fd@1,f7200000
/virtual-memory@0,0
/memory@0,0
/sbus@1,f8000000
/auxiliarry-io@1,f7400003
/interrupt-enable@1,f5000000
/memory-error@1,f4000000
/counter-timer@1,f3000000
/eeprom@1,f2000000
/audio@1,f7201000
/zs@1,f0000000
/zs@1,f1000000
/openprom
/packages
/sbus@1,f8000000/cgsix@2,0
/sbus@1,f8000000/SUNW,bpp@1,200000
/sbus@1,f8000000/SUNW,lpvi@1,300000
/sbus@1,f8000000/le0@0,c00000
/sbus@1,f8000000/esp@0,800000
/sbus@1,f8000000/dma@0,400000
/sbus@1,f8000000/esp@0.800000/st
/sbus@1,f8000000/esp@0.800000/sd
/packages/obp-tftp
/packages/deblocker
/packages/disk-label
ok ▮
```

It is possible to become familiar with the device information tree structure by using the OpenBoot monitor *show-devs* command. At the root of the tree is a node that describes the entire system.

The leaf (child) nodes are indented (with respect to their parent node) to show the hierarchical relationships between nodes.

Nodes on the device information tree that have children generally represent system buses and the associated controllers on the buses. Each parent node is allocated a physical address space, and a major device number that distinguishes the devices on this node from one another. Each child of a parent node is allocated a minor device number, and a physical address within the parent node's address space.

The physical address of nodes is generally assigned in relation to the device's physical characteristics, or the bus slot in which the controller is installed. This is done in an attempt to keep device addresses from changing as systems are upgraded and new devices are installed on the system. Each device node on the device information tree can have the following attributes.

- Properties or data structures that describe the node and associated devices.

- Methods or software procedures used to access the devices.

- Children (devices or other nodes) that are attached to the node and lie directly below it in the tree.

- A parent, or the node that lies directly above the current node in the device tree.

Accessing System Devices with the OpenBoot Monitor

The OpenBoot monitor refers to devices by their points of connection to the system. The operating system refers to devices by mnemonic names and device special files. Most of the device names system administrators deal with are actually aliases that point to some physical device on the system.

For example, the *root* file system may reside on a disk referred to as either */dev/dsk/cot3dos0* or *sd3a* by the administrator. This same disk may be known as *disk3* or */sbus@1,f8000000/esp@0,4000/ sd@3,0:a* to the OpenBoot monitor. On a SPARCStation 2 system, these names are all references to the device the operating system refers to as */devices/sbus@1,f8000000/esp@0,4000/sd@3,0:a*. On an UltraSparc 60 system, the *root* file system would reside on */dev/ dsk/cot0dos0,* which is also known as *sd3a*! On the UltraSparc 60, these names are references to the device the operating system refers to as */devices/pci@1f,4000/scsi@3/sd@0,0:a.*

Examination of the OpenBoot monitor addressing scheme will shed some light on this (at first glance, archaic) device naming system.

➥   *NOTE: On the UltraSparc systems, the root file system is located on* /dev/dsk/cot0dos0 *(SCSI disk 0, slice 0) instead of* /dev/dsk/ cot3dos0 *(SCSI disk 3, slice 0). On IDE-based UltraSparc systems, the default boot device is* /dev/dsk/cot0dos6 *(IDE disk 0, slice 6).*

OpenBoot Access Methods

The OpenBoot monitor allows two methods of accessing devices: by use of an alias, and by physical address references. Until the operating system is up and running, these are the only methods available for accessing system devices.

The device physical addresses are determined as part of the system Power On Self Test (POST). As the system runs the power-on diagnostics, the firmware probes all system buses, and builds a table of devices. The device aliases are hardcoded into the Open-Boot monitor firmware.

EEPROM Device Mappings

The OpenBoot monitor employs an electronically erasable, programmable read-only memory (EEPROM) chip to store certain vital system information. Among the many parameters stored in this chip are the mappings between the device aliases and the physical device addresses. For example, the alias *disk3* refers to the SCSI disk drive connected to the on-board SCSI controller at target address 3. This mapping is controlled by the value of the *scsi-probe-list* variable within the OpenBoot monitor.

The *eeprom* device mappings may be viewed using the OpenBoot monitor *printenv* command. The alias to physical address mappings can be altered by the OpenBoot monitor *setenv* command.

➦   *NOTE: Because* eeprom *mappings differ between SPARC systems, it is best to consult the system hardware reference manuals for more information on the content of the OpenBoot monitor* eeprom.

OpenBoot Monitor Physical Names

The OpenBoot monitor's physical name for a device is constructed by concatenating the physical names of the nodes above the device in the device information tree with a forward slash character (/). This concatenation of nodes yields the generic form of a device physical name: *name@address:arguments*. The following are descriptions of fields in the device name.

- *name:* A name is a text string that generally has a mnemonic value. For example, *sd* represents an SCSI disk. Many controller names use some portion of the manufacturer's name or model number.

- *@:* The "at" sign must precede the address parameter.

- *address:* The address is a hexadecimal numeral string representing the physical address of the device on the system.

The address is usually in the following form: *hex_number,hex_number*.

- "*:*": A colon must precede the arguments parameter.

- *arguments:* The arguments constitute a text string whose format depends on the device. The arguments may be used to pass additional information to the device's driver software.

OpenBoot Monitor Aliases

The generic OpenBoot monitor address for *disk3* may be used to illustrate how the *disk3* device alias is mapped to the physical name for this device. By default, the *sbus-probe-list* variable contains the value *0123*. The *pcia probe-list* variable contains the value *1,2,3,4*.

The *eeprom* defines four disk aliases: *disk0*, mapped to the first address in the probe list; *disk1*, mapped to the second address in the probe list; *disk2*, mapped to the third address in the probe list; and *disk3*, mapped to the last address in the probe list.

On SCSI-based Sparc systems, the *eeprom* maps the *disk3* alias to the SCSI disk connected to the on-board SCSI controller as target 3. This device is mapped to the physical address */sbus@1 f8000000/esp@0,4000/sd@3,0:a*.

On IDE-based Sparc systems, the *eeprom* maps the *disk0* alias to the IDE disk connected to the on-board IDE controller as target 0. This device is mapped to the physical address */pci@1f,0/pci@ 1,1/ ide@3/dad@0,0:g*.

- ➥ NOTE: *The previously described mappings may be altered by changing the value of the* eeprom sbus-probe-list *and* pci{a,b} probe-list *variables.*

Determining Device Connectivity

Examining the physical address of a device allows the system administrator to determine the point of connection for that device. To determine the point of connection for the *disk3* device, simply break down the physical address */pci@1f,0/pci@1,1/ide@3/dad@0,0:g* into the following components.

- */pci@:* Device is connected to a PCI adapter.

- *1f,0:* An address on the main system bus. Therefore, the *pci* bus bridge in question is an on-board device.

- *pci@:* The *pci* interface is the on-board *pci* adapter.

- *1,1:* A *pci* slot number and an address offset within that slot. This means that the *pci* device is in *pci* slot 0 with a hexa-decimal address offset of 1 from the parent node (main system bus) address.

- *ide@:* An IDE adapter connected to the PCI.

- *3:* An IDE adapter address offset from the *pci* bus adapter.

- *dad@:* Device in question is an IDE disk.

- *0,0:g:* Denotes IDE target 0, logical unit number 0, partition *g* of the disk.

Operating System Device Access

The operating system provides the administrator with access to the content of the OpenBoot *eeprom* via the *eeprom* utility. The *eeprom* variables may be examined or changed with this utility. The operating system provides the administrator with device access methods similar to the methods available under the Open-Boot monitor. Before examining operating system device access the administrator should have an understanding of how the UNIX operating system services input/output requests.

Device Special Files

The UNIX operating system treats input/output devices as "special" files, example entries of which are shown in the following illustration. When the operating system references one of these device special files, it is actually invoking a device driver routine to handle the input/output request.

```
┌──────────────────────────────── Console ─────────────────────────────────┐
│ Window  Edit  Options                                              Help    │
├────────────────────────────────────────────────────────────────────────── │
│ # ls -lsa /devices                                                         │
│ total 16                                                                   │
│     2 drwxrwxr-x   4 root    sys        512 Dec 18 15:49 .                  │
│     2 drwxr-xr-x  25 root    root      1024 Feb  6 19:31 ..                 │
│     0 crw-------   1 root    other    28,  0 Dec 18 14:30 audio@1,f7201000:sound,audio │
│     0 crw-------   1 root    other    28,128 Dec 18 14:30 audio@1,f7201000:sound,audioctl │
│     0 crw-------   1 root    sys      68, 11 Dec 18 14:30 eeprom@1,f2000000:eeprom │
│     0 brw-rw-rw-   1 root    sys      36,  0 Dec 18 14:30 fd@1,f7200000:a    │
│     0 crw-rw-rw-   1 root    sys      36,  0 Dec 18 14:30 fd@1,f7200000:a,raw │
│     0 brw-rw-rw-   1 root    sys      36,  1 Dec 18 14:30 fd@1,f7200000:b    │
│     0 crw-rw-rw-   1 root    sys      36,  1 Dec 18 14:30 fd@1,f7200000:b,raw │
│     0 brw-rw-rw-   1 root    sys      36,  2 Dec 18 14:30 fd@1,f7200000:c    │
│     0 crw-rw-rw-   1 root    sys      36,  2 Jan 22 20:23 fd@1,f7200000:c,raw │
│     0 crw-------   1 root    sys      74,  0 Dec 18 14:30 profile:profile    │
│    10 drwxr-xr-x   2 root    sys       4608 Dec 18 15:52 pseudo             │
│     2 drwxr-xr-x   3 root    sys        512 Dec 18 14:30 sbus@1,f8000000    │
│     0 crw-rw-rw-   1 root    sys      29,  0 Dec 18 14:30 zs@1,f1000000:a    │
│     0 crw-------   1 uucp    uucp     29,131072 Dec 18 14:30 zs@1,f1000000:a,cu │
│     0 crw-rw-rw-   1 root    sys      29,  1 Dec 18 14:30 zs@1,f1000000:b    │
│     0 crw-------   1 uucp    uucp     29,131073 Dec 18 14:30 zs@1,f1000000:b,cu │
│ #                                                                          │
└────────────────────────────────────────────────────────────────────────── ┘
```

Typical device special file entries.

The device special files have two unique characteristics not found in typical data files. First, the device special files are associated with "major" and "minor" device numbers. The major device number identifies the "device class," and the minor device number identifies a specific device within a class. These major and minor device numbers are used as entry points into the device driver routines.

The operating system kernel contains a "jump table," which cross-references each device and the major/minor numbers. When the system needs to communicate with a device, it consults the jump table to locate the appropriate device major/minor numbers for this transaction.

The system invokes the device driver routine by referencing the appropriate device special file. The device driver routine determines which to open by extracting the major/minor device number of the device special file referenced by the operating system.

Another unique characteristic of device special files is the concept of "mode bits." Mode bits determine the type of data transfer supported by the device. Types of device special files include character and block mode devices. *Character mode devices*, also known as raw mode devices, transfer data in single byte quantities. *Block mode devices*, also known as cooked mode devices, transfer data in larger quantities (typically disk sectors, or blocks).

Some UNIX variants place all device special files in the */dev* directory. Under Solaris, the device special files are contained in the */devices* directory. Solaris uses the entries in the */dev* directory as device aliases. These alias files are symbolically linked to the device special files in the */devices* directory.

Viewing the Device Information Tree

Under Solaris the physical device interface to the nodes on the device information tree lies in the */devices* directory, shown in the following illustration. This directory contains subdirectories that store device special files for audio devices, floppy disk drives, the *eeprom*, pseudo devices, SBus adapters, *zs* (serial port) devices, and other system devices.

```
total 20
    2 drwxrwxr-x  4 root    sys         512 May  7 20:43 .
    2 drwxr-xr-x 23 root    root        512 Oct  5 19:40 ..
    0 crw-------  1 curt    nobody   28,  0 Apr  9  1994 audio@1,f7201000:sound,audio
    0 crw-------  1 curt    nobody   28,128 Apr  9  1994 audio@1,f7201000:sound,audioctl
    0 crw-------  1 root    sys      68, 11 Apr  9  1994 eeprom@1,f2000000:eeprom
    0 brw-rw-rw-  1 root    sys      36,  0 Apr  9  1994 fd@1,f7200000:a
    0 crw-rw-rw-  1 root    sys      36,  0 Apr  9  1994 fd@1,f7200000:a,raw
    0 brw-rw-rw-  1 root    sys      36,  1 Apr  9  1994 fd@1,f7200000:b
    0 crw-rw-rw-  1 root    sys      36,  1 Apr  9  1994 fd@1,f7200000:b,raw
    0 brw-rw-rw-  1 root    sys      36,  2 Apr  9  1994 fd@1,f7200000:c
    0 crw-rw-rw-  1 root    sys      36,  2 Apr  9  1994 fd@1,f7200000:c,raw
    0 crw-------  1 root    sys      74,  0 Apr  9  1994 profile:profile
   14 drwxr-xr-x  2 root    sys        6656 May  7 20:34 pseudo
    2 drwxr-xr-x  3 root    sys         512 Apr  9  1994 sbus@1,f8000000
    0 crw-------  1 lp      sys      29,  0 Apr  9  1994 zs@1,f1000000:a
    0 crw-rw-rw-  1 root    sys      29,131072 Apr  9  1994 zs@1,f1000000:a,cu
    0 crw-rw-rw-  1 root    sys      29,  1 Apr  9  1994 zs@1,f1000000:b
    0 crw-rw-rw-  1 root    sys      29,131073 Apr  9  1994 zs@1,f1000000:b,cu
~
```

A typical listing of the /devices directory.

Within each subdirectory of */devices* there are more entries, which may be special device files or directories containing special device files. The directory hierarchy simplifies the task of locating a specific device by grouping device entries for that type of device in a single directory, an example of which is shown in the following illustration.

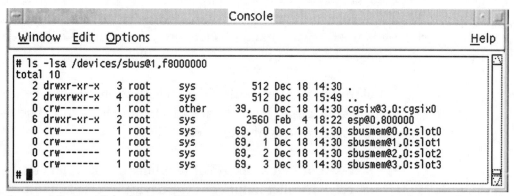

A typical listing of the /devices/SBus directory.

✓   TIP: *A simple method of identifying the address of any device on the system is to use the command* ls –lsaR /devices/*. *The output of this command will list the device special files for all devices on the system.*

Logical Device Names

As seen in the previous section, physical device names can be difficult to decipher, and even more difficult to remember. Fortunately, Solaris provides an easy-to-use, easy-to-remember method of accessing devices that does not require the memorization of several eight-digit hexadecimal addresses.

Alphabet Soup Device Names

Like the OpenBoot monitor, Solaris allows multiple names for each device on the system. As discussed in the section on device special files, the operating system's link to the physical device is through a file in the */devices* directory. However, Solaris also includes aliases for these */devices* entries. Such alias entries reside in the */dev* directory, and are symbolically linked to the entries in the */devices* directory.

For example, the */dev/dsk* directory contains entries with names such as */dev/dsk/c0t3d0s0*. This form of device name is often referred to as the "alphabet soup" designator for the device. These logical names for devices are much easier for most users to remember than names such *as /devices/sbus@1,f8000000/esp@ 0,800000/sd@3,0:a,raw.*

An alphabet soup name for a device consists of four fields: controller (cN), target (tN), device (dN), and slice (sN). The following are descriptions of these fields.

- cN represents controller number *N*. This refers to the logical controller number of the device interface. For instance, a system with a single SCSI interface would have all SCSI devices connected to controller *c0*. A system with two SCSI interfaces may have SCSI devices connected to both the *c0* and the *c1* SCSI interfaces.

- tN represents the target number. This is the SCSI target ID (or SCSI address) of a device connected to the controller.

- dN represents the unit number of the device connected to target controller tN, which is in turn connected to bus controller cN. Some peripherals allow the connection of several devices to a single target controller on the SCSI bus. The Solaris device-naming scheme uses the DN field of the device name to refer to these child devices. One such example is an ESDI disk controller. A typical ESDI disk controller may connect to the SCSI bus as target 2. The ESDI controller may in turn allow the connection of two disk drives. These disk drives may be referred to as */dev/dsk/cot2dos2* and */dev/dsk/cot2d1s2*.

- sN represents the slice or partition number of the device. See Chapter 9, "Disk Subsystem Hardware," for more information on disk slices and partitions.

- ⟶ *NOTE 1: Under Solaris, disk drives are the only devices that use alphabet soup designators. Other System V release 4 operating systems may use similar designators for tape drives and other system devices.*

- ⟶ *NOTE 2: Under Solaris X86, the disk device names are typically of the form cntnsn, representing Controller, Disk, and Slice. This is due to the use of EIDE disk controllers, which do not allow the use of subunits such as the ESDI interfaces on SCSI-based systems.*

Legacy Device Names

Solaris also includes mappings for certain legacy device aliases. Many UNIX variants name SCSI disk devices with the following format: *sd#P*, where # refers to the SCSI target address, and *P* refers to a file system partition on that drive. For example, */dev/sd3a* on a SunOS 4.1.3 system refers to the "a" partition of the disk connected to the on-board SCSI controller at address 3.

These legacy device aliases are also contained in the Solaris */dev* directory hierarchy. Symbolic links connect the aliases to the appropriate device special files under the */devices* directory. For example, the */dev* directory contains an entry for the *sd3a* device. The link for this device points to */dev/rdsk/cot3dos0*. This file is symbolically linked to the device special file */devices/sbus@1, f8000000/esp@0,800000/sd@3,0:a*. Resolving *sd3a* to the device name */devices/SBus@1,f8000000/esp@0,800000/sd@3,0:a* is shown in the following illustration.

```
# ls -lsa /dev/sd3a
  2 lrwxrwxrwx  1 root     root        12 Oct  3 08:31 /dev/sd3a -> dsk/c0t0d0s0

# ls -lsa /dev/dsk/c0t0d0s0
  2 lrwxrwxrwx  1 root     root        73 Jul  1 08:48 /dev/dsk/c0t0d0s0 -> ../../devices/io-unit@f,e0200000/sbi@0,0/dma@0,81000/esp@0,80000/sd@0,0:a

# ls -las /devices/io-unit@f,e0200000/sbi@0,0/dma@0,81000/esp@0,80000/sd@0,0:a
  0 brw-r-----  1 root     sys    32,   0 Jul  1 08:15 /devices/io-unit@f,e0200000/sbi@0,0/dma@0,81000/esp@0,80000/sd@0,0:a

# ls -las /devices/io-unit@f,e0200000/sbi@0,0/dma@0,81000/esp@0,80000/sd@0,0:a,raw
  0 crw-r-----  1 root     sys    32,   0 Jul  1 08:16 /devices/io-unit@f,e0200000/sbi@0,0/dma@0,81000/esp@0,80000/sd@0,0:a,raw
```

Resolving sd3a *to the device name.*

Disk drives are not the only devices that map to legacy device names. For example, to see how the serial port *ttyb* is connected to the system, use *ls -lsa /dev/ttyb*. The output of the *ls* command shows that */dev/ttyb* is a symbolic link to */dev/term/b*. The */dev/term/b* entry is actually a symbolic link to */devices/zs@1,f1000000:b*. Similarly, */dev/rst0* is a symbolic link to */dev/rmt/0*. This file will point to */devices/sbus@1,f8000000/esp@0,800000/st@4,0:*.

➥ *NOTE: Under Solaris the* /device *entries for optional equipment (tape drives, additional disks, CD-ROMs, and so on) will only exist if these devices have been connected to the system and a reconfiguration boot has been performed. The reconfiguration boot is discussed in a later section.*

Displaying System Configurations

It is frequently necessary to examine the configuration of the devices connected to a system. Whenever a new peripheral is to be added, the administrator will need to know how to address the device and where to install it. One way of displaying the system configuration is to use the OpenBoot monitor *show-devs* command. This command causes the OpenBoot monitor to display information about all devices in the device information tree.

The Solaris operating system also provides utilities that may be used to examine the system configuration information. Two utilities are described in the following sections.

prtconf

The Solaris *prtconf* utility allows the administrator to print the system configuration to a terminal or line printer. This utility provides a list of devices and drivers loaded in the system. The device information tree data is preserved by indenting child node information with respect to the parent nodes in the *prtconf* output.

When used with the *-p* flag, *prtconf* will display the OpenBoot *prom* information about the devices. The *-v* flag tells *prtconf* to display the output in verbose format. An example of *prtconf* command output is shown in the following illustration.

Typical output of the
prtconf *command.*

```
System Configuration: Sun Microsystems  sun4c
Memory size: 28 Megabytes
System Peripherals (Software Nodes):

4_75
    packages (driver not attached)
        disk-label (driver not attached)
        deblocker (driver not attached)
        obp-tftp (driver not attached)
    options, instance #0
    aliases (driver not attached)
    openprom (driver not attached)
    zs, instance #0
    zs, instance #1
    audio (driver not attached)
    eeprom (driver not attached)
    counter-timer (driver not attached)
    memory-error (driver not attached)
    interrupt-enable (driver not attached)
    auxiliary-io (driver not attached)
    sbus, instance #0
        dma, instance #0
        esp, instance #0
            sd (driver not attached)
            st (driver not attached)
            sd, instance #0
            sd, instance #1 (driver not attached)
            sd, instance #2 (driver not attached)
            sd, instance #3
            sd, instance #4 (driver not attached)
            sd, instance #5 (driver not attached)
            sd, instance #6 (driver not attached)
        le (driver not attached)
        cgsix, instance #0
    memory (driver not attached)
    virtual-memory (driver not attached)
    fd, instance #0
    pseudo, instance #0
```

sysdef

The Solaris *sysdef* utility also allows the administrator to examine
system configuration information. The *sysdef* utility provides a list
of devices defined on the system and loadable driver modules, as
well as the state of many tunable kernel variables. The device
information tree information is preserved by indenting child
node information with respect to the parent nodes in the *sysdef*
output. An example of *sysdef* command output is shown in the
following illustration.

Typical output of the sysdef command.

```
┌─────────────────────────────────────────────────────┐
│                      Console                          │
├─────────────────────────────────────────────────────┤
│  Window   Edit  Options                      Help     │
├─────────────────────────────────────────────────────┤
│ # sysdef -d                                           │
│                                                       │
│ Node 'TWS,Tatung 4_75', unit #-1                      │
│        Node 'packages', unit #-1 (no driver)          │
│                Node 'disk-label', unit #-1 (no driver)│
│                Node 'deblocker', unit #-1 (no driver) │
│                Node 'obp-tftp', unit #-1 (no driver)  │
│        Node 'options', unit #0                         │
│        Node 'aliases', unit #-1 (no driver)           │
│        Node 'openprom', unit #-1 (no driver)          │
│        Node 'zs', unit #0                             │
│        Node 'zs', unit #1                             │
│        Node 'audio', unit #0                          │
│        Node 'eeprom', unit #-1 (no driver)            │
│        Node 'counter-timer', unit #-1 (no driver)     │
│        Node 'memory-error', unit #-1 (no driver)      │
│        Node 'interrupt-enable', unit #-1 (no driver)  │
│        Node 'auxiliary-io', unit #-1 (no driver)      │
│        Node 'sbus', unit #0                           │
│                Node 'dma', unit #0                     │
│                Node 'esp', unit #0                     │
│                        Node 'sd', unit #-1 (no driver)│
│                        Node 'st', unit #-1 (no driver)│
│                        Node 'sd', unit #3             │
│                        Node 'st', unit #4             │
│                Node 'le', unit #0                      │
│                Node 'cgsix', unit #0                   │
│        Node 'memory', unit #-1 (no driver)            │
│        Node 'virtual-memory', unit #-1 (no driver)    │
│        Node 'fd', unit #0                             │
│        Node 'pseudo', unit #0                          │
│                Node 'clone', unit #0                   │
│                Node 'ip', unit #0                      │
│                Node 'tcp', unit #0                     │
│                Node 'udp', unit #0                     │
│                Node 'icmp', unit #0                    │
│                Node 'arp', unit #0                     │
│                Node 'sad', unit #0                     │
│                Node 'consms', unit #0                  │
│                Node 'conskbd', unit #0                 │
│                Node 'wc', unit #0                      │
│                Node 'iwscn', unit #0                   │
│                Node 'ptsl', unit #0                    │
│                Node 'tl', unit #0                      │
│                Node 'cn', unit #0                      │
│                Node 'mm', unit #0                      │
│                Node 'openeepr', unit #0                │
│                Node 'kstat', unit #0                   │
│                Node 'log', unit #0                     │
│                Node 'sy', unit #0                      │
│                Node 'pm', unit #0                      │
│                Node 'vol', unit #0                     │
│                Node 'ptm', unit #0                     │
│                Node 'pts', unit #0                     │
│                Node 'ksyms', unit #0                   │
│                Node 'winlock', unit #0                 │
│                                                       │
│ #                                                     │
└─────────────────────────────────────────────────────┘
```

sysinfo

Although Solaris provides commands for probing and displaying the system configuration, the output format and information available from each command is somewhat limited. Users desired a command that would provide the information of both *prtconf* and *sysdef*. The *sysinfo* command was developed by MagniComp, Incorporated, as an alternative to the *prtconf* and *sysdef* commands.

Sysinfo allows the administrator to display system information in several formats, and also allows the user to determine system information that is not available from the other commands. *Sysinfo* is available for several UNIX operating systems, including AIX, HP-UX, and Linux. *Sysinfo* is available from *http://www. magnicomp.com/sysinfo*. Output from *sysinfo* is shown in the facing illustration.

Device Instance Names

In the sample *prtconf* and *sysdef* output, many of the devices have an instance number associated with them. These instance numbers refer to the device address or ID on the system. For example, the disk with instance number 3 refers to disk */dev/dsk/ cot3doso*, or *sd3a*. The *esp* adapter with an instance number of 0 is the on-board SCSI adapter on this system. Other adapters and devices follow these conventions, thereby allowing the administrator to determine the structure of the device information tree by looking at the instance numbers of the attached devices.

Reconfiguring Device Information

Sometimes when a new peripheral is installed in a system, it becomes necessary to reconfigure other peripherals. For instance, it may be necessary to move an SBus or PCI adapter card from one slot to another to allow the installation of a new adapter card.

```
                                   xterm
         G E N E R A L    I N F O R M A T I O N
Host Name is              enterprise
Host Address(es) is       192.168.44.55
Host ID is                809e7c71
Serial Number is          2157870193
Manufacturer is           Sun (Sun Microsystems)
System Model is           Ultra 5/10
Main Memory is            64 MB
Virtual Memory is         296 MB
ROM Version is            OBP 3.17.1 1998/11/10 19:10
App Architecture is       sparc
Kernel Architecture is    sun4u
OS Name is                SunOS
OS Version is             5.7

         D E V I C E    I N F O R M A T I O N
SUNW_Ultra-5_10 is a "Sun 375-0009"
    openprom1 is a "Sun Open Boot PROM" device
    options0 is a "PROM Settings"
    aliases1 is a "PROM Device Aliases"
    pci108e,a000(0) is a "Sun sabre" PCI
        simba0 is a "Sun PCI" PCI bus bridge
            ebus0 is a "Sun Microsystems PCI<->OnBoard" bus bridge
                auxio is a "Auxiliary I/O"
                se0 is a "Siemens 82532 ESCC" serial device
                su0 is a "82510/8250/16450/16550AF" serial device
                    kbd is a "Sun Type-5 US Unix" Keyboard
                su1 is a "82510/8250/16450/16550AF" serial device
                ecpp1 is a "IEEE 1284 ECP, Nibble, and Centronics compatible" parallel device
                fd0 is a "Floppy Three" floppy disk drive
                eeprom1 is a "EEPROM" device
                flashprom1 is a "Sun Flash PROM"
                audiocs is a "Sun 4231" audio device
            hme0 is a "Sun" 100-Mb Fast Ethernet network interface
            m64(0) is a "ATI PGX PCI 8-bit [m64]" frame buffer
                monitor0 is a "15-inch/17-inch Entry-Level Color CRT" video monitor
            pci1095,646(0) is a "CMD 646" ATA/IDE/EIDE/UIDE controller
                c0t0d0 (dad0) is a "Seagate Medalist 34342A" 4.0 GB disk drive
                sd17 is a disk drive
        simba1 is a "Sun PCI" PCI bus bridge
            pci1011,24(0) is a "DEC Sun" SunSwift SCSI/Ethernet Multi Function Card
                pci108e,1000 is a "Sun Microsystems Onboard<->PCI" bus bridge
                hme1 is a "Sun cheerio" 100-Mb Fast Ethernet network interface
                SUNW_isptwo0 is a "Q-Logic Ultra" SCSI controller
    cpu0 is a "Sun UltraSPARC-IIi" 270 MHz CPU
```

Sysinfo *output.*

Because the device information tree contains information on the slot number of each device on the SBus, the device information tree must be rebuilt whenever devices are moved or new devices are installed on the system. Solaris provides the following three methods for rebuilding the device information tree.

- *boot -r:* This command tells the OpenBoot monitor to reconfigure the device information tree before starting the operating system. Whenever a device is added or removed, the system should be restarted with the *boot -r* command. This forces the operating system to rebuild the device information tree and load the correct drivers.

- *touch /reconfigure; reboot:* The Solaris start-up code looks in the *root* directory for the */reconfigure* file. If this file exists, Solaris will rebuild the device information tree as part of the boot process.

- A more dangerous way of reconfiguring the system on the fly is to execute the *drvconfig*, *disks*, *tapes*, and *devlinks* commands. These four commands cause the system to probe for new disk and tape devices, and add them to the device information tree while the system is up and running. Although these commands are available and perform the task, it is always best to use the *boot -r* command to let the system find and install the devices at reboot time.

Summary

This chapter examined how devices on Solaris systems are named and how they are addressed. Several examples detailed the relationship between device special files, logical device names, device aliases, physical device names, and addresses of the devices. The examples also showed how to track down the device physical address information for several typical device aliases.

The example output also showed how to identify the devices and driver modules loaded on a system, how to use this information to determine where to install new devices on the system, and how to force the system to reconfigure the device information when devices are added to or removed from the system.

CHAPTER 9
Disk Subsystem Hardware

Previous chapters explored the naming conventions of devices under Solaris. The next few chapters focus on specifics of the devices system administrators deal with every day: disks, tapes, terminals, modems, and printers.

One of the most important peripherals on any computer system is the mass storage subsystem consisting of disk drives, tape drives, and optical media (e.g., CD-ROM drives). Proper system optimization requires understanding of how the mass storage subsystem is organized. This chapter is focused on the disk subsystem on systems running the Solaris operating system.

Understanding Disk Geometry

Formatted disk drives consist of data storage areas called *sectors*. Each Solaris data sector is capable of storing 512 bytes of data. A UNIX file system consists of a particular number of sectors that have been bound together by the *newfs* command. But how are these sectors organized on a disk drive? Which sectors should be

229

bound into a file system? Can the size of a file system be changed? To answer these questions, an examination of the physical disk media is the first step. A typical internal disk layout is shown in the following illustration.

Typical internal disk layout.

Physical Disk Media

A disk drive contains several magnetic surfaces, called *platters*, which are coated with a thin layer of magnetic oxide. A typical SCSI disk may contain nine platters. Depending on the disk drive, there are one or more read/write heads positioned over each platter surface.

Disk sectors are arranged across the surface of the platter into concentric horizontal rings called *cylinders*. You may also find that cylinders are sometimes referred to as "tracks." On the smaller, innermost cylinders, the platter may be divided into a small number of sectors. On the outermost cylinders, the platter could be divided into a larger number of sectors. This situation is referred to as variable sectoring, or zone sectoring. Typically, sectors are logically grouped together (geographically) into clusters.

In order to transfer data to and from these sectors, the read/write head is passed over the magnetic surface of the platter. The read/write head converts electrical impulses from the system into magnetic patterns on the platter surfaces. The magnetic patterns represent the data stored on the platter.

Drive sectors are numbered sequentially beginning with the first sector on the first head of the first cylinder of the drive. On a drive with 55 sectors per track, the first 55 sectors will be numbered 0 through 54. The 55th sector will be the first sector on the second platter of the drive. The 110th sector on the drive will be the first sector on the third platter of the drive, and so on. Upon reaching the last sector on the last platter, the heads are moved to the next cylinder, and sector numbering continues at head 0 of the new cylinder.

➡   *NOTE: Because of the incorporation of alternate sectors, sector numbering may not always be as previously described. Many factors may affect the sequence of sectors. For instance, a bad sector may be mapped to a spare sector such that there would be a sector out of order in the numbering sequence.*

Disk Interfaces

Historically, most UNIX workstations incorporated disk drives that used the Small Computer System Interconnect (SCSI) interface, whereas personal computers (PCs) incorporated disks that used the Integrated Drive Electronics/Advanced Technology Attachment (IDE/ATA) disks. Early versions of these interfaces gave SCSI a performance advantage, whereas IDE drives were less expensive and easier to configure. Recent revisions of both standards have improved the capabilities of both interfaces, and Sun has introduced several "low-end" systems that incorporate Enhanced IDE (EIDE) disk subsystems.

The IDE interface has found its way into the workstation market because of its ease of configuration, and low price. The IDE interface automatically takes care of many of the details of formatting the drive, and mapping out bad sectors. The SCSI interface requires more attention from the operator during format/repair operations, but the data throughput performance makes it the logical choice for high-performance servers and workstations.

The system administrator needs to be aware of the interface capabilities when recommending systems for particular applications! The administrator also needs to be aware of the different operation/maintenance procedures for both drive types. Table 9-1, which follows, summarizes the differences between these two disk interfaces.

Table 9-1: Characteristics of SCSI and EIDE Interfaces

Characteristic	SCSI	EIDE
Performance	High	Moderate
Ease of use	Moderate	High
Expansion	High	Low
Types of devices supported	High	Low
Resource utilization	Moderate to high	Low to moderate
Cost	High	Low

Logical Disks and Partitions

In order to use a disk drive under Solaris, the drive must be partitioned, and the sectors must be bound together with *newfs* to form a file system. But why are disks partitioned in the first place? In order to understand the need to partition the disks, it is necessary to review a page of computer history.

At the time UNIX was first released, most machines that ran UNIX used 16-bit hardware. Consequently, a 16-bit unsigned integer number could address 65,536 sectors on the disk drive. As a result, the disk drive could be no larger than 65,536 sectors * 512 bytes/sector, or roughly 32 Mb.

When large disk drives (300 Mb or more) became available, provisions were made to allow their use on UNIX systems. The solution was to divide such drives into multiple "logical drives," each consisting of 32 MB of data, as shown in the following illustration. By creating several logical drives on a single physical drive, the entire capacity of the 300-Mb disk drive could be utilized. These logical drives became known as *partitions*. Each disk drive may have eight partitions. Under Solaris, the partitions are numbered zero (0) through seven (7).

Early system architecture required that large drives be partitioned into smaller logical units.

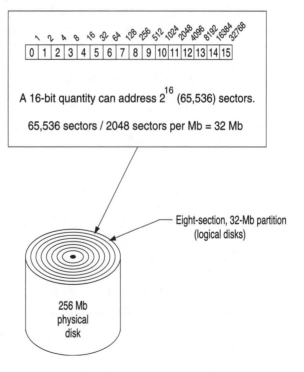

A 16-bit quantity can address 2^{16} (65,536) sectors.

65,536 sectors / 2048 sectors per Mb = 32 Mb

Eight-section, 32-Mb partition (logical disks)

256 Mb physical disk

Because more recent computer hardware employs 32- or 64-bit processors, the system can directly address much larger drives. A 32-bit signed variable can address 2,147,483,648 bytes (2.1 gigabytes or Gb). Under Solaris, the disk driver software is able to address nearly a terabyte of disk space, with files as large as two gigabytes.

Adding a Disk

The general procedure for adding a new disk to a system is very simple. This chapter provides details on the first six steps of physical installation of the new disk, and the entire process is outlined in the following for completeness. Steps 7 through 12 are covered in Chapter 10, "Disk Subsystem Software."

1. Identify the target address the new drive will use. Use *prt-conf* to determine the addresses in use.

2. Use *touch /reconfigure* to cause the system to look for new devices upon reboot.

3. Shut the system down using the *shutdown* command.

4. Set the drive address jumpers, and install the new drive on the system.

5. With the *boot -s* command, boot the system to the single-user state.

6. If necessary, use the *format* utility to format and partition the new drive.

7. Use *newfs* to bind sectors into the partitions.

8. Use *fsck* to check file system integrity.

9. Edit the */etc/vfstab* file to add partitions to the system tables.

10.   Use *mkdir* to create the mount point directories for the partitions.

11.   Use */etc/halt* to bring the system down.

12.   Use *boot* to restart the system.

Pre-format Considerations

The format utility allows the administrator to format, partition, label, and maintain disk drives under Solaris. The command uses the information found in the */etc/format.dat* file to determine the parameters to be applied during the format process.

⮞   *NOTE: Solaris versions 2.0 through 2.5.1 support SCSI, IPI, and SMD disk subsystems in the* format.dat *file. Beginning with Solaris 2.6, only SCSI disk subsystems are supported. EIDE disks do not require* format.dat *entries, as the IDE controller determines the disk geometry, and informs the operating system of the settings for each device.*

The following sections discuss the installation of drives already known to the format software, as well as how to develop a *format.dat* entry for a drive that is not known to the format software. These methods will work for SCSI and IPI disk drives. The procedure for developing *format.dat* entries for SMD disks is similar to the SCSI and IPI procedure.

Installing Drives Known to format.dat

The *format* utility reads a table of available drive types from the */etc/format.dat* file. If the new drive is described in the file, the drive is easily formatted, partitioned, and made available for use.

If an exact description is not available for the drive to be installed, it may be possible to use a Sun-supplied definition. For instance, when attempting to install a 2.4-Gb disk drive, the format utility may be able to use Sun's SUN2.1G entry to format the disk. Although this entry may waste a bit of disk space on the drive, time spent determining an exact *format.dat* entry for the drive will not be necessary.

➥   *NOTE: When formatting SCSI disks, the Solaris format command can assign a default partition table and disk label to unknown drive types. This constitutes a change from the way the format utility worked under SunOS. For IDE disks, Solaris will probe the IDE controller to determine the proper geometry information for the attached devices.*

✓   *TIP: Several disk brands and models may be used in Sun systems. Drives of similar capacity bear the same disk geometry label for purposes of convenience. When installing a drive model known to ship with Sun hardware, take advantage of the* format.dat *entry for the drive.*

Understanding format.dat Entries

If a description of the drive is not contained in *format.dat*, the administrator may need to create an entry that describes the new drive. A brief examination of the fields of a *format.dat* entry will help you understand how to develop such entries. The layout of a typical *format.dat* file appears in the facing illustration.

Listed in the first portion of the *format.dat* file are type, format parameters, and controller combination for each drive. The following fields are mandatory in every *format.dat* entry, and must contain a value.

- *ctlr:* Type of controller the drive emulates. Solaris SCSI drives use the term *SCSI* in this field.

- *ncyl:* Number of data cylinders on the drive according to the operating system.

Layout of a typical format.dat file.

```
┌─────────────────────────────────────────────────────────────────────────┐
│ ▱                              Console                              │▫│□│
├─────────────────────────────────────────────────────────────────────────┤
│ Window  Edit  Options                                              Help   │
├─────────────────────────────────────────────────────────────────────────┤
│ # cat /etc/format.dat                                                  ▲  │
│ #                                                                         │
│ #ident  "@(#)format.dat 1.17     96/06/13 SMI"                            │
│ #                                                                         │
│ # Copyright (c) 1991 by Sun Microsystems, Inc.                            │
│ #                                                                         │
│ # Data file for the 'format' program.  This file defines the known        │
│ # disks, disk types, and partition maps.                                  │
│ #                                                                         │
│ #                                                                         │
│ # This is the list of Sun supported disks for embedded SCSI.              │
│ #                                                                         │
│ disk_type = "SUN1.3G" \                                                   │
│         : ctlr = SCSI : fmt_time = 4 \                                    │
│         : trks_zone = 17 : asect = 6 : atrks = 17 \                       │
│         : ncyl = 1965 : acyl = 2 : pcyl = 3500 : nhead = 17 : nsect = 80 \│
│         : rpm = 5400 : bpt = 44823                                        │
│                                                                           │
│ disk_type = "SUN2.1G" \                                                   │
│         : ctlr = SCSI : fmt_time = 4 \                                    │
│         : ncyl = 2733 : acyl = 2 : pcyl = 3500 : nhead = 19 : nsect = 80 \│
│         : rpm = 5400 : bpt = 44823                                        │
│                                                                           │
│ disk_type = "SUN2.9G" \                                                   │
│         : ctlr = SCSI : fmt_time = 4 \                                    │
│         : ncyl = 2734 : acyl = 2 : pcyl = 3500 : nhead = 21 : nsect = 99 \│
│         : rpm = 5400                                                      │
│ #                                                                         │
│ # This is the list of partition tables for embedded SCSI controllers.     │
│ #                                                                         │
│ partition = "SUN1.3G" \                                                   │
│         : disk = "SUN1.3G" : ctlr = SCSI \                                │
│         : 0 = 0, 34000 : 1 = 25, 133280 : 2 = 0, 2672400 : 6 = 123, 2505120│
│                                                                           │
│ partition = "SUN2.1G" \                                                   │
│         : disk = "SUN2.1G" : ctlr = SCSI \                                │
│         : 0 = 0, 62320 : 1 = 41, 197600 : 2 = 0, 4154160 : 6 = 171, 3894240│
│                                                                           │
│ partition = "SUN2.9G" \                                                   │
│         : disk = "SUN2.9G" : ctlr = SCSI \                                │
│         : 0 = 0, 195426 : 1 = 94, 390852 : 2 = 0, 5683986 : 6 = 282, 5097708│
│ #                                                                      ▼  │
└─────────────────────────────────────────────────────────────────────────┘
```

- *acyl:* Number of alternate cylinders to be used for bad block remapping.

- *pcyl:* Actual physical number of cylinders on the drive (usually derived from the drive technical manual).

- *nhead:* Number of data heads per drive according to the operating system; sometimes referred to as "tracks."

- *nsect:* Number of sectors per track according to the operating system. Note that most current SCSI drives provide a variable number of sectors per track. The *nsect* entry is typically an average number derived by dividing the total number of sectors by the total number of tracks.

- *rpm:* Rotational speed of the drive.

- *bpt:* Number of bytes per track on the drive.

The following *format.dat* fields are optional.

- *cache:* Value used to disable the drive cache during the format procedure.

- *fmt_time:* Value used to tell the format program the amount of time necessary to format the drive.

- *trks_zone:* Typically seen on drives that use a variable number of sectors per track. The drive is divided into zones, with *N* tracks per zone.

- *atrks:* Number of alternate tracks available (per zone) for bad block remapping.

- *asect:* Number of alternate sectors available (per zone) for bad block remapping.

The final section of the *format.dat* file contains partition maps for each drive/controller combination. Fields in the partition map section of the *format.dat* file follow.

- *partition:* ASCII label used to identify the partition map.

- *disk:* ASCII label used to identify the type of disk drive.

- *ctlr:* Type of controller the drive is connected to.

- *0=0,N:* Patterns that define the slice or partition number (0 through 7), the starting cylinder of the partition, and the size (in sectors) of the partition.

- ↝ NOTE: *Because slice 2 is always defined as encompassing the entire drive, the partition starts at cylinder 0, and encompasses all sectors between the first sector and the maximum sector on the drive.*

Creating format.dat Entries for SCSI Disks

What do the numbers in the *format.dat* entry really mean? To the operating system, a SCSI disk appears to be a series of contiguous sectors. The drive controller manages the sector number to cylinder/head/sector address mapping. Consequently, the numbers in the *format.dat* entry of a SCSI disk can be just about anything, as long as the entry does not exceed the formatted capacity of the drive. To calculate a valid *format.dat* entry, you need to know either (1) the number of sectors available on the formatted drive or (2) the drive's formatted capacity (in megabytes).

➡ *NOTE: For SMD drives, it is also necessary to know the number of sectors per track and the number of bytes per sector.*

Knowing the number of bytes per track and rotational speed of the drive is also useful. However, these items of information can be derived, as explained in the following.

When the number of sectors on the drive is known, *format.dat* entries can be developed in two steps. First, use */bin/factor* to factor the number of sectors to prime numbers. Next, use the prime factors to develop the *format.dat* entry.

➡ *NOTE: The /bin/factor program is a "game" that ships with Solaris. The program will factor the numeric input argument into a series of prime numbers, and print those prime numbers in the output.*

For example, assume a drive with 10,000 sectors. The */bin/factor* program reports that $10,000 = 2*2*2*2*5*5*5*5$. Upon grouping these factors into cylinder/head/sector groupings, you can develop a *format.dat* entry as follows: $5*5*5*2$ cylinders, $2*2$ heads, $2*5$ sectors (250 cylinders * 4 heads * 10 sectors).

Assume that the formatted capacity of a drive (in megabytes) is known (512 Mb). Additional steps in the calculations follow.

1.   Divide the capacity by the sector size (512 bytes/sector): 512,000,000/512 = 1,000,000 sectors.

2.   Feed the resulting number to */bin/factor*: */bin/factor* 1,000,000 = 2*2*2*2*2*2*5*5* 5*5* 5*5.

3.   Use the resulting prime factors to develop the *format.dat* entry by organizing them into cylinder/head/sector groupings: 5*5*2*2 sectors (1,250 cylinders * 8 heads * 100 sectors/ track).

Because the number of sectors on real drives never factor easily, the time has come to apply the method on a real drive. One particularly popular drive had 1,658 data cylinders, 15 heads, and a formatted capacity of 1,079.1 Mb. The drive manual lists the drive capacity as 50,910 bytes per track with a rotational speed of 3,600 RPM. Based on this information, the second method would be used to develop the *format.dat* entry that follows.

•   1,079,100,000 bytes / 512 bytes = 2,107,617 sectors.

•   */bin/factor* 2107617 = 3 * 702,539. The result does not constitute adequate prime factors to develop a *format.dat* entry. Subtract a few sectors from the total, and try again. Subtract 17 sectors, and try factoring 2,107,600 sectors. The result is */bin/factor* 2107600 = 2 * 2 * 2 * 2 * 5 * 5 * 11 * 479.

•   By organizing the factors into cylinder/head/sector groupings, develop a *format.dat* entry: 479*4 cylinders, 2*2*5 heads, 5*11 sectors (1,916 cylinders, 20 heads, 55 sectors).

If the speed of the drive is not known, you can use 3,600 RPM for drives that transfer at 5 Mb/second or less. If the drive transfers at 5 Mb/second or more, use 5,400 or 7,200 RPM. High-performance drives typically spin at 10,000 RPM. If the drive manual does not specify the number of bytes per track, calculate an entry by multiplying the number of bytes/sector by the number of sectors/track.

✓   *TIP: Most manufacturers have bulletin boards or technical support phone numbers that users can contact to obtain the manufacturer's format.dat information for most of the popular systems on the market. Because the manufacturers know their drives better than anyone, following their recommendations whenever possible is recommended.*

Using the format Utility

Once a *format.dat* entry exists for a drive, it is time to begin the format process. Invoke the *format* utility by logging in as *root*, and typing *format* at the prompt, as shown in the following illustration.

Starting the format utility.

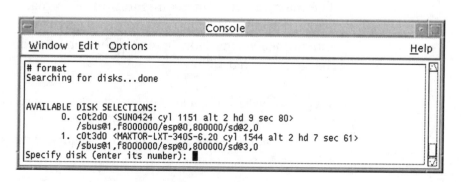

The *format* utility, the main menu of which is shown in the following illustration, will respond by listing all drives it can find attached to the system. Select the new drive by entering the number for the desired disk drive and pressing <Enter>. If the drive is already formatted, *format* will display this information and then display the main format menu. If the drive is not formatted, the *format* command will determine a default configuration (determined by probing the device) and ask if it is okay to use a default label that matches the current drive configuration.

```
┌─┐─────────────────────── Console ───────────────────┌─┐┌□┐
│─│                                                    │◄││ │
│ Window  Edit  Options                           Help│ └─┘
├────────────────────────────────────────────────────┬─┐
│Specify disk (enter its number): 0                  │▲│
│selecting c0t2d0                                    ├─┤
│[disk formatted]                                    │ │
│Warning: Current Disk has mounted partitions.       │ │
│                                                    │ │
│                                                    │ │
│FORMAT MENU:                                        │ │
│        disk       - select a disk                  │ │
│        type       - select (define) a disk type    │ │
│        partition  - select (define) a partition table │
│        current    - describe the current disk      │ │
│        format     - format and analyze the disk    │ │
│        repair     - repair a defective sector      │ │
│        label      - write label to the disk        │ │
│        analyze    - surface analysis               ├─┤
│        defect     - defect list management         │ │
│        backup     - search for backup labels       │ │
│        verify     - read and display labels        │ │
│        save       - save new disk/partition definitions │
│        inquiry    - show vendor, product and revision │
│        volname    - set 8-character volume name    │ │
│        !<cmd>     - execute <cmd>, then return      ├─┤
│        quit                                        │ │
│format> █                                           │▼│
└────────────────────────────────────────────────────┴─┘
```

The main menu allows configuration of format operation, defect list extraction and management, drive partitioning, drive labeling, and surface verification. Each of the menu items is examined in the sections that follow.

✗   *WARNING: When used to format a disk or to perform surface analysis,* format *is a destructive procedure. Verify that file system backups have been performed before formatting a drive that has been in operation.*

disk Command

The *disk* menu, shown in the following illustration, is used to select the disk drive to be modified. This menu is a near duplicate of the *format* command menu. When formatting several disks it is possible to simply select a new disk via the disk menu instead of exiting and restarting *format* for each drive. The disk menu item displays the current controller and drive geometry information for each drive available on the system.

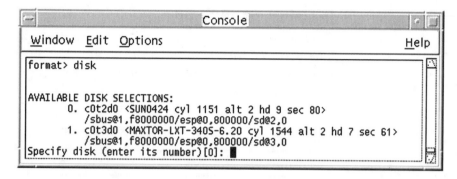

The format *utility* disk *menu.*

type Command

The *type* menu, shown in the following illustration, allows the selection of the appropriate drive type, and geometry information for the drive to be formatted. The geometry and type information are read in from the entries in the *format.dat* file.

The format *utility* type *menu.*

```
                              Console
    Window  Edit  Options                              Help

format> type

AVAILABLE DRIVE TYPES:
        0. Auto configure
        1. Quantum ProDrive 80S
        2. Quantum ProDrive 105S
        3. CDC Wren IV 94171-344
        4. SUN0104
        5. SUN0207
        6. SUN0327
        7. SUN0340
        8. SUN0424
        9. SUN0535
        10. SUN0669
        11. SUN1.0G
        12. SUN1.05
        13. SUN1.3G
        14. SUN2.1G
        15. SUN2.9G
        16. MAXTOR-LXT-340S-6.20
        17. other
Specify disk type (enter its number)[8]: █
```

partition Command

The *partition* menu, shown in the following illustration, is used to examine and modify the partition information for the drive about to be formatted.

The format *utility* partition *menu.*

```
┌─────────────────────────────────────────────────────────────────┐
│ ─                         Console                          ▸ □   │
├─────────────────────────────────────────────────────────────────┤
│ Window   Edit  Options                                  Help     │
├─────────────────────────────────────────────────────────────────┤
│ format> part                                                     │
│                                                                  │
│ PARTITION MENU:                                                  │
│        0        - change `0´ partition                           │
│        1        - change `1´ partition                           │
│        2        - change `2´ partition                           │
│        3        - change `3´ partition                           │
│        4        - change `4´ partition                           │
│        5        - change `5´ partition                           │
│        6        - change `6´ partition                           │
│        7        - change `7´ partition                           │
│        select - select a predefined table                        │
│        modify - modify a predefined partition table              │
│        name   - name the current table                           │
│        print  - display the current table                        │
│        label  - write partition map and label to the disk        │
│        !<cmd> - execute <cmd>, then return                       │
│        quit                                                      │
│ partition> █                                                     │
└─────────────────────────────────────────────────────────────────┘
```

The numbered entries in the *partition* menu refer to available drive partitions. The partitions are always numbered 0 through 7. Remaining entries in the partition menu are submenus that allow customization of the partition tables.

The *select* menu, shown in the following illustration, allows selection of a predefined partition map as supplied in */etc/format.dat*. It is also possible to "copy" the partition map of another drive of the same type on the system.

The partition
select
submenu.

The *modify* menu, shown in the following illustration, allows modification of the current partition map. This menu is used to set the size and name of the file systems on the drive. Within this menu, commands are provided to allow the partition map to be examined, printed, or saved to a file.

The partition
modify *submenu.*

The *name* menu, shown in the following illustration, allows selection of a partition name for the current partition map. The name is saved in the */etc/format.dat* file for future reference.

*The partition
name submenu.*

The *print* submenu, shown in the following illustration, makes it possible to view the partition map that is being modified. The information provided in the printout allows the determination of the size of the partition(s), starting cylinder for the partition(s), and other information about the partition(s) on the disk drive.

*The partition
print
submenu.*

The *label* menu, shown in the following illustration, includes the current partition size, and name information to be written into the volume label. This command performs in the same way as the label menu item in the *format* utility main menu.

The partition label submenu.

current Command

The *current* menu, shown in the following illustration, displays information about the currently selected drive. Information provided includes the addressing information for the controller for the drive and the drive geometry information.

The format utility current menu.

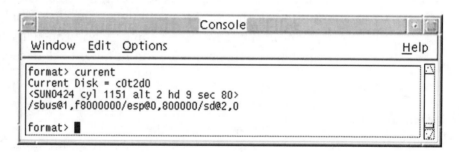

repair Command

The *repair* menu, shown in the following illustration, provides the ability to add and subtract entries to and from the current drive defect list. Bad spot locations are specified via an absolute block number (for SCSI and IPI disks), or by giving the cylinder/head/sector information for the bad spots (on SMD drives).

➥   *NOTE: The IDE controller handles all repair operations automatically; therefore, the format program will not allow the operator to "repair" an IDE disk drive.*

The format utility repair menu.

label Command

The *label* menu, shown in the following illustration, writes a label on the current drive. This selection operates the same as the label selection in the *partition* submenu. The disk label contains information on the disk drive geometry and the size of the partitions, including starting cylinder and number of data sectors.

The format *utility* label *menu.*

analyze Menu

The *analyze* menu, shown in the following illustration, is used to set the surface analysis parameters for the verification portion of the *format* utility. This menu also allows various surface analysis programs to be executed independent of the *format unit* operation.

The format *utility* analyze *menu.*

```
+------------------------------------------------------------------+
|  -                        Console                         | · | □ |
+------------------------------------------------------------------+
|   Window  Edit  Options                               Help        |
+------------------------------------------------------------------+
| format> analyze                                               [▒] |
|                                                                   |
| ANALYZE MENU:                                                     |
|         read      - read only test    (doesn't harm SunOS)        |
|         refresh   - read then write   (doesn't harm data)         |
|         test      - pattern testing   (doesn't harm data)         |
|         write     - write then read        (corrupts data)        |
|         compare   - write, read, compare (corrupts data)          |
|         purge     - write, read, write   (corrupts data)          |
|         verify    - write entire disk, then verify (corrupts data)|
|         print     - display data buffer                           |
|         setup     - set analysis parameters                       |
|         config    - show analysis parameters                      |
|         !<cmd>    - exeucte <cmd> , then return                    |
|         quit                                                      |
| analyze> ▮                                                        |
+------------------------------------------------------------------+
```

The *read, refresh, test, write, compare,* and *purge* menu items allow control of respective test sequences.

✗   *WARNING: The* write, compare, *and* purge *options will destroy data stored on the disk. The* read, refresh, *and* test *items do not destroy data stored on the disk drive.*

The *print* item prints out a disk data buffer to the screen, which permits the operator to view the pattern currently being written to the drive. The *setup* option in the *analyze* menu, shown in the following illustration, allows configuration of the surface analysis portion of format.

The analyze setup
submenu.

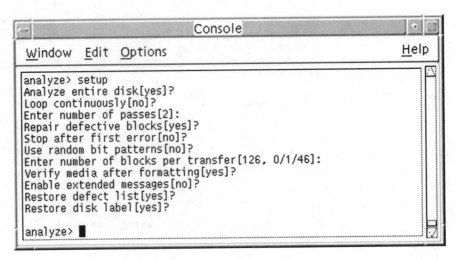

```
                         Console
 Window   Edit   Options                              Help

 analyze> setup
 Analyze entire disk[yes]?
 Loop continuously[no]?
 Enter number of passes[2]:
 Repair defective blocks[yes]?
 Stop after first error[no]?
 Use random bit patterns[no]?
 Enter number of blocks per transfer[126, 0/1/46]:
 Verify media after formatting[yes]?
 Enable extended messages[no]?
 Restore defect list[yes]?
 Restore disk label[yes]?

 analyze> ▮
```

The *setup* menu allows selection of the portion of the disk to be analyzed, as well as the transfer size to use. This menu also provides control over whether bad spots are mapped out or merely reported to the operator. The *analyze config* menu item, shown in the following illustration, allows verification of current settings before continuing with the format operation.

Using the analyze
config *command
to show current
setup.*

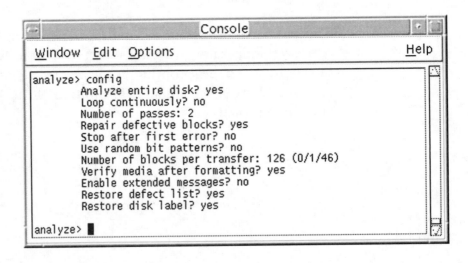

```
                         Console
 Window   Edit   Options                              Help

 analyze> config
          Analyze entire disk? yes
          Loop continuously? no
          Number of passes: 2
          Repair defective blocks? yes
          Stop after first error? no
          Use random bit patterns? no
          Number of blocks per transfer: 126 (0/1/46)
          Verify media after formatting? yes
          Enable extended messages? no
          Restore defect list? yes
          Restore disk label? yes

 analyze> ▮
```

defect Menu

The *defect* menu of *format*, shown in the following illustration, is used to read and manipulate the manufacturer defect list from a drive. It also allows defects to be added or removed from the defect list.

➡ *NOTE: The IDE controller handles all repair operations automatically; therefore, the format program will not allow the operator to use the* defect *menu when working with an IDE disk drive.*

The format *utility* defect *submenu.*

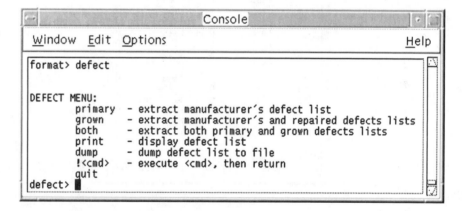

```
                              Console
  Window  Edit  Options                              Help

 format> defect

 DEFECT MENU:
         primary  - extract manufacturer's defect list
         grown    - extract manufacturer's and repaired defects lists
         both     - extract both primary and grown defects lists
         print    - display defect list
         dump     - dump defect list to file
         !<cmd>   - execute <cmd>, then return
         quit
 defect>
```

The *primary* menu item, shown in the following illustration, allows the vendor defect list to be read from the disk media. The *grown* and *both* menu items allow the vendor defect list and the repaired defect list to be extracted from the disk media.

Using the defect *menu* primary *command to extract the vendor defect list.*

```
                              Console
  Window  Edit  Options                              Help

 defect> primary
 Extracting primary defect list...Extraction complete.
 Defect List has a total of 14 defects.

 defect>
```

Once the defect list(s) have been extracted from the media, this information can be saved in a file with the *dump* menu item. The *print* menu allows the defect list to be viewed on the screen, as shown in the following illustration.

Viewing and saving the defect list with the print and dump commands.

```
                            Console
 Window   Edit  Options                                    Help
defect> print
  num     cyl    hd      bfi     len    sec    blk
   1        0     0      867      0
   2       44     6     3179      0
   3      568     0    18207      0
   4      965     1    17051      0
   5     1174     0    10693      0
   6     1216     0    11849      0
   7     1225     0    30923      0
   8     1267     5    18207      0
   9     1289     0    21675      0
  10     1374     0    23409      0
  11     1414     0    25721      0
  12     1451     0    23409      0
  13     1465     0    11849      0
  14     1477     0     7225      0
total of 14 defects.

defect> dump
Enter name of defect file: /tmp/424.bads
```

backup Menu

The *backup* menu, shown in the following illustration, is used to look for backup copies of the disk label. This can be particularly useful for drives on which the primary label has somehow been damaged. If the primary label is intact, *format* will display this information, and ask if the search for backup labels should continue. Once a valid label is found, the *backup* command will write this label to the disk.

Using the format
utility backup
command to
restore the drive
label.

```
┌─────────────────────────── Console ───────────────────────────┐
│ Window  Edit  Options                                     Help │
├────────────────────────────────────────────────────────────────┤
│ format> backup                                                  │
│ Disk has a primary label, still continue? y                     │
│ Searching for backup labels...found.                            │
│ Restoring primary label.                                        │
│                                                                 │
│ format>                                                         │
└────────────────────────────────────────────────────────────────┘
```

verify Menu

The *verify* menu, shown in the following illustration, searches the disk for a valid label. Once the label is found, *verify* displays the information on the screen.

Using the format
utility to verify the
disk label.

```
┌─────────────────────────── Console ───────────────────────────┐
│ Window  Edit  Options                                     Help │
├────────────────────────────────────────────────────────────────┤
│ format> verify                                                  │
│                                                                 │
│ Primary label contents:                                         │
│                                                                 │
│ Volume name = <       >                                         │
│ ascii name  = <SUN0424 cyl 1151 alt 2 hd 9 sec 80>              │
│ pcyl        = 2500                                              │
│ ncyl        = 1151                                              │
│ acyl        =    2                                             │
│ nhead       =    9                                             │
│ nsect       =   80                                             │
│ Part      Tag     Flag   Cylinders      Size         Blocks     │
│   0 unassigned     wm    0              0      (0/0/0)       0   │
│   1 unassigned     wm    0              0      (0/0/0)       0   │
│   2     backup     wm    0 - 1150   404.65MB   (1151/0/0) 828720 │
│   3 unassigned     wm    0 -  438   154.34MB   (439/0/0)  316080 │
│   4 unassigned     wm    0              0      (0/0/0)       0   │
│   5 unassigned     wm    0              0      (0/0/0)       0   │
│   6        usr     wm  439 - 1150   250.31MB   (712/0/0)  512640 │
│   7 unassigned     wm    0              0      (0/0/0)       0   │
│                                                                 │
│ format>                                                         │
└────────────────────────────────────────────────────────────────┘
```

save Menu

The *save* menu, shown in the following illustration, saves the current setup and partitioning information to the system */etc/format.dat* file. The user is given the option of selecting another file to which the information will be saved.

Using the format *utility to save partition information.*

```
format> save
Saving new partition definition
Enter file name["./format.dat"]: /tmp/format.dat
format> █
```

inquiry Menu

The *inquiry* menu, shown in the following illustration, probes the disk drive and displays the vendor name, product, and firmware revision level of the current disk.

Using the format *utility* inquiry *command.*

```
format> inquiry
Vendor:    SEAGATE
Product:   ST1480    SUN0424
Revision: 6266
format> █
```

volname Menu

The *volname* menu, shown in the following illustration, allows the creation of a volume name for the drive. Once the volume name is entered, the *volname* command writes the volume name to the current drive.

Using the format
utility volname
command.

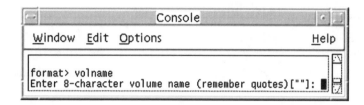

Using the format
utility volname
command.

format Command

Once all setup and customization are completed, it is time to format the drive. This is accomplished with the *format* menu selection on the main menu.

The *format* menu, shown in the following illustration, is used to start the format process. Starting the process will cause the system to issue the *format-unit* command (for IPI and SCSI drives), or, for SMD disk drives, to begin writing format information to the disk drive via program control. Once the formatting is complete, the *format* utility will run a few passes of surface analysis on the drive. When the surface analysis is complete, a geometry label is written to the disk and control switches back to the *format* utility main menu.

The format *utility*
format *menu.*

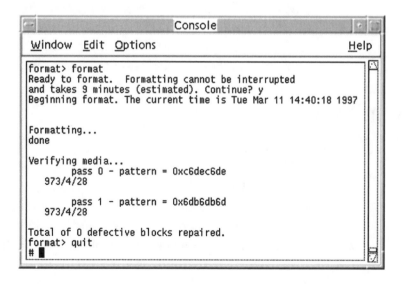

quit Menu

The *quit* menu, shown in the following illustration, exits the *format* utility, and returns you to the *root* shell prompt.

Using the format *utility* quit *command to return to the shell.*

```
                          Console
  Window  Edit  Options                              Help

format> ?
Expecting one of the following: (abbreviations ok):
          disk       - select a disk
          type       - select (define) a disk type
          partition  - select (define) a partition table
          current    - describe the current disk
          format     - format and analyze the disk
          repair     - repair a defective sector
          label      - write label to the disk
          analyze    - surface analysis
          defect     - defect list management
          backup     - search for backup labels
          verify     - read and display labels
          save       - save new disk/partition definitions
          inquiry    - show vendor, product and revision
          volname    - set 8-character volume name
          !<cmd>     - execute <cmd>, then return
          quit

format> quit
# █
```

✓   *TIP: One of the simplest ways of breathing new life into a sluggish system is to add disk space and I/O channels. These procedures allow the I/O load of the system to be dispersed across more channels and devices, thereby reducing the load on existing devices.*

Summary

This chapter focused on the mass storage subsystem, one of the most critical Solaris subsystems. Topics included how to install a new disk drive, and formatting and partitioning a disk drive. In the following chapters, selected topics will be revisited and refined as more advanced troubleshooting and storage subsystem management techniques are presented.

Disk Subsystem Software

Once new disks have been formatted and before they are put into operation, strategic planning questions are in order. For instance, you should determine the size of the file systems on the disk, how the space will be optimized, and how the file systems are to be mounted. This chapter focuses on software issues associated with disk subsystem management, including how to determine partition sizes, mounting file systems, and the use of Volume Manager.

Calculating Disk Slice Sizes

What happens if the company purchases a disk drive that is not formatted and partitioned for Solaris systems? How do you change the formatting and partitioning such that the drive is useful for a particular application? As shown in Chapter 9, "Disk Subsystem Hardware," the *format* utility provides the capabilities of formatting the drive, and writing a Solaris partition map on the drive.

The partition maps define how the drive is sliced into logical units. The partitions are sometimes referred to as *disk slices*. You can use a default partition map or develop a partition map that customizes the file system(s) for the local environment. To ensure that operations are performed on the correct entity, you need a method of referring to a specific disk slice.

Disk Slice References

Using block numbers to describe the partition start and end points is cumbersome. Most UNIX operating systems define a symbolic system, which is used to describe the partitions on a hard disk drive. Under Solaris, different regions of the disk are called *slices*. The drive label contains partition information that details the blocks encompassed by each slice of the drive. Under Solaris, the disk slices are numbered as shown in table 10-1, which follows.

Table 10-1: Solaris Disk Partitions

Disk Slice	Typical Usage
0	Root file system (bootable partition)
1	Swap space (*swapfs* file system that is not mounted)
2	Entire drive (sometimes called the overlap partition)
3	/var (optional)
4	/usr/openwin (optional)
5	/opt (optional)
6	/usr filesystem (default)
7	/export/home partition (default)

➥   *NOTE: The floppy diskette device uses an older convention of naming slices, with the letters "a" through "h." This convention held that the third or "c" slice would be sized to encompass the entire disk.*

Every disk drive stores the partition information for the file systems in the disk label on the drive. The disk label contains the volume table of contents (VTOC) and other information about the disk drive. The next section describes the VTOC and the information it stores.

Volume Table of Contents (VTOC)

A label including the VTOC is written on the disk as part of the format process. The VTOC contains information about the drive geometry (number of cylinders, heads, and sectors) and the partition map for the drive, as well as information about how each partition is being used by the system. Much of this information is summarized by the tags and flags fields of the VTOC.

Slice Tags and Flags

The information about mount points, and mount options, is stored in a portion of the VTOC called the "tags and flags" fields. These fields identify how the partitions are used and mounted. The tag field is used to identify how the slice is being used. Valid entries in this field are listed in table 10-2, which follows. Note that the tag field value is not the same as the slice number.

Table 10-2: Solaris VTOC Tag Fields

Tag Field Value	Description
0	Unassigned partition
1	Boot partition
2	Root partition
3	Swap partition
4	User partition
5	Backup partition

Tag Field Value	Description
6	Stand partition
7	Home partition

The flag field is used to determine how the partition is to be mounted. Valid entries are summarized in table 10-3, which follows.

Table 10-3: Solaris VTOC Flag Fields

Flag Field Value	Description
00	Mounted read/write
01	Not mounted
10	Mounted read-only

Viewing the VTOC with the prtvtoc Utility

It is possible to view the content of the VTOC using the */etc/prtvtoc* utility, as shown in the following illustration. The *prtvtoc* utility reads the disk label, and lists the geometry of the drive, the size of each partition, the mount point for the partition (if known), and other information stored in the VTOC and drive label. The syntax for the *prtvtoc* command follows.

```
# prtvtoc /dev/rdsk/c#t#d#s2
```

Using prtvtoc to view the volume table of contents (VTOC).

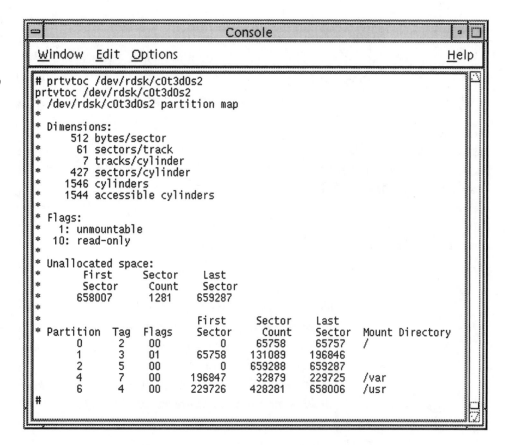

```
# prtvtoc /dev/rdsk/c0t3d0s2
prtvtoc /dev/rdsk/c0t3d0s2
* /dev/rdsk/c0t3d0s2 partition map
*
* Dimensions:
*      512 bytes/sector
*       61 sectors/track
*        7 tracks/cylinder
*      427 sectors/cylinder
*     1546 cylinders
*     1544 accessible cylinders
*
* Flags:
*    1: unmountable
*   10: read-only
*
* Unallocated space:
*       First     Sector      Last
*       Sector     Count     Sector
*       658007      1281     659287
*
*                             First    Sector    Last
* Partition  Tag  Flags      Sector     Count   Sector   Mount Directory
          0    2   00             0     65758    65757   /
          1    3   01         65758    131089   196846
          2    5   00             0    659288   659287
          4    7   00        196847     32879   229725   /var
          6    4   00        229726    428281   658006   /usr
#
```

Over-allocation Overhead

As mentioned in Chapter 2, "Solaris Concepts and Terminology," a typical rule of thumb when partitioning disks is to allow for 10 percent of the sectors in a file system to be consumed by *over-head*. The simplest way to illustrate overhead is by examining the output of the */bin/df* command.

In the following output, it is obvious that the *used* and *avail* numbers provided by *df* do not add up to the *kbytes* amount. The difference between these quantities points out the existence of file system overhead. Part of this difference may be attributed to the way the kernel reserves space on a file system. The Berkeley Fast File system used by Solaris reserves a certain number of sectors for use in short-term over-allocation situations. The amount of space reserved for such emergencies is controlled by the parameters used when the file system is created.

```
# df -k
Filesystem          kbytes    used     avail     capacity    Mounted on
/dev/dsk/c0t3d0s0   32431     32262    0         100%        /
/dev/dsk/c0t3d0s6   240463    77640    38783     36%         /usr
/proc               0         0        0         0%          /proc
fd                  0         0        0         0%          /dev/fd
/dev/dsk/c0t1d0s6   30651     12436    15155     46%         /var
swap                267916    4        267912    1%          /tmp
```

➾ *NOTE: When a file system reaches the 100% full level, the system will generate a message on the system console. The administrator should examine the file system and remove any temporary or unused files. If this operation does not free enough space to allow for normal operation, you will need to take more drastic measures (such as moving files to another file system) to ensure proper system operation.*

Super-block Overhead

Another source of file system overhead is the support structure required to keep track of the files in the file system. One of these support structures is called the *super-block*. The super-block can be considered the index of the file system content. Any references to a file or directory on the partition must first reference the super-block to determine which sectors are part of that file or directory.

The primary super-block is stored in one of the first few sectors of the partition. Backup or alternate copies of the super-block are stored in several places on the partition in case of failure of the primary super-block. On small file systems there are few copies of the alternate super-blocks. On large file systems, there are many copies of the alternate super-blocks.

General Rules for Determining Partition Sizes

The following general rules help simplify the task of determining the appropriate size for a partition.

1. Determine how much space the application currently occupies.

2. Add 50% to the number from step 1 for growth and expansion of the file system.

3. Add 10% more space for file system overhead.

4. Divide the number from the previous step by the sector size (512).

5. Divide the number from the previous step by *Nheads* * *Nsects* to determine how many cylinders are required. Round up to the nearest whole number.

For example, an application that requires 100 Mb of storage space, and a disk drive with a geometry of 20 heads and 97 sectors/track, would yield the following calculations.

- Original package size: N = 100 Mb

- Add space for growth: M = N + (N * .5) = 150 Mb

- Add 10% overhead space: O = M + (M * .1) = 165 Mb

- Divide by 512 bytes per sector: P = 173,015,040 bytes / (512 bytes/sector) = 337,920 sectors required

- Ncyl = 337,920 sectors / (20 heads * 97 sectors per head per cylinder) = 175 cylinders required for the partition

✓   *TIP: The following formula is a quick method of estimating disk space requirements: 2,048 sectors * 512 bytes/sector is equivalent to 1 Mb of disk space.*

Repartitioning Disk Drives

As file systems outgrow current partitions or as file systems become unnecessary, you may have occasion to repartition a disk. It should be apparent from the content of the VTOC that repartitioning a disk is a destructive process.

✗   *WARNING: Before repartitioning a disk drive, it is wise to save all data stored on the target disk drive. Failure to do so will result in data loss.*

If it becomes necessary to repartition a disk, save all files from the disk drive by dumping them to tape. Once all information from the disk has been saved, you can use the *format* utility to reformat/repartition the drive. Once the drive has been repartitioned, restore the files from the backup tape.

Types of File Systems

Once a disk is partitioned, you need to create a file system on each partition such that the disk space is available to the system. Solaris supports a variety of file system types that can be mounted as part of the overall file tree. File system types must be understood in order for you to create the proper type for use within the local environment. Table 10-4, which follows, describes the file systems that may be used with Solaris.

Table 10-4: File System Types Supported by Solaris

File System Type	Hardware Medium	Description
ufs	Hard disk	Standard UNIX file system using the BSD fast file system.
hsfs	CD-ROM	High Sierra or ISO 9660 file system; can be mounted read-only.
pcfs	Floppy disk	MS-DOS file system.
nfs	Network file system	Network-based file system.
tmpfs	Memory/swap area	Special-purpose, high-performance file system for temporary data.
cachefs	Hard disk and nfs file systems	Special-purpose, high-performance file system to cache CD-ROM and NFS disks.
lofs	Loopback	Special-purpose file system that allows multiple mount points for a single file system.
procfs	Memory	Special-purpose memory-based file system for use by debuggers and other development tools.
udf	CD-ROM, DVD, hard disk, diskette	Universal Disk Format – used by Digital Video disks, CD-ROM media. Available under Solaris 7 11/99 and Solaris 8.
vxfs	RAID disk arrays	Veritas File System.
swapfs	Hard disk	Special-purpose file system used to extend virtual memory on the system.

The first five file system types listed—*ufs*, *hsfs*, *pcfs*, *nfs*, and *tmpfs*—are the most commonly used under Solaris. In the following diagram showing a sample UNIX file tree and file system mount points, the */tmp* file system is a *tmpfs* file system, whereas the other file systems are *ufs* file systems. This is a fairly common situation for a standalone workstation. A workstation on a network may employ many types of file systems.

Sample Solaris file tree.

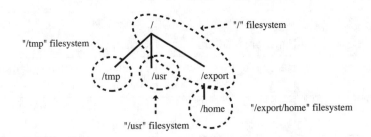

ufs File Systems

The *ufs* file system is the standard UNIX file system. Under Solaris, the *ufs* file system is based on the Berkeley Fast File system. This file system implemented several modifications to early UNIX file systems in order to improve disk subsystem performance and reliability. Many system utilities will work only on *ufs* file systems. The most notable of these utilities are the *ufsdump* and *ufsrestore* commands.

Beginning with Solaris 7, a new UFS option has been added: *ufs* logging. A file system that employs *ufs* logging is more fault tolerant than a standard *ufs* file system. The *ufs* logging feature allocates 1 megabyte of file system space per gigabyte of file system for use as a transaction log. The system disk driver writes all transactions to the log. Once the *ufs* log has been written, the information in the *ufs* log is written to the file system. Consult the manual page for *mount_ufs* for more information on *ufs* logging.

hsfs File Systems

The High Sierra file system (HSFS, also known as the ISO 9660 file system) is a special-purpose file system developed for use on CD-ROM media. The geometry and structure of CD-ROM media are different from those of a hard disk drive. The Solaris *hsfs* file system also supports the Rock-Ridge extensions to the HS file system. Consult the manual page for *hsfs* for more information about HS file systems.

pcfs File Systems

The *pcfs* file system is a UNIX implementation of the DOS file attribute table (FAT) file system. This file system allows Solaris systems to mount PC-DOS formatted file systems. The *pcfs* file system allows UNIX users direct read/write access to PC-DOS files using standard UNIX utilities (e.g., *cat*, *ls*, and *rm*). Files in the *pcfs* file system are limited to the standard 8.3 file name conventions (eight characters followed by a dot, followed by three characters). File names are created using all upper-case characters. Consult the manual page for *pcfs* for more information on *pcfs* file systems.

nfs File Systems

Network file systems (NFS) allow users to share files between many types of systems on the network. NFS provides a method of making a disk drive on one system (a server) appear as though it were connected to another system (a client). Extra kernel-level code is required to ensure that valuable information is not lost from client machines if and when a file server is down. Consult the manual page for *nfs* for more information about *nfs* file systems.

tmpfs File Systems

The *tmpfs* file systems are a memory-based, special-purpose file system created and destroyed every time a system is rebooted.

The memory-based file system provides much better performance than a disk-based file system. The *tmpfs* file systems are often used to provide temporary scratch storage (which may be used by system utilities, and users) while using the system. Consult the manual pages for *tmpfs* for more information on *tmpfs* file systems.

cachefs File Systems

The *cachefs* file systems are an offshoot of the *nfs* file systems. Modern disk subsystem technology provides much better throughput than exists on many networks. Network bottlenecks typically occurred when an *nfs* client required large quantities of data from a server. The server could get the information from its local disk very quickly, but had to "feed it through a funnel" to get it to the client.

The *cachefs* is a disk-based cache on the client machine. The *cachefs* software uses predictive algorithms to determine which disk blocks the client machine will request. These blocks are copied to the disk-based *cachefs* file system on the client machine. When the client machine needs the data, it can access it from its own disk drive. If the server changes the cached data, the *cachefs* is informed that the data is "dirty," which causes the client to request a fresh copy from the server. Any changes to the cached data made by the client machine are written directly to the server's disk via NFS.

lofs File Systems

The *lofs* file systems are virtual file systems that may be used to create new file systems that allow alternate paths to the files in existing file systems. Consult the manual page for *lofs* for more information.

procfs File Systems

The *procfs* file system is similar in concept to the device access methods provided by the file entries in the */dev* directory. The files found in a *procfs* file system are entries that map the address space of a process into a file. Opening one of these files allows you to read directly from the address space of a running process. This file system is used by certain debugging tools as a method of observing and changing the behavior of a process while it is running.

udf File Systems

The *udf* file system provides support for the file system format used on digital video disk/digital versatile disk (DVD) systems. The *udf* file system is available for use with DVD and CD-ROM drives, as well as with diskettes and SCSI hard drives.

vxfs File Systems

The *vxfs* file system is a virtual file system employed with RAID disk arrays. The *vxfs* file system allows the administrator to concatenate several disks into a single large file system, and to create mirrored or striped file systems within the RAID array. Consult the documentation for the Sun StorEdge system for more information on the *vxfs* file system.

swapfs File Systems

The *swapfs* is a special file system used by the SunOS kernel for paging and swapping. SunOS, like many modern versions of UNIX, manages memory through a system of fixed size units called *pages*. Program instructions and data are broken up into pages and loaded into memory wherever free pages exist. The operating system, by means of a page table and special memory management hardware, presents a linear memory map to each process even though the instructions and data that comprise the process may be scattered throughout memory.

Virtual Memory

Virtual memory is the process by which physical memory is extended by moving pages that have not been recently referenced out of memory and on to disk. This process of moving program instructions and data in and out of memory is called *paging*. The *swapfs* file system is used by SunOS to store pages that are not currently in use. Because of its special usage, a *swapfs* file system does not appear in the UNIX file tree, and the disk space consumed by a *swapfs* file system does not appear in the output of commands such as *df*.

➡   NOTE: *Although the* df *command lists the swap partition, the space utilization numbers do not always reflect an accurate view of swap usage. This is due to the fact that the swap partition is also used by the* /tmp *file system. In order to determine the swap space utilization, use the* swap *command, as detailed in the next section.*

Using the swap Command

It is possible to determine the disk space in use by *swapfs* file systems with the */usr/sbin/swap* command. This command can read and interpret the content of a *swapfs* file system and report back on its size and current usage. An example appears in the following illustration.

The swap *command's two reporting options.*

The first form of the *swap* command, using the *-l* option, gives the size of the *swapfs* file system in 512-byte disk blocks, along with its current usage, offset, SunOS device name, and major and minor device numbers. The second form is more frequently used. The *-s* option lists current usage by detailing the amount of space allocated, the space claimed for usage by SunOS but not currently being used, the total of the two previous values, and the amount of unused space. This form of the *swap* command is often used to help determine how much *swapfs* space is needed by a system when running certain programs.

✓   *TIP: If a system requires additional swap space, it is possible to create swap files or swap partitions, and to direct Solaris to add them to the total system virtual memory using the* swap *command with the -a option.*

Creating File Systems

Except in special cases, a file system must be created on a partition before the partition can be used under Solaris. The act of creating a file system puts a framework on the partition that enables Solaris to store files there. The file system framework provides index information for the disk blocks initialized and mapped when the disk was formatted.

Creation of a new file system can be accomplished by invoking the */usr/sbin/newfs* utility. The *newfs* utility allows you to create file systems with default (or custom) file system parameters, as shown in the following illustration. In effect, *newfs* is just a user-friendly front end to the */etc/mkfs* program. The *mkfs* command creates the file system super-blocks, divides the file system into several cylinder groups, and takes care of creating all file system sub-structures required by the operating system.

➡   *NOTE: In order to create a* udf *file system, use the command* # mkfs -F udfs /dev/rdsk/device-name. *To display* udf *file system parameters, use the command* # mkfs -F udfs -m /dev/rdsk/device-name.

Using newfs *to create file systems.*

Working with the newfs Command

The default type of file system created by *newfs* is a *ufs* file system. The *newfs* command consults the */etc/format.dat* file used by format and the disk label to identify the specifications for the disk, and build the appropriate file system. The syntax for invoking the *newfs* command follows.

```
# newfs [mkfs options] /dev/rdsk/c#t#d#s#
```

Invoking *newfs* with the *-N* option causes *newfs* to display the basic parameters of the file system without actually creating it. An example is shown in the following illustration.

Viewing file system parameters.

The *newfs* command will display information about the new file system as it operates. The first line printed by *newfs* describes the basic disk geometry. The previous illustration shows information regarding a disk that is a two-sided (two heads) floppy diskette with 18 sectors and 80 cylinders for a total of 2,880 sectors (or blocks).

The second line of output from *newfs* describes the *ufs* file system created in this partition. In the example, the file system is 1.4 Mb in size, containing five cylinder groups of 16 cylinders per group. Cylinder groups are a concept introduced in the Berkeley UNIX Fast File system. Cylinder groups provide a method by which file blocks are allocated close together on the disk to improve performance.

✓ TIP: *Varying the size of the cylinder groups is one way to tune the performance of a* ufs *file system.*

The final two lines of *newfs* output list the locations of the super-block backups. The super-block is the head of the file index node or *inode* information used by the *ufs* file system. The *inodes* are "used" by routines in the Solaris kernel to allocate, read, write, and delete files. Backup copies of these super-blocks are created in the event that a system crash, power failure, or other problem occurs that leaves the main copy of the super-block damaged.

How newfs Connects to mkfs

As previously mentioned, the *newfs* command is a front end to *mkfs*. The *newfs* command invokes *mkfs* to perform the actual creation of the file system on the target device. The *mkfs* command requires a long list of parameters. For most applications a detailed knowledge of the *mkfs* parameters is unnecessary; *newfs* makes proper choices for the parameters it passes to *mkfs*. However, it is useful to know the parameters so that modifications may be performed when applications require specialized adjustments to a file system.

For example, it is possible to build a file system tuned to store a large quantity of small, frequently changing files such as those found in the *Usenet net-news* system. Such a file system would require more index nodes or *inodes* than are provided by the default options used by *newfs*. Situations that require manual modification of *mkfs* parameters are covered in more detail in Chapter 11, "Disk and File System Maintenance."

✓   TIP: *Use of the -v option flag with* newfs *will allow the administrator to see which parameters it passes to* mkfs. *The example shown in the following illustration uses the -N option, as well as to display the parameters without actually making the file system.*

Viewing the mkfs parameters with the -v and -N options.

```
# newfs -N -v /dev/rdiskette
mkfs -F ufs -o N /dev/rdiskette 2880 18 2 8192 1024 16 10 5 2048 t 0 -1 8 1
/dev/rdiskette: 2880 sectors in 80 cylinders of 2 tracks, 18 sectors
        1.4MB in 5 cyl groups (16 c/g, 0.28MB/g, 128 i/g)
super-block backups (for fsck -F ufs -o b=#) at:
 32, 640, 1184, 1792, 2336,
#
```

Some of the parameters in the preceding example should look familiar. Many of these parameters may be altered by using an option flag to *newfs* instead of manually invoking *mkfs*. Table 10-5, which follows, lists each *newfs*/*mkfs* option, reading from left to right across the *mkfs* command line. The first column lists the *mkfs* parameter, the second column lists the *newfs* option flag used to modify the parameter, and the third column describes the parameter.

Table 10-5: Options Cross Reference for newfs to mkfs

mkfs Parameter	newfs Option Flag	Description
-F ufs	–	Indicates the type of file system to be created.
-o N	-N	No change flag; shows what would be done, but does not actually create the file system.
/dev/rdiskette	–	Raw device on which the file system should be created (in this case, a floppy diskette).
2880	-s	Total size of the file system (in blocks).
18	–	Number of sectors per track.
2	-t	Number of tracks per cylinder.
8192	-b	Size of the allocation block.
1024	-f	Size of a file fragment.
16	-c	Number of cylinders per cylinder group.
10	-m	Minimum free space required on the file system.
5	-r	Rotational speed of disk specified in rpm for *newfs*, and in revolutions per second for *mkfs*.
2048	-I	Number of bytes allocated per *inode*.
T	-o	Optimization method (space or time).
0	-a	Alternate blocks per cylinder.

When *newfs* is invoked, it prompts the operator for confirmation before proceeding. The creation of a file system is a destructive process. The administrator must ensure that the data on the partition has been copied to tape before allowing *newfs* to operate. Creation of a file system is shown in the following illustration.

Creating a file system.

```
# newfs -v /dev/rdiskette
newfs: construct a new file system /dev/rdiskette: (y/n)? y

mkfs -F ufs /dev/rdiskette 2880 18 2 8192 1024 16 10 5 2048 t 0 -1 8 15

/dev/rdiskette: 2880 sectors in 80 cylinders of 2 tracks, 18 sectors
        1.4MB in 5 cyl groups (16 c/g, 0.28MB/g, 128 i/g)
super-block backups (for fsck -F ufs -o b=#) at:
 32, 640, 1184, 1792, 2336,
#
```

After asking the operator for confirmation, *newfs* invokes *mkfs* to create the file system. The *mkfs* command reports the location of each of the super-block backups as it creates them. It creates a set of index nodes (*inodes*) and the linked data structure used to access them. Typically, about 10 percent of the total data blocks available are also set aside to avoid the greatly reduced performance a *ufs* file system encounters as the last few data blocks are allocated.

If *mkfs* reports an error, it usually means that the disk has been improperly partitioned, or formatted incorrectly. The *mkfs* command expects a disk partition to contain a number of blocks that is an integer multiple of the number of blocks in a single disk cylinder. Consult Chapter 9, "Disk Subsystem Hardware," for help in correctly formatting and partitioning a disk.

Tuning Space Usage with the newfs Command

The default *newfs* parameters will typically result in the loss of 10 percent of the usable disk space on the partition. Part of the space loss is due to the amount of reserved over-subscription space. The *newfs* command allows alterations in the amount of space reserved for over-subscription storage. The *newfs* utility also allows tuning of the number of copies of alternate super-blocks, and several other parameters that affect the amount of storage space consumed by the partitioning process.

➥ *NOTE: See Chapter 11, "Disk and File System Maintenance," for more information on using* newfs *for file system tuning to reduce overhead and improve disk subsystem performance.*

Working with Local File Systems

Once a file system has been created, it must be added to the UNIX file tree. At selected spots called mount points, the file tree spans from one file system to the next. File systems are often referred to by the names of their mount points. One could say the system shown in the following illustration contains the */, /tmp, /usr,* and */export/home* file systems. Before mounting a new file system, the administrator should determine which file systems are already in use on the system.

A file tree spanning four file systems.

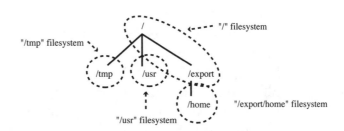

Identifying Mounted File Systems

Solaris provides several ways for the administrator to identify the mounted file systems. The */etc/mount* command invoked with no arguments, the *df* command, and examination of the */etc/mnttab* file are the most common methods.

df *Command*

The *df* command offers several different ways of displaying information about file systems mounted on a system. Running the command with no argument results in the display of information for all mounted file systems. As shown in the following, it is possible to restrict the *df* output to a single file system by giving a directory in that file system as the last argument to the *df* command.

```
glenn% df
/               (/dev/dsk/c0t3d0s0):         11690 blocks      9292 files
/usr            (/dev/dsk/c0t3d0s6):        786776 blocks    322309 files
/proc           (/proc):                        0 blocks       204 files
/dev/fd         (fd):                           0 blocks         0 files
/tmp            (swap):                    218816 blocks      3288 files
/opt            (/dev/dsk/c0t3d0s5):        91236 blocks     73801 files
/var/mail       (mail.astro.com:/var/mail): 309376 blocks    96776 files
glenn%
```

The output produced by *df* shows the mount point, the device mounted (in parentheses), the number of free 512-byte disk blocks, and the number of free files that can be created on the file system. Note that *nfs*-mounted file systems list the name of the server exporting the file system, and the name of the file system on the server. In the previous entry example, it is evident that the *glenn* machine obtains the */var/mail* file system from the *mail.astro.com* machine. When *df* is invoked with a directory argument, the output format changes, as seen in the following.

```
glenn% df /
File system          kbytes      used       avail      capacity    Mounted on
/dev/dsk/c0t3d0s     20664       14819      3785       80%         /
glenn%
```

This listing shows the device name first, followed by the total, used, and available space in 1024-byte (kbyte) units; the proportion of the file system already in use; and the mount point. To see this listing for all mounted file systems, use the *-k* option flag with no directory argument.

The most complete *df* listing is obtained by using the *-g* option, an example of which is shown in the following entry. This option formats and displays the *statvfs* structure, which details some of the parameters used to create the file system and provides a complete description of the available space. The *-g* option flag can be used with or without a directory argument to display all file systems or a single file system.

```
glenn% df -g /
/                      (/dev/dsk/c0t3d0s0):8192 block size  1024 frag size
41328 total blocks 11690 free blocks    7570 available    11520 total files
9292 free files     8388632 filesys id
ufs fstype           0x00000004 flag       255 file name length
glenn%
```

> ➥ *NOTE: It is important to remember the differences among the three block sizes discussed in this chapter. A disk block is the basic unit of storage on the physical medium. This unit is often 512 bytes long, but not always. An allocation block is the basic unit of space in a file system. It is either 8,192 or 4,096 bytes in length. A fragment, often 1,024 bytes in length, is a portion of an allocation block.*

/etc/mnttab

Another way to determine which file systems are mounted is by examination of the */etc/mnttab* file. This file is created and maintained by the */etc/mount* command. Every time a file system is mounted, an entry is added to the *mnttab* file. When a file system is unmounted, the entry is removed from the *mnttab* file. A typical *mnttab* file might resemble the following example.

```
# cat /etc/mnttab
/dev/dsk/c0t3d0s0 / ufs rw,suid,dev=800018 869622929
/dev/dsk/c0t3d0s6 /usr ufs rw,suid,dev=80001e 869622929
/proc /proc proc rw,suid,dev=2740000 869622929
fd /dev/fd fd rw,suid,dev=27c0000 869622929
/dev/dsk/c0t1d0s6 /var ufs rw,suid,dev=80000e 869622929
swap /tmp tmpfs dev=0 869622932
mail.astro.com:/var/mail /var/mail nfs soft,bg,rw,dev=2700002 869764852
```

mount *Command*

The *mount* command allows the administrator to view which file systems are mounted, as well as providing a method of mounting file systems. When invoked without arguments, the *mount* command lists mounted file systems by their mount points, showing the device mounted at each mount point, the mount options used, and the time the file system was mounted.

```
glenn% /etc/mount
/ on /dev/dsk/c0t3d0s0 read/write/setuid on Sat Apr 1 1:23:45 2000
/usr on /dev/dsk/c0t3d0s6 read/write/setuid on Sat Apr 1 1:23:45 2000
/proc on /proc read/write/setuid on Sat Apr 1 1:23:45 2000
/dev/fd on fd read/write/setuid on Sat Apr 1 1:23:45 2000
/tmp on swap read/write on Sat Apr 1 1:23:45 2000
/opt on /dev/dsk/c0t3d0s5 setuid/read/write on Sat Apr 1 1:23:45 2000
/var/mail on mail.astro.com:/var/mail soft/bg/read/write/remote on Sat Apr
    1 1:23:45 2000
glenn%
```

Mounting File Systems

File systems can be mounted via several methods: manual mounting by invoking the */etc/mount* command, the system can automatically mount the file system at boot time after consulting the */etc/vfstab* file, or the volume manager may be used to mount the file system.

General Procedure for Mounting a File System

The general procedure for mounting a file system is as follows.

1. Format the disk drive.

2. Partition the disk drive.

3. *newfs* the partition.

4. Create a mount point for the file system.

5. Determine which method will be used to mount the file system (manual mount, *vfstab* mount, automounter, Volume Manager).

6. Mount the file system.

mount *Command Revisited*

The *mount* command has several options that control the way in which a file system is mounted. The default action is to mount the file system and base its read and write access on the permissions of the files in the file system. The syntax of the mount follows.

```
# mount -o option,option,...,option disk_device mount_point
```

Options are given as a comma-separated list following the *-o* flag (e.g., *mount -o rw,quota*). Some of the more common options for the *mount* command are summarized in table 10-6, which follows. The illustration that follows shows an example of mounting a file system.

Table 10-6: mount Command Options

Option	Description
quota	Activates the user disk quota system when the file system is mounted.
ro	Mount the file system read-only.
-F fstype	Mount a file system of fstype.
-a	Mount all file systems with "mount at boot" (in the vfstab file) set to yes.
-o	Use specific fstype options when mounting file system.
-p	Provide a list of mounted file systems in vfstab format.
-v	Provide verbose output when mounting file system(s).
-O	Overlay the file system on an existing mount point.
-m	Mount file system without making an entry in /etc/mnttab.
nosuid	Disallow setUID execution for files in this file system.

Mounting a file system.

```
# cd /
# mkdir /mnt
# mount /dev/rdiskette /mnt
# df -k /mnt
Filesystem          kbytes    used    avail capacity  Mounted on
/dev/diskette         1263       9     1253      1%    /mnt
#
#
```

Automatic Mounting

If the administrator needed to type in the *mount* command every time it was necessary to mount a disk, it would quickly become a tiresome chore. Solaris provides several methods for automating the process of mounting file systems.

- The *vfstab* file provides the ability for file systems to be mounted automatically at boot time.

- The *automounter* allows network file systems to be mounted on demand. The *automounter* is discussed in detail in Chapter 20, "Automating NFS with Automount."

- The volume manager provides a means for users to mount/ unmount removable media without granting them *root* privileges.

vfstab *file*

One method of automating the process of mounting file systems is to add the file system to the */etc/vfstab* file. This file is read by the */etc/mountall* command when it is run as part of the system boot sequence. The *vfstab* file lists file systems to be mounted when the system is booted. A quick way to add items to *vfstab* is to use the - *p* option to the *mount* command.

An example using a floppy disk file system follows. First, the file system is mounted, then the current *vfstab* is saved, and finally, the *mount* and *grep* commands are used to append the mount instructions to the */etc/vfstab* file. This is shown in the following illustration.

✓   *TIP: Saving a copy of the current* vfstab *is good practice, as it provides an easy way to recover from any mistakes made when adding new file systems.*

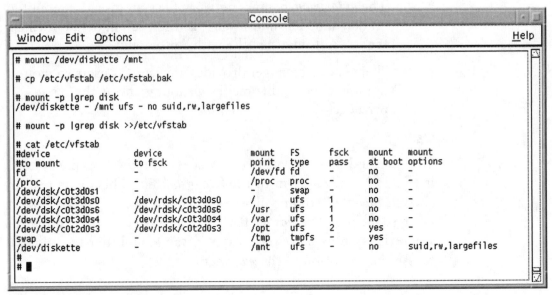

```
                                    Console
 Window   Edit   Options                                              Help

 # mount /dev/diskette /mnt

 # cp /etc/vfstab /etc/vfstab.bak

 # mount -p |grep disk
 /dev/diskette - /mnt ufs - no suid,rw,largefiles

 # mount -p |grep disk >>/etc/vfstab

 # cat /etc/vfstab
 #device                                mount   FS     fsck   mount   mount
 #to mount            to fsck           point   type   pass   at boot options
 fd                   -                 /dev/fd fd     -      no      -
 /proc                -                 /proc   proc   -      no      -
 /dev/dsk/c0t3d0s1    -                 -       swap   -      no      -
 /dev/dsk/c0t3d0s0    /dev/rdsk/c0t3d0s0 /      ufs    1      no      -
 /dev/dsk/c0t3d0s6    /dev/rdsk/c0t3d0s6 /usr   ufs    1      no      -
 /dev/dsk/c0t3d0s4    /dev/rdsk/c0t3d0s4 /var   ufs    1      no      -
 /dev/dsk/c0t2d0s3    /dev/rdsk/c0t2d0s3 /opt   ufs    2      yes     -
 swap                 -                 /tmp    tmpfs  -      yes     -
 /dev/diskette        -                 /mnt    ufs    -      no      suid,rw,largefiles
 #
 # ▮
```

Adding a file system to /etc/vfstab.

It is also possible to add new file systems by editing the *vfstab* file and entering the required information manually. The *vfstab* file format is described in the comment lines at the top of the file. The fields are separated by a tab or spaces. The following is a typical *vfstab* file. The fields and respective functions of those fields appear in table 10-7, which follows.

#device	device	mount	FS	fsck	mount	mount
#to mount	to fsck	point	type	pass	at boot	options
#						
#/dev/dsk/cld0s2	/dev/rdsk/cld0s2	/usr	ufs	1	yes	-
/proc	-	/proc	proc	-	no	-
fd	-	/dev/fd	fd	-	no	-
swap	-	/tmp	tmpfs	-	yes	-
/dev/dsk/c0t3d0s0	/dev/rdsk/c0t3d0s0	/	ufs	1	no	-
/dev/dsk/c0t3d0s6	/dev/rdsk/c0t3d0s6	/usr	ufs	2	no	-
/dev/dsk/c0t3d0s5	/dev/rdsk/c0t3d0s5	/opt	ufs	5	yes	-
/dev/dsk/c0t3d0s1	-	-	swap	-	no	-
/dev/dsk/c1t3d0s1	-	-	swap	-	no	-
mail.astro.com:/var/mail	-	/var/mail	nfs	no	yes	soft,bg,rw

Table 10-7: Fields and Functions for vfstab

Field Name	Description
Device to mount	Name of device to be mounted.
Device to *fsck*	Raw device to be checked by the *fsck* utility.
Mount point	Directory where the device should be added to the UNIX file tree.
FS type	File system type.
fsck pass	Number indicates the ordering method by which the file system will be checked.
Mount at boot	Yes to cause file system to mount at boot. No to prevent file system mount at boot.
Mount options	Options to be passed to the mount command.

fsck *Pass Field*

UNIX file systems (*ufs*) are the only file systems on which the file system checker (*fsck*) operates. If the *fsck* pass field contains a minus sign (-) or a zero (0), no file system integrity checking is performed. A file system with an *fsck* pass of one (1) indicates that the file system is to be checked sequentially (in the order it is listed in the *vfstab* file). Note that the / (*root*) file system is always checked first.

File systems with an *fsck* pass number greater than one (1) are checked in parallel (simultaneously). For systems with a large number of disks spanning multiple disk controllers or disk busses, parallel file system checking is generally faster than sequential checking. For efficiency, use *fsck* on file systems of similar size on different disks simultaneously. Once the appropriate information has been entered in the *vfstab* file, the administrator can cause the system to mount all file systems by invoking the following command.

```
# /etc/mount -a
```

Unmounting a File System

The complement to mounting a file system is to unmount it by using the */etc/umount* command. Exercise care when unmounting file systems. Some file systems are required for proper system operation. Other file systems may be unmounted while allowing the system to continue to operate, but the system will not perform as expected. The syntax for the *umount* command follows.

```
# umount mount_point (i.e. umount /mnt)
```

or

```
# umount device (i.e. umount /dev/dsk/c0t2d0s1)
```

If you are uncertain about the effect unmounting a file system may have on the system, it is best to bring the system to the single-user state before invoking the *umount* command. In the case of a file server, unmounting a file system may also have an effect on the *nfs* client machines on which the file system is mounted. An example of unmounting a file system is shown in the following illustration.

Unmounting a file system.

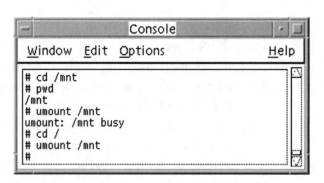

Volume Manager

. . . .

Adding entries to the *vfstab* file works well for hard disks but is not suitable for removable media such as floppies and CD-ROMs. These devices tend to be mounted and unmounted much more frequently than hard disks. To handle such devices, use the *Volume Manager*. The Volume Manager automatically detects the presence of removable media, determines the file system type, and then mounts the media appropriately.

Automatic detection of the media is limited to the CD-ROM device. The Solaris File Manager accomplishes this with a combination of a daemon (*vold*) and a configuration file (*/etc/vold.conf*), which specifies the actions to be taken for various removable devices and file system types. For floppy disks, the Volume Manager is unable to detect the presence of new disks and must be informed that a new disk is present. This is accomplished by invoking the */bin/volcheck* command, as shown in the following illustration.

The Volume Manager always mounts floppy disks under the */floppy* directory it controls. CD-ROMs are likewise mounted under the */cdrom* directory. Note that in the example the *volcheck* command was issued from an ordinary user account and not the super-user. The Volume Manager removes the need for users to have *root* privileges in order to mount removable media.

```
┌─────────────────────────────────────────────────────────────────┐
│ ─                          Console                          ▪ □ │
├─────────────────────────────────────────────────────────────────┤
│  Window  Edit  Options                                     Help  │
├─────────────────────────────────────────────────────────────────┤
│ # df -k                                                          │
│ Filesystem          kbytes    used    avail capacity  Mounted on │
│                                                                   │
│ /dev/dsk/c0t3d0s0    30702   23310     7362     76%   /          │
│ /dev/dsk/c0t3d0s6   200515  181176    19139     91%   /usr       │
│ /proc                    0       0        0      0%   /proc      │
│ fd                       0       0        0      0%   /dev/fd    │
│ /dev/dsk/c0t3d0s4    15342    5705     9622     38%   /var       │
│ /dev/dsk/c0t2d0s3   148167   10969   137050      8%   /opt       │
│ /dev/dsk/c0t2d0s6   240463  201981    38242     85%   /usr/openwin│
│ swap                 56920     196    56724      1%   /tmp       │
│ # volcheck                                                       │
│                                                                   │
│ # df -k                                                          │
│                                                                   │
│ Filesystem          kbytes    used    avail capacity  Mounted on │
│ /dev/dsk/c0t3d0s0    30702   23313     7359     77%   /          │
│ /dev/dsk/c0t3d0s6   200515  181176    19139     91%   /usr       │
│ /proc                    0       0        0      0%   /proc      │
│ fd                       0       0        0      0%   /dev/fd    │
│ /dev/dsk/c0t3d0s4    15342    5705     9622     38%   /var       │
│ /dev/dsk/c0t2d0s3   148167   10969   137050      8%   /opt       │
│ /dev/dsk/c0t2d0s6   240463  201981    38242     85%   /usr/openwin│
│ swap                 56864     200    56664      1%   /tmp       │
│ /vol/dev/diskette0/unnamed_floppy                                │
│                       1423       0     1423      0%   /floppy/unnamed_floppy │
│ #                                                                │
│ # ■                                                              │
└─────────────────────────────────────────────────────────────────┘
```

Mounting a floppy disk using the Volume Manager.

The Solaris File Manager communicates with the Volume Manager and will automatically open a new File Manager window when a removable device is mounted. The Volume Manager may also activate the File Manager. Under the File menu on the main File Manager window is a menu item that allows the user to check for new floppies. This allows the File Manager to perform the same action as typing the *volcheck* command, as shown in the following illustration. The newly created File Manager window also has an Eject button, which will unmount the file system and eject the disk from the drive.

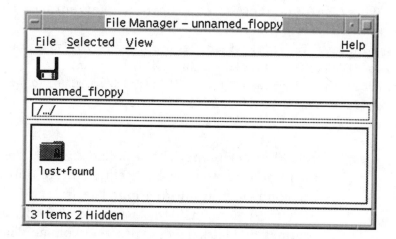

*File Manager
detection of
Volume Manager
mounted disks.*

Configuring the Volume Manager

Changing the actions taken and devices under control of the Volume Manager is a simple matter of modifying the */etc/vold.conf* file. The configuration information in the file is formatted by section. The following is a sample */etc/vold.conf* file.

```
#
# @(#)VOLD.CONF = 693/05/17 SMI
#
# Volume Daemon Configuration file
#
# Database to use (must be first)
db db_mem.so
# Labels supported
label dos label_dos.so floppy
label cdrom label_cdrom.so cdrom
label sun label_sun.so floppy
# Devices to use
use cdrom drive /dev/dsk/c0t6 dev_cdrom.so cdrom0
use floppy drive /dev/diskette dev_floppy.so floppy0
# Actions
```

```
insert /vol*/dev/diskette[0-9]/* user=root /usr/sbin/rmmount
insert /vol*/dev/dsk/* user=root /usr/sbin/rmmount
eject /vol*/dev/diskette[0-9]/* user=root /usr/sbin/rmmount
eject /vol*/dev/dsk/* user=root /usr/sbin/rmmount
notify /vol*/rdsk/* group=tty /usr/lib/vold/volmissing -c
# List of file system types unsafe to eject
unsafe ufs hsfs pcfs
```

The lines beginning with a pound sign (#) are considered comments and are used here to help delineate the various sections of the configuration information. The first two sections, beginning with the comments *Database* and *Labels*, describe the database routines and disk label types the Volume Manager recognizes. These two sections should not be modified.

Devices to use *Section*

The third section, marked by the comment *Devices to use*, lists the names and types of the removable media devices the Volume Manager should monitor. Each line in this section starts with the keyword *use*.

```
use cdrom drive /dev/dsk/c0t6 dev_cdrom.so cdrom0
```

The *use* keyword is followed by the type of device, either CD-ROM or floppy, and the keyword *drive*.

```
use cdrom drive /dev/dsk/c0t6 dev_cdrom.so cdrom0
```

Following the device type is the Solaris name for the device. Note that the CD-ROM device name specifies only the first five characters of the full special device name. Because the Volume Manager will monitor and mount all available slices it finds on a CD-ROM disk, the only information items needed are the specific controller and target portions of the device name.

```
use cdrom drive /dev/dsk/c0t6 dev_cdrom.so cdrom0
```

Following the special device name is the name of the shared object used to manage the device. This must match up with the device type specified (e.g., if the device type is *cdrom*, the shared object must be *dev_cdrom.so*). Finally, the symbolic name used in the */device* directory is listed. The first device of a given type has a 0 (zero) appended to its name, whereas the second is appended with a 1, and so on. For instance, a second CD-ROM drive located at target 5 on the built-in SCSI controller would be placed under Volume Manager control by adding the following line to the devices section of the *vold.conf* file.

```
use cdrom drive /dev/dsk/c0t5 dev_cdrom.so cdrom1
```

Actions *Section*

The next section, which begins with the comment *Actions*, specifies the actions to be taken when certain events occur. The basic events are the insertion of media into a drive (*insert*), removal of media from a drive (*eject*), and notification of problems (*notify*). An example entry in the *Actions* section follows.

```
eject /vol*/dev/diskette[0-9]/* user=root /usr/sbin/rmmount
```

Entry Content

Each line lists an event followed by a regular expression. When an event occurs, each line that begins with the name of the event (*insert*, *eject*, or *notify*) is checked. If the volume on which the event occurs matches the regular expression, the remainder of the action line comes into play. The remainder of this line includes the name of the user or group identification to be used to run the listed command with the listed arguments. In the previous example line, when the *eject* event occurs and the volume matches the regular expression */vol*/dev/diskette[0-9]/*, the command */usr/sbin/rmmount* would be run with the *root* user permissions.

rmmount.conf *Configuration File*

The */usr/sbin/rmmount* command has its own configuration file named */etc/rmmount.conf.* Although not often modified, this file allows the specification of additional actions to occur when a disk is mounted. A common use of this feature is to allow CD-ROMs mounted by the Volume Manager to be automatically shared, or made accessible to other workstations on the network via NFS. To accomplish this, a share line is added to the bottom of the */etc/ rmmount.conf* file as follows.

```
share cdrom*
```

This line would share any CD-ROM mounted by the Volume Manager without any restrictions. To control access the administrator can add options to the share line in a form similar to the *share* command. The *share* command is covered in more detail in Chapter 19, "Accessing Network Resources with NFS."

Mounting Non-ufs File Systems with the Volume Manager

The Volume Manager is also able to handle file system types other than *ufs*. For instance, inserting an MS-DOS formatted floppy and using the File Manager or the *volcheck* command results in the disk being mounted under the */floppy* directory on a mount point that bears the floppy volume name. Thus, inserting an MS-DOS floppy whose volume name is "transfer" and selecting Check For Floppy from the File menu in the File Manager will cause the window shown in the following illustration to appear. Note the DOS label on the floppy disk icon that indicates the file system type.

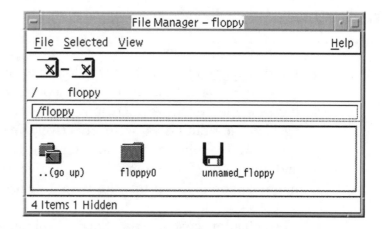

File Manager mounting an MS-DOS floppy.

Starting and Stopping the Volume Manager

The CD-ROM and floppy disk drives on a Solaris system are usually under the control of the Volume Manager. Some CD-ROM and floppy operations require that these devices not be under the influence of the Volume Manager. Solaris makes it possible to disable the Volume Manager and work directly with the CD-ROM and/or floppy drive. To disable the Volume Manager, as shown in the following illustration, invoke the */etc/init.d/volmgt* script with the *stop* option. To restart the Volume Manager, invoke the */etc/init.d/volmgt* script with the *start* option.

Stopping the Volume Manager.

Summary

Local file systems and their relationships to the UNIX file tree constitute a basic building block of a Solaris system. Understanding how file systems are created, mounted, checked, and tuned enables the administrator to effectively manage corporate disk space. More importantly, understanding how to mount, create, and unmount file systems on an active system can result in minimization of down-time for corporate users.

This chapter covered the Solaris Volume Table of Contents (VTOC), types of file systems available under Solaris, the creation of file systems, as well as a number of methods that may be employed to make the file system available for users. These methods include the *automounter*, the Volume Manager, and the *vfstab*. The next chapter discusses file system tuning and the maintenance operations required to keep the basic file system structures healthy.

CHAPTER 11
Disk and File System Maintenance

Previous chapters explored the naming conventions of devices under Solaris, disk subsystem hardware, and disk subsystem software. These chapters covered the information the system administrator needs in order to install and manage system disk drives. Unfortunately, disk drives break down. Disk drives are also throughput-limited devices. The bits can only be written to, and retrieved from, the magnetic platters at a finite speed.

This chapter presents information to help the system administrator recognize disk problems, and methods that can be used to avoid or recover from disk failures. In addition, the discussion focuses on file system performance tuning and alternate disk subsystem technology that can be employed for high-throughput file servers.

File System Repair

Although UNIX file systems are fairly rugged, problems may occur that cause damage to the index structure. Solaris keeps some of the file system structure in memory and periodically updates that structure to disk as a means of improving file system performance. However, an inadvertent shutdown, such as a power failure, might result in the index structure on the disk not being completely updated. This type of file system damage requires repair before the file system can be mounted.

Automatic File Checking

The integrity of file systems automatically mounted at boot time is checked by the *fsck* command. The Volume Manager will also use the *fsck* command to check any *ufs* file systems it mounts. In both of these cases, *fsck* is run using the *-p* (*preen*) option. The *preen* option, shown in the following illustration, automatically fixes minor file system problems, but will exit if major file system damage is encountered.

Using the fsck *preen option to check a file system.*

```
                    Console
  Window   Edit   Options              Help
 # fsck -o p /dev/rdsk/c0t1d0s2
 /dev/rdsk/c0t1d0s2: is clean.
 #
```

In the event of major damage to the index structure, *fsck* must be run manually without the *preen* option. Running *fsck* in this mode provides more detail about the checks performed by *fsck*. The following illustration is typical of *fsck* output for an intact file system.

Sample fsck *output for a clean file system.*

```
                            Console
  Window   Edit   Options                          Help
 # fsck /dev/rdsk/c0t1d0s2
 ** /dev/rdsk/c0t1d0s2
 ** Last Mounted on /book
 ** Phase 1 - Check Blocks and Sizes
 ** Phase 2 - Check Pathnames
 ** Phase 3 - Check Connectivity
 ** Phase 4 - Check Reference Counts
 ** Phase 5 - Check Cyl groups
 2 files, 9 used, 95412 free (20 frags, 11924 blocks,  0.0% fragmentation)
 #
```

Recoverable File System Damage

Typically, file system damage occurs when a system is shut down incorrectly (e.g., power failure or system crash). As a result, the file system index information in memory may not have been written to disk. A disk with this type of file system damage can usually be repaired with *fsck*.

The example in the following illustration is a file system damaged when information was written over the primary super-block. To recover from this problem, *fsck* needs to know where to find an alternate super-block. To specify the alternate super-block, invoke *fsck* with the *-o b=#* command line options. This allows *fsck* to reconstruct the primary super-block by using the information stored in an alternate super-block.

Disk with a bad super-block.

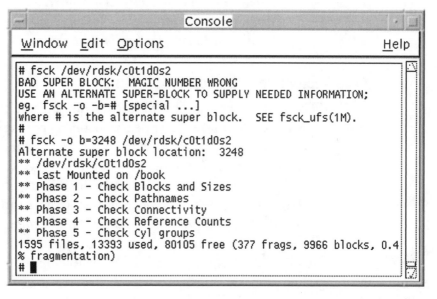

```
                              Console

  Window   Edit   Options                          Help

 # fsck /dev/rdsk/c0t1d0s2
 BAD SUPER BLOCK:   MAGIC NUMBER WRONG
 USE AN ALTERNATE SUPER-BLOCK TO SUPPLY NEEDED INFORMATION;
 eg. fsck -o -b=# [special ...]
 where # is the alternate super block.   SEE fsck_ufs(1M).
 #
 # fsck -o b=3248 /dev/rdsk/c0t1d0s2
 Alternate super block location:   3248
 ** /dev/rdsk/c0t1d0s2
 ** Last Mounted on /book
 ** Phase 1 - Check Blocks and Sizes
 ** Phase 2 - Check Pathnames
 ** Phase 3 - Check Connectivity
 ** Phase 4 - Check Reference Counts
 ** Phase 5 - Check Cyl groups
 1595 files, 13393 used, 80105 free (377 frags, 9966 blocks, 0.4
 % fragmentation)
 #
```

✓   *TIP: The alternate super-block locations are listed by the* newfs *command when the file system is created. The locations can be listed again with* newfs -N *on the raw disk device, as shown in the following illus-tration.*

```
# newfs -N /dev/dsk/c0t3d0s0
/dev/rdsk/c0t3d0s0:     65758 sectors in 154 cylinders of 7 tracks, 61 sectors
        32.1MB in 10 cyl groups (16 c/g, 3.34MB/g, 1600 i/g)
super-block backups (for fsck -F ufs -o b=#) at:
 32, 6928, 13824, 20720, 27616, 34512, 41408, 48304, 54688, 61584,
#
```

Invoking newfs *with the -N flag to identify alternate super-block locations.*

➥   *NOTE: The "FILE SYSTEM WAS MODIFIED" warning given by* fsck *should serve to remind the operator to run* fsck *again. Sometimes a single invocation of* fsck *does not correct all problems present in the index structure of a disk. Using* fsck *a second time catches any problems not corrected during the first invocation.*

Example of a Recoverable Disk

The disk shown in the following illustration contains extensive file system damage. It appears to be recoverable, but successful recovery will take some work. The first problem is a damaged directory. The *fsck* command will reconstruct the directory struc-ture by replacing mandatory parts of the directory file, but it may not be able to recover the pointers to the individual files. The result is files (and possibly directories) no longer contained in any directory. The complete sequence of *fsck* actions needed to recon-struct the file is illustrated in the following sections.

Using fsck *to repair a heavily damaged disk, part 1.*

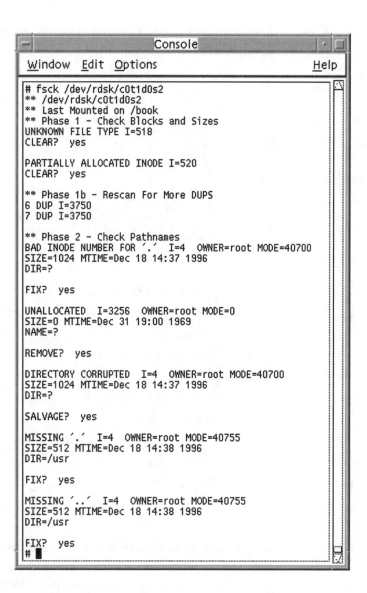

```
┌─┬──────────────────────── Console ─────────────────────┬─┬─┐
│ ─ │                                                    │ ▸ │□│
├───┴────────────────────────────────────────────────────┴───┤
│ Window  Edit  Options                               Help   │
├─────────────────────────────────────────────────────────┬──┤
│ # fsck /dev/rdsk/c0t1d0s2                               │▲│
│ ** /dev/rdsk/c0t1d0s2                                   │ │
│ ** Last Mounted on /book                               │ │
│ ** Phase 1 - Check Blocks and Sizes                    │ │
│ UNKNOWN FILE TYPE I=518                                 │ │
│ CLEAR?  yes                                            │ │
│                                                        │ │
│ PARTIALLY ALLOCATED INODE I=520                        │ │
│ CLEAR?  yes                                            │ │
│                                                        │ │
│ ** Phase 1b - Rescan For More DUPS                     │ │
│ 6 DUP I=3750                                           │ │
│ 7 DUP I=3750                                           │ │
│                                                        │ │
│ ** Phase 2 - Check Pathnames                           │ │
│ BAD INODE NUMBER FOR '.'  I=4  OWNER=root MODE=40700   │ │
│ SIZE=1024 MTIME=Dec 18 14:37 1996                      │ │
│ DIR=?                                                  │ │
│                                                        │ │
│ FIX?  yes                                              │ │
│                                                        │ │
│ UNALLOCATED  I=3256  OWNER=root MODE=0                 │ │
│ SIZE=0 MTIME=Dec 31 19:00 1969                         │ │
│ NAME=?                                                 │ │
│                                                        │ │
│ REMOVE?  yes                                           │ │
│                                                        │ │
│ DIRECTORY CORRUPTED  I=4  OWNER=root MODE=40700        │ │
│ SIZE=1024 MTIME=Dec 18 14:37 1996                      │ │
│ DIR=?                                                  │ │
│                                                        │ │
│ SALVAGE?  yes                                          │ │
│                                                        │ │
│ MISSING '.'  I=4  OWNER=root MODE=40755                │ │
│ SIZE=512 MTIME=Dec 18 14:38 1996                       │ │
│ DIR=/usr                                               │ │
│                                                        │ │
│ FIX?  yes                                              │ │
│                                                        │ │
│ MISSING '..'  I=4  OWNER=root MODE=40755               │ │
│ SIZE=512 MTIME=Dec 18 14:38 1996                       │ │
│ DIR=/usr                                               │ │
│                                                        │ │
│ FIX?  yes                                              │▼│
│ # █                                                    │ │
└────────────────────────────────────────────────────────┴──┘
```

lost+found *Directory*

Once the directories are repaired, the repair effort moves on to reconnecting files and directories to appropriate directories. The *lost+found* directory is used by *fsck* as a place to deposit files and directories that will be relinked to the directory tree. In the following illustration, the *lost+found* directory itself has been damaged and is recreated.

File system repair efforts, part 2.

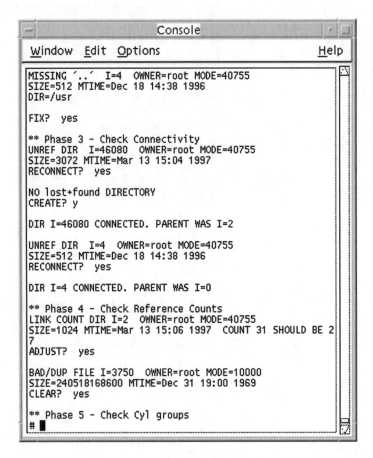

```
                          Console
 Window  Edit  Options                              Help

MISSING '..'  I=4  OWNER=root MODE=40755
SIZE=512 MTIME=Dec 18 14:38 1996
DIR=/usr

FIX?  yes

** Phase 3 - Check Connectivity
UNREF DIR  I=46080  OWNER=root MODE=40755
SIZE=3072 MTIME=Mar 13 15:04 1997
RECONNECT?  yes

NO lost+found DIRECTORY
CREATE?  y

DIR I=46080 CONNECTED. PARENT WAS I=2

UNREF DIR  I=4  OWNER=root MODE=40755
SIZE=512 MTIME=Dec 18 14:38 1996
RECONNECT?  yes

DIR I=4 CONNECTED. PARENT WAS I=0

** Phase 4 - Check Reference Counts
LINK COUNT DIR I=2  OWNER=root MODE=40755
SIZE=1024 MTIME=Mar 13 15:06 1997  COUNT 31 SHOULD BE 2
7
ADJUST?  yes

BAD/DUP FILE I=3750  OWNER=root MODE=10000
SIZE=240518168600 MTIME=Dec 31 19:00 1969
CLEAR?  yes

** Phase 5 - Check Cyl groups
#
```

The files and directories in *lost+found* are listed by the number of the index node (*inode*) they inhabit. File name information is stored in the directory file structure. When a file system sustains damage to the directory file structure, the names of the files in the directory are frequently damaged (or lost). However, the file's content, ownership, and permissions are usually left intact because they are stored in the *inode* for the file. If the *inode* for the file is intact, *fsck* can save the file's content even if the name of the file is lost.

✓ TIP: *It is often faster (and easier) not to reconnect zero-length files. Zero-length files may occur if a program creates a temporary file, opens it, and then removes it while the file is still open. The result is a file that does not belong to any directory, but is still available to be written and read by the program. Under normal operations the file is cleaned up when the program exits.*

With the files and directories reconnected to the file tree in the *lost+found* directory, *fsck* repairs the reference count information in the directories and updates free block information in the super-block. The modified directories and super-block are thereby corrected with respect to the actions *fsck* took earlier, as shown in the following illustration.

File system repair efforts, part 3.

```
┌─────────────────────────────────────────────────────────────────┐
│ ─                          Console                          · ▣  │
├─────────────────────────────────────────────────────────────────┤
│ Window  Edit  Options                                      Help  │
├─────────────────────────────────────────────────────────────────┤
│ ** Phase 4 - Check Reference Counts                              │
│ LINK COUNT DIR I=2  OWNER=root MODE=40755                        │
│ SIZE=1024 MTIME=Mar 13 15:06 1997  COUNT 31 SHOULD BE 27         │
│ ADJUST?  yes                                                     │
│                                                                  │
│ BAD/DUP FILE I=3750  OWNER=root MODE=10000                       │
│ SIZE=240518168600 MTIME=Dec 31 19:00 1969                        │
│ CLEAR?  yes                                                      │
│                                                                  │
│ ** Phase 5 - Check Cyl groups                                   │
│ CG 0: BAD MAGIC NUMBER                                          │
│ SUMMARY INFORMATION BAD                                          │
│ SALVAGE?  yes                                                    │
│                                                                  │
│ CG 1: BAD MAGIC NUMBER                                          │
│ CG 2: BAD MAGIC NUMBER                                          │
│ CG 3: BAD MAGIC NUMBER                                          │
│ CG 4: BAD MAGIC NUMBER                                          │
│ CG 5: BAD MAGIC NUMBER                                          │
│ CG 6: BAD MAGIC NUMBER                                          │
│ CG 7: BAD MAGIC NUMBER                                          │
│ CG 8: BAD MAGIC NUMBER                                          │
│ CG 9: BAD MAGIC NUMBER                                          │
│ 1595 files, 13393 used, 80105 free (377 frags, 9966 blocks,  0.4% fragmentatio│
│                                                                  │
│ ***** FILE SYSTEM WAS MODIFIED *****                             │
│ # █                                                              │
└─────────────────────────────────────────────────────────────────┘
```

Finally, *fsck* is run again to verify that no problems remain from the first run of *fsck*. With the file system structure intact, the file system is mounted. Now comes the tough job of looking through the files in the *lost+found* directory. This is not easy because the file names are missing. Fortunately, in this case the only items in the *lost+found* directory are directories, and their content provides clues to respective names. The directories can be returned to their proper places and the file system can be put back into service.

Restoring Files from Backup Versus Repairing a File System

Fortunately, repairing file systems is not a common task for the system manager. When confronted with repairing a file system, keep in mind the balance between restoration of the file system from backups versus the repair effort. If a recent backup is available, it may be quicker to restore the file system than attempt repair. Repairing a file system can be very time consuming, and it is not always easy to determine what to do with the remnants *fsck* leaves behind.

For example, the damaged file system in the preceding illustration generated over 5,000 lines of *fsck* output, and required almost an hour to manually repair the damage. A full restore of the 14 Mb of data on this file system would have been a much more productive use of the administrator's time. The fourth part of the file system repair effort is shown in the following illustration.

*File system
repair efforts,
part 4.*

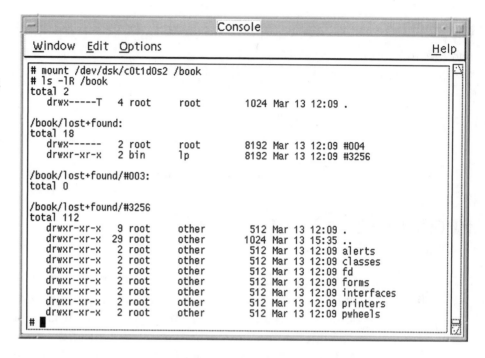

```
┌─                              Console                            ◄ ─│□│
│  Window  Edit  Options                                         Help  │
├──────────────────────────────────────────────────────────────────┬──┤
│ # mount /dev/dsk/c0t1d0s2 /book                                    │▲│
│ # ls -lR /book                                                     │ │
│ total 2                                                            │ │
│     drwx-----T   4 root      root         1024 Mar 13 12:09 .      │ │
│                                                                    │ │
│ /book/lost+found:                                                  │ │
│ total 18                                                           │ │
│     drwx------   2 root      root         8192 Mar 13 12:09 #004   │ │
│     drwxr-xr-x   2 bin       lp           8192 Mar 13 12:09 #3256  │ │
│                                                                    │ │
│ /book/lost+found/#003:                                             │ │
│ total 0                                                            │ │
│                                                                    │ │
│ /book/lost+found/#3256                                             │ │
│ total 112                                                          │ │
│     drwxr-xr-x   9 root      other         512 Mar 13 12:09 .      │ │
│     drwxr-xr-x  29 root      other        1024 Mar 13 15:35 ..     │ │
│     drwxr-xr-x   2 root      other         512 Mar 13 12:09 alerts │ │
│     drwxr-xr-x   2 root      other         512 Mar 13 12:09 classes│ │
│     drwxr-xr-x   2 root      other         512 Mar 13 12:09 fd     │ │
│     drwxr-xr-x   2 root      other         512 Mar 13 12:09 forms  │ │
│     drwxr-xr-x   2 root      other         512 Mar 13 12:09 interfaces│
│     drwxr-xr-x   2 root      other         512 Mar 13 12:09 printers│ │
│     drwxr-xr-x   2 root      other         512 Mar 13 12:09 pwheels│▼│
│ # █                                                                │░│
└──────────────────────────────────────────────────────────────────┴──┘
```

Unrecoverable File System Damage

Some of the problems in a damaged file system can be repaired, whereas others cannot. Disk drives that have suffered media damage may contain unrecoverable data sectors. The *fsck* command will notify the user about blocks that are no longer readable, and then exit. Disks that have sustained media damage are often beyond repair.

It may be possible to copy the undamaged files to another disk (or tape) and manually select critical files that have not been backed up. Damaged floppy diskettes can often be reused once they are reformatted, but it is best to replace hard disks if you suspect media damage. The next section explores a few typical disk failure modes. The following illustration shows a disk with unrecoverable errors.

Disk with unrecoverable errors.

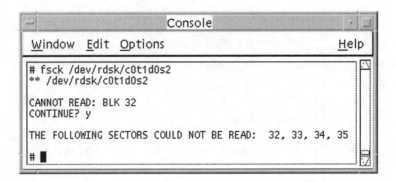

```
# fsck /dev/rdsk/c0t1d0s2
** /dev/rdsk/c0t1d0s2

CANNOT READ: BLK 32
CONTINUE? y

THE FOLLOWING SECTORS COULD NOT BE READ:   32, 33, 34, 35

#
```

Recognizing Failing Disks

Over time, all disk drives will experience damage of one sort or another. Damage is usually limited to repairable file system problems due to unexpected shutdowns. Hardware failure is also a possible (albeit less frequent) cause of disk subsystem failure. What are the symptoms of a failing disk drive? Can anything be done to prevent such failures? Once the telltale signs of an impending failure are present, what can be done to avoid catastrophe?

Disk Failure Modes

Most disk drive failures are not characterized by immediate catastrophic failure. More typically, bad spots will appear gradually until the system can no longer read/write major portions of the drive. The system will usually print messages on the console informing the user that something unexpected has occurred.

Soft Errors

In many cases, bad spots on the disk surface can be read without losing data because the disk subsystem contains logic to detect and correct multi-bit information errors. As the system encounters these bad spots, the *syslog* utility will be informed. The *syslog* facility will create console error messages informing the operator that a retryable error has been encountered.

These messages mean that the system had to try several times to read a portion of the disk. System administrators should not be concerned if they notice occasional retry errors, such as that shown in the following illustration. A few soft errors are to be expected from time to time. If the errors become more frequent, you should study the problem and take corrective action before it becomes critical.

Typical retryable disk error message.

Hard Errors

Sometimes the system cannot read a bad spot. The drive logic that detects these errors fails to correct them. In this case the system will report that it has encountered a fatal disk error, such as that shown in the following illustration. Whenever a fatal disk error is reported, the administrator should be very concerned about the integrity of the files on the disk drive. Steps must be taken to rectify the problem immediately, or data may be destroyed.

Typical hard disk error message.

Electronics Failures

A less typical disk subsystem failure involves problems with disk drive electronics. When a drive electronics failure occurs, the system may not be able to open the drive for read/write access. This type of failure is not easily repaired by the end user. In the unfortunate event that the drive had not been backed up to tape, it is sometimes possible to replace the drive circuit boards so the content of the failing drive may be copied to tape. This operation usually voids the manufacturer's warranty, and should not be attempted by the system administrator. Most disk manufacturers and third-party disk repair centers will perform this service for a fee.

✓   *TIP: With SCSI disk systems, a failure resulting from loose connections or faulty terminators is common. Checking all cable connections, power supplies, and terminators before replacing a drive is recommended.*

✗   *WARNING: The administrator should always ensure that there is a usable, recent backup of the data on a disk before attempting to reformat a failing drive. If a backup has not been performed, contact your disk vendor or data recovery service before attempting to work on the disk. Data recovery services can often salvage a large percentage of the data from a failing disk drive. If the administrator has written format information or surface test patterns over the data, the task of recovering the data is much more difficult and expensive.*

Repairing Damaged Disks

There are several courses of action that may be followed when the administrator suspects a disk is failing. The following are basic procedures for repairing bad disks.

- Reboot the system to determine if the failure is due to corrupted system information.

- If the block number of a bad spot is known, use the *repair* portion of the *format* utility to map out the spot on the drive.

➥ *NOTE: The repair option does not work with IDE disk drives. The IDE subsystem should handle bad-block mapping on its own.*

- If there are several bad spots on the drive, use the *analyze* portion of the *format* utility to perform non-destructive surface analysis of the drive to search for and optionally repair bad spots.

- Reformat the defective disk. Remember to perform a file system backup of all file systems on the drive before you reformat it. A simple reformat will often allow the drive to be placed back into operation.

- Replace the drive with a known good drive. Copy all data to the new drive. Note that this option can be time consuming and expensive.

Using format to Repair Bad Spots

If a disk drive is generating a large number of error messages on a few disk blocks, consider repairing the bad spots with the *format* utility *repair* command. The bad spot will be marked as unusable so that the system will not attempt to access this sector in the future. The *repair* facility will also "map" the bad spot to a spare sector on the disk drive, and attempt to copy the data from the bad spot to the replacement sector.

If the *analyze* facility locates a large number of bad spots on the disk, it may be prudent to reformat the entire disk. Before formatting the disk, verify that current backups are available for file systems on the failing disk. Once the drive has been reformatted, reload the file systems from the dump tapes. The following illustration shows *repair* being used to correct a bad spot on a disk.

Using repair to fix a bad spot on a disk.

Cases of Fatal Error

In some cases, *format* cannot recover the data from a bad spot on a disk. In this case, the system will print a warning message stating that it has encountered a fatal error. If an attempt is made to repair such a sector, the bad spot will be mapped out, but the file that contained the bad sector may be destroyed by the mapping process. If this occurs, you will need to restore the damaged file from a backup tape. The following illustration shows a fatal error being corrected with the *repair* command.

```
Console
Window  Edit  Options                                          Help

format> repair
Enter block number of defect: 1488457
Ready to repair defect, continue? y
Block 1488460 (680/16/4), Fatal media error (header not found) during read
Warning: unable to save track data.
Block 1488526 (680/16/70), Fatal media error (header not found) during write
Warning: unable to save track data.
Repairing block 14888457 (680/16/1)...done

format>
```

Repairing a fatal error with the repair *command.*

If the block number of the bad spot is not known, or if there appear to be multiple bad spots on the disk, configure the *format* utility *analyze* routine to scan the entire drive to find and repair all bad spots, as shown in the following illustration.

```
Console
Window  Edit  Options                                          Help

analyze> read
Ready to analyze (won't harm SunOS), This takes a long time.
but is interruptable with CTRL-C.  Continue? y

        pass 0
Block 1425054  (651/16/21), Fatal media error (header not found) during read
Repairing...succeeded.

Block 1433798  (655/16/17), Fatal media error (header not found) during read
Warning: unable to save track data
Repairing...succeeded.

Block 1436028  (656/16/60), Corrupted non-media error (soft retry) during read
succeeded.

Block 1435979  (656/16/11), Fatal media error (header not found) during read
Repairing...Warning: unable to save defective data for blkno 1435979.
succeeded.

Total of 4 defective blocks repaired
analyze>
```

Using analyze *to locate and repair bad spots on a disk.*

If *analyze* continues to find errors on successive passes, it may be time to replace the disk drive with a new one. If another drive (of the same model and geometry) is available, a simple shell script may be used to copy files from a failing drive to a new drive, as shown in the following illustration.

```
# for i
> in 0 6
> do
> dd if=/dev/rdsk/c0t1d0s"$i" of=/dev/rdsk/c0t2d0s"$i" bs=100b
> fsck /dev/rdsk/c0t2d0s"$i"
> done
read: I/O error
162+0 records in
162+0 records out
** /dev/rdsk/c0t2d0s0
** Last Mounted on /usr/tools
** Phase 1 - Check Blocks and Sizes
** Phase 2 - Check Pathnames
** Phase 3 - Check Connectivity
** Phase 4 - Check Reference Counts
** Phase 5 - Check Cyl groups
1669 files, 4946 used, 2562 free (26 frags, 317 blocks,  0.3% fragmentation)
read: I/O error
1603+0 records in
1603+0 records out
** /dev/rdsk/c0t2d0s6
** Last Mounted on /usr/lang
** Phase 1 - Check Blocks and Sizes
** Phase 2 - Check Pathnames
** Phase 3 - Check Connectivity
** Phase 4 - Check Reference Counts
** Phase 5 - Check Cyl groups
1420 files, 12788 used, 61934 free (222 frags, 7714 blocks,  0.2% fragmentation)
#
```

Using dd *to copy files between disks.*

If the new disk drive is a different model, or if the geometry of the failing disk differs from that of the new one, use *ufsdump* to copy the bits from the failing drive to another disk or directly to tape. Once the failing disk has been backed up to tape, remove it from the system and install the replacement drive. With the new drive installed and formatted, use *ufsrestore* to restore the file system content to the new drive, as shown in the following illustration.

```
┌─────────────────────────────────────────────────────────────────┐
│  ─                              Console                      ▲  ▫ │
├─────────────────────────────────────────────────────────────────┤
│  Window  Edit  Options                                     Help   │
├─────────────────────────────────────────────────────────────────┤
│ # ufsdump 0f - /dev/rdsk/c0t2d0s6 | (cd /mnt ; ufsrestore rf -) ▒│
│   DUMP: Writing 32 Kilobyte records                             ▒│
│   DUMP: Date of this level 0 dump: Sat Mar 15 23:44:06 1997      │
│   DUMP: Date of last level 0 dump: the epoch                    │
│   DUMP: Dumping /dev/rdsk/c0t2d0s6 (grumpy:/usr/lang) to standard output.│
│   DUMP: Mapping (Pass I) [regular files]                        │
│   DUMP: Mapping (Pass II) [directories]                         │
│   DUMP: Estimated 27952 blocks (13.65MB).                       │
│   DUMP: Dumping (Pass III) [directories]                        │
│   DUMP: Dumping (Pass IV) [regular files]                       │
│ Warning: ./lost+found: File exists                              │
│   DUMP: 27902 blocks (13.62MB) on 1 volume at 97 KB/sec          │
│   DUMP: DUMP IS DONE                                             │
│ # sync                                                          │
│ # umount /mnt                                                   │
│ # fsck /dev/rdsk/c0t1d0s6                                       │
│ ** /dev/rdsk/c0t1d0s6                                           │
│ ** Last Mounted on /mnt                                         │
│ ** Phase 1 - Check Blocks and Sizes                            │
│ ** Phase 2 - Check Pathnames                                   │
│ ** Phase 3 - Check Connectivity                                │
│ ** Phase 4 - Check Reference Counts                            │
│ ** Phase 5 - Check Cyl groups                                  │
│ 1420 files, 12788 used, 57326 free (230 frags, 7137 blocks,  0.3% fragmentation)│
│ # █                                                             ▒│
└─────────────────────────────────────────────────────────────────┘
```

Using ufsdump/ufsrestore *to copy files between disks.*

Improving Disk Subsystem Performance

The input/output (I/O) subsystem of any computer is one of the major pieces of the system performance puzzle. An inadequate I/O subsystem can severely limit the performance of the system. On personal computers, this I/O limitation is not (usually) a significant problem because only one user is accessing the system at any given time. On the other hand, such limitations in large file server systems can bring a corporation to its knees.

Because of the importance of high-performance I/O in the corporate computing environment, several new technologies have emerged to address this need. Not all performance problems, however, are the result of insufficient I/O subsystems. Sometimes the file system may be tweaked, or tuned, to provide significantly better system performance. The following sections touch upon selected aspects of file system tuning and technologies available to improve I/O subsystem performance.

➡️   *NOTE: Due to the performance characteristics of SCSI and IDE disk subsystems, it would be advisable to incorporate SCSI-based disk drives in any high-performance disk subsystems. The high-end SCSI drives outperform their IDE counterparts in such applications.*

File System Optimization

The file system created by *newfs* will provide acceptable performance for most applications. There are a few situations in which adjusting or tuning the basic file system parameters will result in better performance. One such situation involves applications that have very specific file system needs.

For instance, file system tuning usually improves performance of special disks, disk controllers, and applications that write files of a particular size (e.g., many small files, few large files). The various parameters that affect file system performance can be set using option flags when the file system is created, or they can be adjusted later using the *tunefs* command.

tunefs *Command*

The *tunefs* command provides a means of adjusting various file system parameters after a file system has been created. The *tunefs* command can also be used to correct mistakes made when the file system was created with *newfs*, or to adjust for changing disk usage. Some file system parameters cannot be adjusted without recreating the file system. In these cases it is necessary to dump the file system to tape and recreate the file system using different parameters.

Similar to tuning a musical instrument or a small engine, the best approach to file system tuning is to adopt a cycle of test, adjust, and retest. The testing should resemble the application that will be in use on the file system. In fact, the application itself should be used as the test if possible. File system performance measurements are helpful, such as the amount of time it takes for a basic operation to complete (to judge the effect of tuning efforts). With the measurement defined and a test method developed, it is possible to evaluate file system changes to determine whether they improve disk subsystem performance.

➥   *NOTE: A growing number of modern SCSI disks are "smart" in that they have a sophisticated internal controller and memory buffer that hide certain performance characteristics. These disks provide high performance, but are not readily tunable. The internal controller and memory buffer mask the effects of selected adjustments such as rotational delay factors.*

What tunefs Can and Cannot Do

File system tuning can only provide limited performance improvements. As servers became faster and jobs more demanding, the roles of the file server and file server disk subsystem have come under increasing scrutiny.

Some file servers provide large file systems for users to access over the network. These file systems are full of current information (files) and are frequently accessed. Other file servers are used for archival storage; files consist of historical information and are not accessed very often. Both types of file servers could be referred to as "disk farms."

Performance tuning is critical to the first type (active files), and probably not very important for the archive server. Unfortunately, *tunefs* can only do so much to improve performance for either type of server. When *tunefs* is no longer sufficient, you may want to investigate other available technologies.

In most instances, disk farms are set up to provide high-speed access to important data. Typical applications include database systems, medical images, and other space-intensive information services. In a few instances, the disk farm is used for long-term archival storage. Because the information on these systems is not accessed very often, high performance is not always the driving criterion in file system tuning.

Although *tunefs* is capable of changing many characteristics of the disk subsystem, it is not a do-all–save-all I/O subsystem fix-it tool. Certain performance parameters are beyond repair with utilities such as *tunefs*. A few of these parameters are the rate at which the disk can transfer data, rotational speed of the drive, and the amount of storage space available on the drive.

On the other hand, *tunefs* can change several important file system parameters. The following sections outline several typical changes made with the *tunefs* command. Many of these changes can be made without the need for backups and restoration of files in the file system. In order to implement a few types of changes, the file system will need to be dumped to tape, then reloaded from the backup media once the changes are implemented. In all cases, adjustment of file system parameters requires that the file system be unmounted.

➥   *NOTE: It is always a good idea to verify that the file system has been backed up to tape before making changes. If something goes wrong, the backup media could be used to reload the files once a new file system is created with the desired parameters.*

Tuning for Special Disks and Disk Controllers

By default, *newfs* uses the information in the disk label to calculate several factors related to the ability of the disk, controller, and CPU to read or write information. As the disk rotates, each block in the track moves past the disk heads. If the controller and CPU are fast enough, blocks may be written or read in the order they come under the heads.

Slower controller/CPU combinations must skip one or more blocks between read/writes in order to keep up with the I/O demands of the system. The *newfs* and *tunefs* commands allow the rotational interleave aspect of the file system to be adjusted by calculating the required time for a single block read to be processed. The proper ordering of read/write operations and skip operations is calculated from this value in combination with the rotational speed of the disk.

Some disk controllers are capable of writing or reading multiple blocks in a single operation due to high-speed buffer memory located in the controller. The *tunefs* command allows specification of the number of buffers the controller hardware provides. This improves file system performance by avoiding unnecessary block skipping. The option flags used to adjust these parameters are listed in table 11-1, which follows.

➤ *NOTE: Setting the proper parameter values requires detailed knowledge of the disk hardware. These values are generally set by* newfs.

Table 11-1: tunefs Parameters Associated with High-performance Disk Controllers

newfs Option	tunefs Option	Parameter
[-d gap	-d gap	The "gap" is the time (in milliseconds) required for a disk service interrupt to be completed and a new transfer started. This value will be used along with rotational speed of the drive, and number of sectors per track, to determine the number of blocks to skip between disk transfers.
-C maxcontig	-a maxcontig	The maximum number of blocks that can be written contiguously before a block must be skipped. This number depends on disk and controller characteristics.

Tuning for Small and Large Files

Applications that consistently read and write very small (or very large) files can often benefit from file system tuning. Large files are often split between cylinder groups on the disk. A cylinder group is a collection of cylinders used by the disk I/O routines in the SunOS kernel to improve disk performance by grouping a file's data blocks close together on the disk.

When a large file is split over two or more cylinder groups, the file's data blocks are spread across the disk. Consequently, extra seek time is required when reading or writing large files. Adjusting the maximum number of blocks that can be allocated to a single file within a cylinder group may help to reduce this problem.

It is also possible to adjust basic allocation block and fragment sizes on the file system. Larger allocation block and fragment sizes favor large files by reducing the time required for file allocation at the expense of reduced space efficiency. If an application stores small files (exclusively), a smaller allocation block will improve speed, and a smaller fragment size will improve space utilization efficiency by avoiding the allocation of blocks that are much larger than the data to be stored.

Tuning for Usenet-news

Applications that use many small files, such as *Usenet-news*, typically require more index nodes (*inodes*) than are normally created by the default *newfs* operation. This value can be changed by adjusting the parameters that control the number of bytes per *inode*. (See the following illustration.)

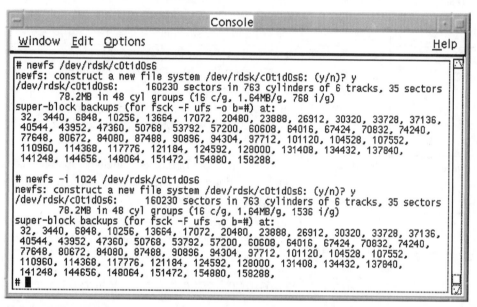

```
# newfs /dev/rdsk/c0t1d0s6
newfs: construct a new file system /dev/rdsk/c0t1d0s6: (y/n)? y
/dev/rdsk/c0t1d0s6:      160230 sectors in 763 cylinders of 6 tracks, 35 sectors
        78.2MB in 48 cyl groups (16 c/g, 1.64MB/g, 768 i/g)
super-block backups (for fsck -F ufs -o b=#) at:
 32, 3440, 6848, 10256, 13664, 17072, 20480, 23888, 26912, 30320, 33728, 37136,
 40544, 43952, 47360, 50768, 53792, 57200, 60608, 64016, 67424, 70832, 74240,
 77648, 80672, 84080, 87488, 90896, 94304, 97712, 101120, 104528, 107552,
 110960, 114368, 117776, 121184, 124592, 128000, 131408, 134432, 137840,
 141248, 144656, 148064, 151472, 154880, 158288,

# newfs -i 1024 /dev/rdsk/c0t1d0s6
newfs: construct a new file system /dev/rdsk/c0t1d0s6: (y/n)? y
/dev/rdsk/c0t1d0s6:      160230 sectors in 763 cylinders of 6 tracks, 35 sectors
        78.2MB in 48 cyl groups (16 c/g, 1.64MB/g, 1536 i/g)
super-block backups (for fsck -F ufs -o b=#) at:
 32, 3440, 6848, 10256, 13664, 17072, 20480, 23888, 26912, 30320, 33728, 37136,
 40544, 43952, 47360, 50768, 53792, 57200, 60608, 64016, 67424, 70832, 74240,
 77648, 80672, 84080, 87488, 90896, 94304, 97712, 101120, 104528, 107552,
 110960, 114368, 117776, 121184, 124592, 128000, 131408, 134432, 137840,
 141248, 144656, 148064, 151472, 154880, 158288,
#
```

Using newfs *to create a* Usenet-news *file system with more* inodes *than default file systems.*

To find the number of *inodes* created for the file system, divide the size of the file system by the number of bytes per *inode*. The default size number of bytes that can be referenced by an *inode* is 2,048. This assumes that the average file size will be at least 2,048 bytes. A smaller value may be needed in some cases. When in doubt, err on the side of more *inodes* (a smaller number of bytes per *inode*). If a file system runs out of *inodes*, additional files cannot be created even if additional storage blocks are available.

Table 11-2, which follows, lists the parameters used to tune the number of *inodes* created by *newfs*.

Table 11-2: newfs and tunefs Options for Adjusting Drive Block Sizes

newfs Option	tunefs Option	Parameter
-b size	N/A	Basic allocation block size is 8,192 bytes. Only 4,096 or 8,192 are valid entries for this parameter. Use 4,096 only for file systems that will contain very small files.
-f frag	N/A	Size of a block fragment. Acceptable values are 512, 1,024, 2,048, 4,096, and 8,192. (8,192 is valid only if the basic allocation block value is also 8,192.) The smaller values use disk space efficiently at the expense of speed. Larger values provide better read/write speed at the expense of storage space efficiency. The default is 1,024.
N/A	-e maxbpg	Maximum number of blocks a single file may be allocated within a single cylinder group. The default is approximately 25% of the blocks in the cylinder group.
-i nbpi	N/A	Number of bytes per *inode*. Lower values result in more *inodes*, and larger values result in fewer *inodes*. The default is 2,048 bytes.

Tuning for Storage Space Efficiency Versus Speed

The disk storage routines in the SunOS kernel have two strategies available for disk allocation: time and space. Time efficiency refers to the time required to allocate space and write files. Optimizing time is wasteful of space because transfers often result in gaps being created in lieu of long, continuous disk writes.

Space efficiency refers to the efficient use of scattered blocks on the disk. Optimizing space wastes time because a file is allocated to blocks scattered around the disk, and the disk heads must move more frequently to read or write a file. The administrator can use *tunefs* to set the preferred method on each file system.

Table 11-3, which follows, lists the options available for tuning the disk subsystem storage strategies.

Table 11-3: newfs and tunefs Options Related to Space Versus Time Argument

newfs Option	tunefs Option	Parameter
-o strategy	-o strategy	Strategy is either space or time.
-m free	-m free	Minimum free space percentage. A reserve of less than 10% will adversely affect performance.

➥ *NOTE: The space versus time storage strategy routines are automatically set by* newfs *depending on the allocated minimum free space. The defaults are to use space optimization when the minimum free space is less than 10% of the disk blocks, and to use time optimization when the minimum free space is 10% or more of the disk blocks.*

Using the space strategy on a small disk may be desirable when efficient utilization of disk space is more important. On a large file system that has a small value for the minimum free space, it may desirable to use the time strategy to improve performance. Another factor related to the strategy routines is the free space reserve.

The SunOS file system routines operate most efficiently when at least 10% of the basic allocation blocks are free. Routines must do much more work to allocate space for a file when the free space falls below the 10% threshold. File systems are normally created with a 10% reserve, but *tunefs* can be employed to adjust the reserve, as shown in the following illustration.

Using tunefs to change the file system optimization method.

```
# tunefs -o space /dev/rdsk/c0t1d0s6
optimization preference changes from time to space
should optimize for time with minfree >= 10%

# tunefs -o time -m 15 /dev/rdsk/c0t2d0s3
optimization preference changes from space to time
minimum percentage of free space changes from 10% to 15%
#
```

Alternate Disk Subsystem Technologies

For most applications, the standard disk subsystem would provide adequate performance and storage space efficiency. However, in some cases I/O subsystem demands require more throughput or higher availability than is available with standard interfaces. The following sections explore a few of these high-throughput, high-availability technologies.

Fibre Channel

Although current SCSI subsystems can transfer data at 160+ megabytes per second, this was not always the case. Designers wanted a faster channel between the host and the disk farm than SCSI could provide. One technology that provided a high-speed channel already existed: FibreChannel.

In order to provide maximum I/O subsystem performance, many RAID arrays include a fibre channel host interface. This interface provides a 25-megabit-per-second (or faster) interface between the disk array and the file server. High-end RAID arrays may include a 100-megabit-per-second FibreChannel interface, or a Fibre Channel Arbitrated Loop (FCAL) interface. The noise immunity and improved signal quality of the fiber optic based interconnect provide a virtually error-free I/O channel for high-traffic file servers. For high-end file servers, such fiber-based links can provide excellent disk performance.

For a small to medium file server, the standard SCSI interface may be used to connect a RAID array to the system. The 40 to 160 megabit/second capabilities of current SCSI interfaces on many of today's workstations provide very good performance in all but the most demanding environments.

Redundant Arrays of Inexpensive Disks (RAID)

Servers that provide disk farm service are prime candidates for the use of alternate disk subsystem technology. One increasingly popular technology is Redundant Array of Inexpensive Disks (RAID). RAID disk arrays consist of a high-performance host-to-array interface, specialized controllers (typically SCSI based), and an array of disk drives. RAID systems ship with specialized software to control and administer the operation of the disk array. All standard Solaris commands work for the RAID disk array because the driver software takes care of the interface to the RAID subsystem.

RAID disk systems allow for high-throughput, high-availability disk I/O subsystems. The data is broken up into small segments. Each small segment is written to a disk via separate controllers. The throughput improvement is similar to the improvement noticed when the highway department adds more traffic lanes to a major highway.

RAID technology offers several new options to consider in I/O subsystem design and implementation. In particular, RAID I/O subsystems provide several "levels" of service. Each RAID level describes a different method of data storage on the disk drive(s). Each level has unique characteristics, and some of these levels are more suitable to certain applications than other I/O implementations. The most frequently encountered RAID implementations are discussed in the following sections.

RAID Level 0

RAID level o implements a striped disk array. This array contains several disk drives, each attached to a separate disk controller. The data is broken down into small segments, and each data segment is written to a separate disk drive. Disk I/O performance is improved by spreading the I/O load across many channels and drives.

RAID level o, an example of which is shown in the following illustration, does not offer data redundancy, nor does it provide better fault tolerance than a standard I/O subsystem. RAID level o does, however, provide the best I/O throughput of all current RAID implementations.

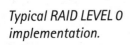

Typical RAID LEVEL 0 implementation.

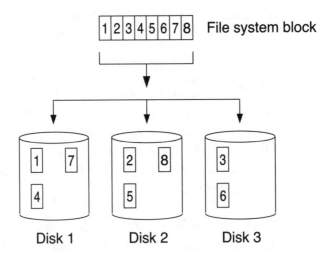

RAID Level 1

RAID level i implements a mirrored disk array. The information written to one disk drive is simultaneously written (mirrored) to another disk drive. This ensures that the data is available even if one of the disk drives would happen to fail.

Due to the mirroring, RAID level 1, an example of which is shown in the following illustration, offers better data availability, at the cost of high throughput. The primary disadvantage of using RAID level 1 is the cost of purchasing duplicate disk drives.

Typical RAID level 1 implementation.

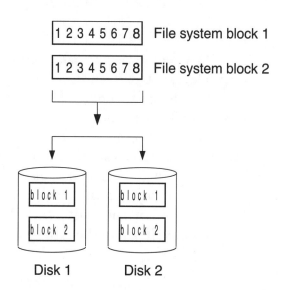

It is possible to create a RAID array that implements a mirror of a striped disk array. This configuration will provide moderate I/O performance as well as high data availability. This striped-mirrored configuration is referred to as RAID 0+1.

✗   *WARNING: When implementing a mirrored disk array, take care not to place both halves of the mirror on the same disk. A common mistake occurs when some portion of the "original" data and the "copy" of the data resides on a single disk. If this disk were to fail, both copies of the data would be lost!*

RAID Level 2

RAID level 2 interleaves data sectors across many disks. The controller breaks up the blocks and spreads them over a small group of disk drives. An error detection and correction process is built into RAID level 2 systems to provide some measure of fault tolerance. RAID level 2 is not found on many systems because other RAID levels offer better performance and fault tolerance.

RAID Level 3

RAID level 3 interleaves data sectors across many disks much like RAID level 2. However, RAID level 3 uses a single parity disk per group of drives to implement a fault tolerant disk system. All drive spindles are synchronized such that the read/write heads on all drives in the array are active at the same time. This synchronization also ensures that the data is written to the same sector on every drive in the array. If a disk fails, the system continues to operate by recreating the data on the failed disk from the information stored on the parity disk. A typical RAID level 3 implementation is shown in the following illustration.

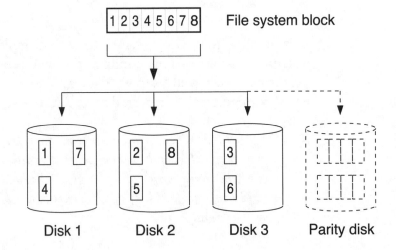

Typical RAID level 3 implementation.

RAID Level 4

RAID level 4 implements striping with the addition of a parity disk. This provides for some of the fault tolerance missing in RAID level 0. RAID level 4 is not found on many systems, because other RAID levels can provide for better throughput and fault tolerance.

RAID Level 5

RAID level 5 implements a large disk storage array. Similar to a level 0 RAID array, the disk array uses striping to improve system performance. RAID level 5 also implements a parity protection scheme. This scheme allows the system to continue operation even if one of the disk drives in the array fails. The data normally stored on the failed drive can be recreated from the parity information stored elsewhere in the array. Once the failed disk drive is replaced, the data is restored automatically by the RAID controller. RAID level 5 provides improved system availability with reasonable I/O subsystem throughput.

The primary disadvantages of RAID level 5 arrays are limited throughput and loss of usable disk space. The performance of a RAID 5 array is limited due to the requirements of creating the parity information. Each time a disk write is performed, the system must calculate the parity information corresponding to the data being saved. The space-loss issue occurs because the RAID 5 array is creating parity information that must be stored in addition to the data. A typical RAID level 5 implementation is shown in the following illustration.

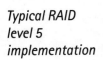

*Typical RAID
level 5
implementation*
.

Special Software Requirements for RAID

Most RAID arrays require special software drivers in order to take advantage of the specialized hardware. This software allows the administrator to create and manage the disk arrays, and provides an interface between the operating system and the disk array's processing system. Many of these packages provide a GUI interface to simplify the management of the array.

One package available for RAID subsystems on a Solaris host is the Veritas file system from Veritas Software. The Veritas package allows the operator to make a large file system out of many small drives, create/manage RAID arrays, migrate data between arrays, set up and control "hot spare" disk drives in the array, and much more.

Software that Emulates RAID Arrays

Hardware solutions are not the only means of implementing disk mirroring and striping. Some software vendors offer applications packages that allow standard SCSI disks connected to standard SCSI controllers to act as a disk array. One such package is Sun Microsystems' Solstice DiskSuite product.

The DiskSuite software allows the administrator to group standard SCSI disk drives into array-like structures. These pseudo arrays may be used to create logical volumes that span the multiple disk drives in the array. For example, it is possible to build a single 10-Gb file system out of five 2.1-Gb disk drives using the DiskSuite software. The performance of these pseudo arrays will not be as good as that of a RAID array, but the cost differential may make the pseudo array more cost effective for some applications.

Summary

The I/O subsystem of any computer system is one of the most important pieces of the system performance puzzle. The system administrator must monitor the system for disk errors and other signs of problems in order to ensure the safety of users' files. At the same time, the administrator must strive to provide file systems that meet the needs of the corporate computing environment.

This chapter covered some typical disk failure modes, and methods the administrator may employ to recover from these failures. Also discussed were file system performance issues, and methods to improve performance. Various current-technology alternative (high-performance/high-availability) disk subsystem technologies were also explored as options available to provide better performance for large file servers.

Managing System Software

Previous chapters of this book focused on disk geometry and file systems. The discussion centered on how disk drives are divided into file systems, and how these file systems track files and directories. With these physical setup tasks complete, the time has come to move on to a discussion of how to install and manage the software that will occupy the file systems.

Software Maintenance Concepts

This chapter introduces the concepts of software packages, software clusters, and software configurations. Under Solaris, application software distributions are divided into three general categories discussed in the following sections.

Packages

Software for Solaris is distributed in what is commonly referred to as software packages. A software package is a collection of files that performs a function, such as the on-line manual pages, or the OpenWindows demonstration programs. There are over 80 software packages in Solaris. Package names are prefixed with the letters SUNW. For example, the package name for the on-line manual pages is *SUNWman*, and the OpenWindows demos, *SUNWowdem*.

The software packages in Solaris conform to a software standard known as the applications binary interface (ABI). ABI-compliant software can be easily managed (i.e., installed and/or removed) with standard systems administration utilities, discussed later in this chapter.

Clusters

Related software packages are grouped into software clusters. Cluster names do not use the *SUNW* prefix. Instead, they use logical names found in the Sun environment. For example, the OpenWindows Version 3 cluster contains 13 software packages related to OpenWindows version 3. Individual packages may be added to or deleted from a cluster depending on the specific needs at each site. *SunInstall* stores a list of the packages that are installed and removed from the system. Other tools in the package maintenance suite allow the administrator to determine which software packages and clusters are installed on the system.

Configurations

The *SunInstall* utility groups software packages and clusters into categories known as software configurations. There are four software configurations in Solaris. Each configuration supports different levels of system sophistication. The four configurations are *core*, *end user*, *developer*, and *entire*.

> ➡ *NOTE: Recent releases of Solaris also include the "entire plus OEM support" configuration. This configuration provides support for third-party SPARC-compatible systems.*

The core software configuration contains the minimum software required to boot and operate a standalone host. The core does not include the OpenWindows software nor the on-line manual pages. It does, however, include sufficient networking software and OpenWindows drivers to run OpenWindows from a server sharing the OpenWindows software. A server would not be built from the core configuration. The core software configuration requires approximately 112 Mb of disk space.

The end user software configuration contains the core configuration software and additional software typically employed by end users. It includes OpenWindows and the end user version of AnswerBook. It does not, however, contain the on-line manual pages. The end user software configuration requires approximately 312 Mb of disk space.

The developer software configuration contains the core and end user configuration software and additional software typically used by systems and software developers. It includes the on-line manual pages and the full implementation of OpenWindows and compiler tools, but does not include compilers nor debugging tools. The developer's software configuration requires approximately 561 Mb of disk space.

> ➡ *NOTE: In Solaris 1 and earlier releases of Sun operating systems, compilers were included (bundled) with the release media. Compilers and debuggers are not bundled with Solaris.*

The entire software configuration contains the complete Solaris release, and under Solaris 8 requires approximately 2,000 Mb of disk space. Unbundled products are not included in the entire release configuration.

It is important to note that the software configurations can be modified. For example, the on-line manual pages cluster could be added to the end user configuration. Likewise, the on-line manual pages could be removed from the developer configuration. These modifications can be made when the system is being built or afterward.

Software Maintenance Utilities

Solaris provides two methods for use during software installation: the command line oriented *pkg* utilities, and tools that provide a graphical interface. Because the command line utilities provide the basis for the graphical tools, they are discussed first.

Command Line Utilities

The command line software maintenance utilities provide the system administrator with the ability to perform software installation, verification, and removal over a terminal-based interface. The command line utilities can also be used from a graphics console if the administrator so desires.

pkginfo *Command*

The *pkginfo* command, shown in the following illustration, is used to identify the packages installed on the system.

Using pkginfo *to identify the packages installed on a system.*

```
                                    Console
 Window  Edit  Options                                            Help
# pkginfo
system        SUNWadmap     System administration applications
system        SUNWast       Automated Security Enhancement Tools
system        SUNWcar       Core Architecture, (Root)
system        SUNWcsr       Core Solaris, (Root)
system        SUNWcsu       Core Solaris, (Usr)
system        SUNWdtab      CDE DTBUILDER
system        SUNWdtbas     CDE application basic runtime environment
system        SUNWdtcor     Solaris Desktop /usr/dt filesystem anchor
system        SUNWdtdmn     CDE daemons
system        SUNWdtdst     CDE Desktop Applications
system        SUNWdtdte     Solaris Desktop Login Environment
system        SUNWdthj      HotJava for Solaris
system        SUNWdticn     CDE icons
system        SUNWdtim      Solaris CDE Image Viewer
system        SUNWdtlog     System boot for Desktop Login
system        SUNWdtwm      CDE DESKTOP WINDOW MANAGER
system        SUNWinst      Install Software
system        SUNWkvm       Core Architecture, (Kvm)
system        SUNWman       On-Line Manual Pages
application SUNWwabi       Wabi Application to run Microsoft Windows Applications
#
```

The *pkginfo* command can also be used to display more detailed information on particular packages on your system. Command line options are summarized in the following.

- *-q*: Do not list any information.

- *-x*: Designate an extracted listing of package information.

- *-l*: Specify long format output.

- *-p*: Display information on partially installed packages.

- *-I*: Display information for fully installed packages.

- *-r*: List the installation base for relocatable packages.

- *-a arch*: Specify the package architecture as *arch*.

- *-v version*: Specify the version of the package as *version*.

- *-c category*: Display all packages that match category.

- *-d device*: Specify the device that contains the package to be checked.

- -*R root_path*: Specify the path to the package to be checked.

For example, to cause *pkginfo* to display more detailed information on a specific package, as shown in the following illustration, use the *-l* option.

Using pkginfo *to display detailed information about packages on a system.*

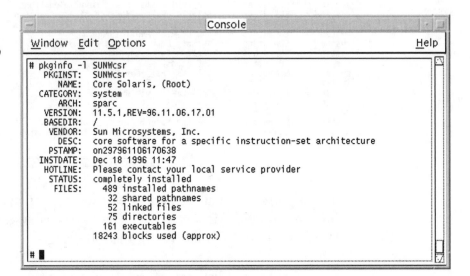

```
Console
Window  Edit  Options                                                    Help
# pkginfo -l SUNWcsr
   PKGINST:  SUNWcsr
      NAME:  Core Solaris, (Root)
  CATEGORY:  system
      ARCH:  sparc
   VERSION:  11.5.1,REV=96.11.06.17.01
   BASEDIR:  /
    VENDOR:  Sun Microsystems, Inc.
      DESC:  core software for a specific instruction-set architecture
    PSTAMP:  on297961106170638
  INSTDATE:  Dec 18 1996 11:47
   HOTLINE:  Please contact your local service provider
    STATUS:  completely installed
     FILES:     489 installed pathnames
                 32 shared pathnames
                 52 linked files
                 75 directories
                161 executables
              18243 blocks used (approx)
#
```

NOTE: *The descriptions of command line options (flags) in this and subsequent sections are condensed synopses of flag operations. Consult the on-line manual pages for more information on these commands.*

pkgadd *Command*

The *pkgadd* utility allows the operator to copy a package from the distribution media to a location on one of the system's disks. Applications are typically distributed on magnetic or optical media such as cartridge tape, floppy diskette, or CD-ROM. The *pkgadd* utility copies the package from the media, and installs it in the specified directory. Some of the more useful command line options for the *pkgadd* command follow.

- *-n*: Non-interactive mode.

- *-a admin*: Employ the user-supplied *admin* file in place of the default *admin* file.

- *-d device*: Define the device that contains the package to be installed.

- *-R root_path*: Define the full path to where the package should be installed.

- *-r response*: Use the response file for answers to interactive mode installations.

- *-s spool*: Install package in the spool directory.

- *pkginst*: Package name and instance of the package to be installed.

The following is a typical invocation of the *pkgadd* command.

```
pkgadd -d device SUNWpkgA
```

This statement instructs *pkgadd* to search the device specified by the name device to locate the *SUNWpkgA* package. Part of the information included with the distributed package is a default installation map. The default installation map specifies installation directory, file modes, and file ownership for all files in the package. Once the system locates the package on the distribution media, the user is prompted for more information. These prompts allow the user to override the default selections.

➤ *NOTE: In releases subsequent to Solaris 2.4, certain unbundled and third-party applications are no longer compatible with the* pkgadd *utility. These applications require interaction throughout the installation process. To install these packages (released prior to 2.4), the NONABI_SCRIPTS environment variable must be defined as TRUE. This can be accomplished with the following commands:*

setenv NONABI_SCRIPTS=TRUE and export
NONABI_SCRIPTS. *(See the following illustration.)*

Using pkgadd *to install a package under Solaris.*

```
┌─────────────────────────────────────────────────────────────┐
│                          Console                            │
├─────────────────────────────────────────────────────────────┤
│ Window   Edit   Options                              Help   │
├─────────────────────────────────────────────────────────────┤
│ grumpy# pkgadd -d `pwd` SUNWsmaud                           │
│                                                             │
│ Processing package instance <SUNWsmaud> from </cdrom/showme_video_2_0_│
│ 1>                                                          │
│                                                             │
│ SunSolutions ShowMe Audio                                   │
│ (sparc) 2.0.1,REV=1.0.1                                     │
│         Copyright 1994 Sun Microsystems, Inc. All Rights Reserved. │
│                Printed in the United States of America.     │
│ 2550 Garcia Avenue, Mountain View, California, 94043-1100 U.S.A. │
│                                                             │
│ Using </opt> as the package base directory.                 │
│ ## Processing package information.                          │
│ ## Processing system information.                           │
│ ## Verifying package dependencies.                          │
│ WARNING:                                                    │
│     The <SUNWsm-ml> package "SunSolutions Motif" is a       │
│     prerequisite package and should be installed.           │
│                                                             │
│ Do you want to continue with the installation of <SUNWsmaud> [y,n,?] y │
│ ## Verifying disk space requirements.                       │
│ ## Checking for conflicts with packages already installed.  │
│ ## Checking for setuid/setgid programs.                     │
│                                                             │
│ Installing SunSolutions ShowMe Audio as <SUNWsmaud>         │
│                                                             │
│ ## Installing part 1 of 1.                                  │
│ /opt/SUNWsunsol/ShowMe/bin/bin-SUNOS/showmeaud              │
│ /opt/SUNWsunsol/ShowMe/bin/bin-SVR4/showmeaud               │
│ /opt/SUNWsunsol/bin/showmeaud <symbolic link>               │
│ [ verifying class <none> ]                                  │
│                                                             │
│ Installation of <SUNWsmaud> was successful.                 │
│ grumpy# █                                                   │
└─────────────────────────────────────────────────────────────┘
```

pkgrm *Command*

The *pkgrm* command is used to undo the operation of the *pkgadd* command, or to remove a software package from the system. Before removing a package, *pkgrm* checks other packages on the system to identify interdependencies. If interdependencies are found, the *pkgrm* command consults the *admin* file to determine the action to be taken. Selected useful command line options for the *pkgrm* command follow.

- **-n**: Non-interactive mode. The command exits if interaction is required.

- **-R**: Defines the full path name to the package to be removed.

- **-admin**: Employs the user-supplied *admin* file in place of the default *admin* file.

- **-spool**: Removes the package from the spool directory.

- **Pkginst**: Specifies the name and instance of the package to be removed.

Using pkgrm to remove a package under Solaris.

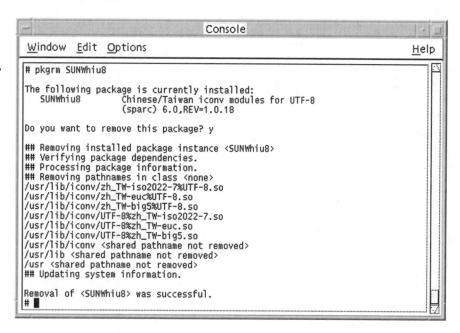

```
                                Console
 Window  Edit  Options                                        Help

# pkgrm SUNWhiu8

The following package is currently installed:
   SUNWhiu8         Chinese/Taiwan iconv modules for UTF-8
                    (sparc) 6.0,REV=1.0.18

Do you want to remove this package? y

## Removing installed package instance <SUNWhiu8>
## Verifying package dependencies.
## Processing package information.
## Removing pathnames in class <none>
/usr/lib/iconv/zh_TW-iso2022-7%UTF-8.so
/usr/lib/iconv/zh_TW-euc%UTF-8.so
/usr/lib/iconv/zh_TW-big5%UTF-8.so
/usr/lib/iconv/UTF-8%zh_TW-iso2022-7.so
/usr/lib/iconv/UTF-8%zh_TW-euc.so
/usr/lib/iconv/UTF-8%zh_TW-big5.so
/usr/lib/iconv <shared pathname not removed>
/usr/lib <shared pathname not removed>
/usr <shared pathname not removed>
## Updating system information.

Removal of <SUNWhiu8> was successful.
#
```

- *NOTE: In releases subsequent to Solaris 2.4, certain unbundled and third-party applications are no longer compatible with the* pkgrm *utility. These applications require interaction throughout the process of removal. To remove these packages (released prior to Solaris 2.4), the NONABI_SCRIPTS environment variable must be defined as TRUE. This can be accomplished with the following commands:*

setenv NONABI_SCRIPTS=TRUE and export NONABI_
SCRIPTS.

pkgchk *Command*

The *pkgchk* utility, shown in the following illustration, checks a
package to determine if it is properly installed, or if it has been
properly removed. The following are command line options for
this utility.

Using pkgchk *to
check the
installation of a
package.*

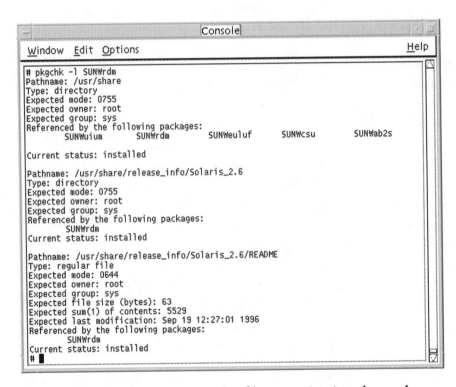

```
# pkgchk -l SUNWrdm
Pathname: /usr/share
Type: directory
Expected mode: 0755
Expected owner: root
Expected group: sys
Referenced by the following packages:
        SUNWuium        SUNWrdm        SUNWeuluf        SUNWcsu        SUNWab2s

Current status: installed

Pathname: /usr/share/release_info/Solaris_2.6
Type: directory
Expected mode: 0755
Expected owner: root
Expected group: sys
Referenced by the following packages:
        SUNWrdm
Current status: installed

Pathname: /usr/share/release_info/Solaris_2.6/README
Type: regular file
Expected mode: 0644
Expected owner: root
Expected group: sys
Expected file size (bytes): 63
Expected sum(1) of contents: 5529
Expected last modification: Sep 19 12:27:01 1996
Referenced by the following packages:
        SUNWrdm
Current status: installed
#
```

- *-l*: Lists information on the files constituting the package.
  This option cannot be used with *a*, *c*, *f*, *g*, and *v* options.

- *-a*: Audits the attributes, but does not check file content.

- *-c*: Audits file content, but does not check attributes.

- *-f*: Corrects file attributes when possible.

- *-q*: Quiet mode.

- *-v*: Verbose mode.

- *-x*: Searches exclusive directories.

- *-n*: Prevents checking of volatile (editable) files.

- *-p path*: Checks only the listed path(s) for accuracy.

- *-i file*: Reads a list of path names from a file.

- *-d device*: Specifies the device containing the spool directory.

- *-R root_dir*: Specifies the path to the package(s).

- *-m pkgmap*: Checks the entire package against the package map.

- *-e envfile*: Uses the *envfile* to resolve parameters listed in the *pkgmap* file.

- *pkginst*: Specifies the package and instance to be checked.

Other pkg *Commands*

Solaris also provides a series of commands that can be used by applications developers to create package distributions. Several of these commands follow.

- *pkgask*: Stores answers to a request script.

- *pkgmap*: Packages content description file.

- *pkgmk*: Produces an installable package.

- *pkgparam*: Displays package parameter values.

- *pkgproto*: Generates prototype file entries for input to *pkgmk*.

- *pkgtrans*: Translates the following package formats: (1) file system to data stream, (2) data stream to file system, or (3) file system type 1 to file system type 2.

- *installf*: Adds a file to the installed software database.

- *removef*: Removes a file from the installed software database.

For more information on the use of these commands, consult the *Solaris AnswerBook*, and the on-line manual pages.

Software Manager Tool (swmtool)

The second method of software maintenance is to use the Software Manager Tool (*swmtool*). The *swmtool* can be used to install, remove, and perform an integrity check on the software on local or remote systems.

swmtool *Interface*

To install or remove packages with *swmtool*, the operator must be operating in a windowing environment, and have *root* permission. To view the installed packages on a system, use the following statement; *root* permission is not required. The illustration that follows shows start-up of *swmtool*.

```
/usr/sbin/swmtool &
```

This statement causes *swmtool* to open a new window on your display. Buttons available within this window allow the operator to set the parameters for package installation or removal, as shown in the following illustration.

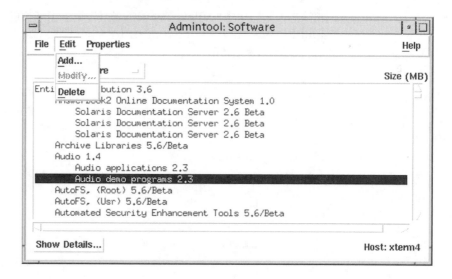

*Setting
parameters
for package
installation.*

By clicking on the Props button, the operator can set the source media, and the directory on the source media that contains the package to be installed, as shown in the following illustration.

Setting the source media and package directory.

NOTE: *Beginning with Solaris 2.6, the* swmtool *buttons have been redefined. The Props button is now called Properties.*

The *swmtool* will read the source volume and display a description of each package available on the media, as shown in the following illustration. Select the package of interest by clicking on the icon to the left of the package description. More information on the package or individual clusters within a package is available by double clicking on the icon, or by selecting the Expand button on the software menu.

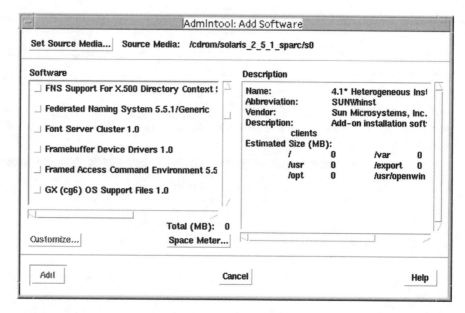

Packages available for installation displayed by the swmtool command.

Once you have chosen the software to be installed, click on the Begin Installation button to start the installation process. The *swmtool* command may request input to allow it to properly install the selected packages.

➡️   *NOTE: Beginning with Solaris 2.6, button names have been redefined. The Begin Installation button is now called the Add button.*

Removing a Package with swmtool

To remove a package with *swmtool*, the operator must first select the Remove mode of operation. The *swmtool* command will display a list of packages available on the system, as shown in the following illustration.

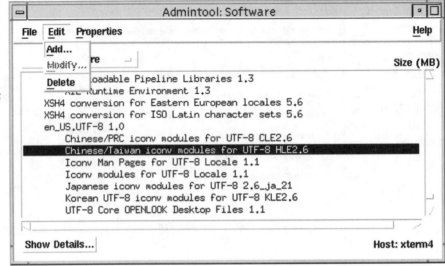

In Remove (Delete) mode, swmtool displays the packages available for

Select the package to be removed by clicking on the icon for the desired package. Once the operator has selected all packages to be removed, click on the Begin Removal button. The *swmtool* will ask for confirmation of the removal operation, as shown in the following illustration.

The swmtool command requests removal confirmation.

➼ *NOTE: Beginning with Solaris 2.6, the names of the buttons have been redefined. The Remove button is now located in the Edit menu, and is labeled as the Delete button.*

Upon completion of *swmtool* tasks, the operator should exit by selecting the box in the upper left corner of the *swmtool* window, and moving the pointer to the Exit item on the menu.

Software Maintenance Log Files

The *pkgadd*, *pkgrm*, and *swmtool* utilities create several files in the */var/sadm* directory to inform the operator of the success or failure of respective operations. These files may be browsed with the *more* command to monitor the steps followed during a package installation or removal.

If the installation of a package fails, the reason for the failure will be written to the appropriate package log file in */var/sadm*. In the event problems are encountered during the installation or removal of a particular package, it may be helpful to peruse the installation log file for the package to determine the source of the problems. The content of a typical installation log file is shown in the following illustration.

Typical content of a /var/sadm/ system package installation log file.

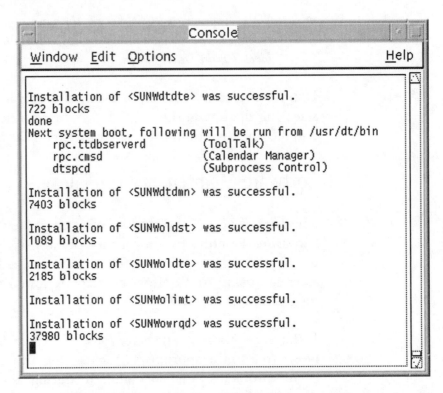

```
                          Console
 Window   Edit   Options                          Help

Installation of <SUNWdtdte> was successful.
722 blocks
done
Next system boot, following will be run from /usr/dt/bin
     rpc.ttdbserverd          (ToolTalk)
     rpc.cmsd                 (Calendar Manager)
     dtspcd                   (Subprocess Control)

Installation of <SUNWdtdmn> was successful.
7403 blocks

Installation of <SUNWoldst> was successful.
1089 blocks

Installation of <SUNWoldte> was successful.
2185 blocks

Installation of <SUNWolimt> was successful.

Installation of <SUNWowrqd> was successful.
37980 blocks
```

▪◆   *NOTE: Although it may appear that log files are superfluous and could be removed, it is generally recommended that these files be left on the system for future reference.*

The Product Registry

Beginning with Solaris 7, Sun introduced a new way of keeping track of the software installed on the system. The Product Registry provides a comprehensive list of all packages and clusters installed on the system. This information may be viewed via a new tool, */usr/bin/prodreg*. The main screen of *prodreg* is shown in the following illustration.

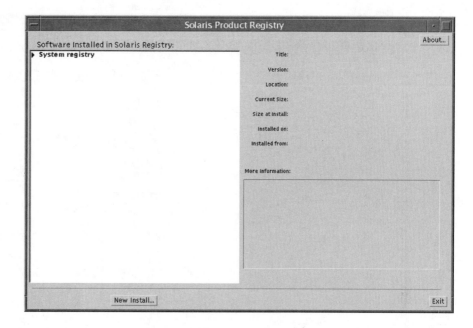

The Solaris Product Registry as shown by /usr/bin/prodreg.

➪ *NOTE: The* prodreg *tool is also available via the CDE control panel under the Application Manager > "System Admin" menu.*

In order to view the content of the registry, click on the small triangle to the left of the "System registry" entry. This will cause the registry to open this folder, allowing the user to view the next level of detail within the registry, as shown in the following illustration.

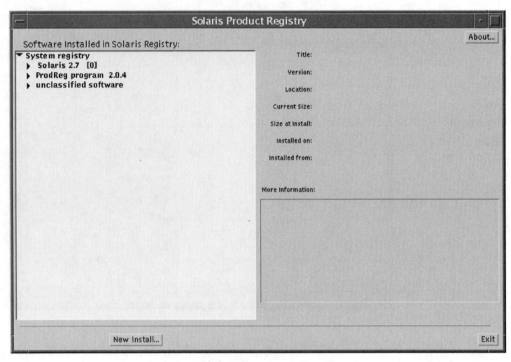

Intermediate level of detail within the product registry.

With the registry opened, the user should have several choices to explore. In order to view the details of an entry, click on the small triangle to the left of the registry entry, as shown in the following illustration.

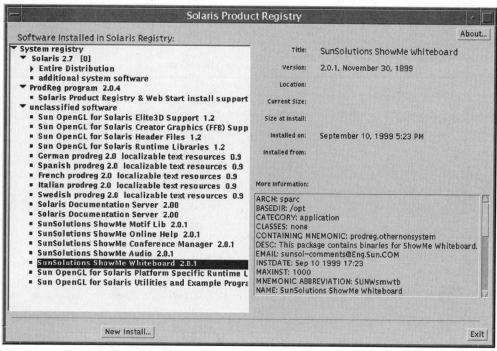

A detailed view within the product registry.

With the registry opened, the administrator may examine the details of each package and cluster installed on the system. In the previous illustration the administrator has selected the "SunSolutions ShowMe Whiteboard 2.0.1" entry by single clicking it. The More Information box shows the details of this package. Scroll bars allow the administrator to scroll through information regarding this package.

The product registry tool (*prodreg*) also allows the administrator to install new packages and clusters on the system. Selecting the New Install button causes *prodreg* to open a software installation window, as shown in the following illustration.

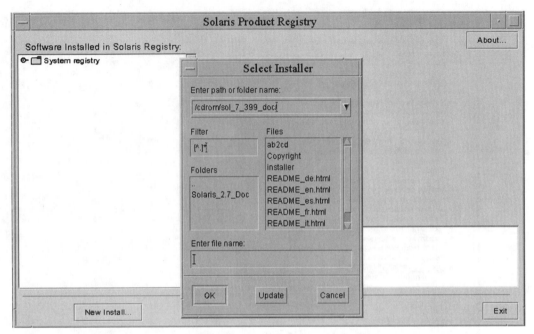

Installing software using the prodreg *New Install feature.*

Once the Select Installer window is available, use the top pull-down menu box to select the source of the package. In this case, the package source is a CD-ROM. Therefore, the administrator has selected the */cdrom/sol_7_399_doc* entry from the menu, as shown in the following illustration. Next, the administrator needs to select the item to install, by double clicking on the appropriate entry in the Files menu. This action will cause the file name to be transferred to the file name box.

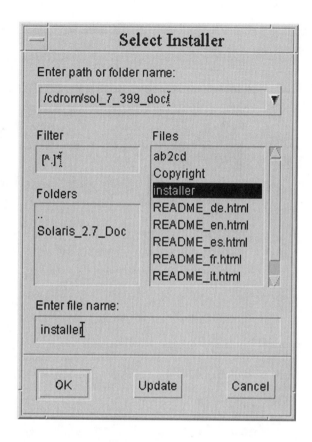

Installing software using the prodreg *installation tool.*

Once the appropriate entries have been selected, clicking on the OK button will start the installation process. A new window will appear to guide the user through the install, as shown in the following illustration.

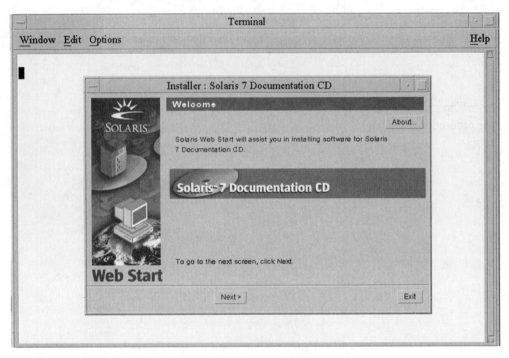

The Installer window will walk the user through the installation.

The installer opens a terminal window, and a GUI installer window. The terminal window displays text messages detailing the package installation. The GUI window walks the administrator through the installation process. The administrator will be prompted to select the files to be installed, the location where they will be installed, and other package-specific information for this installation. Once all of the information has been provided, the installer will begin installing the software, as shown in the following illustration.

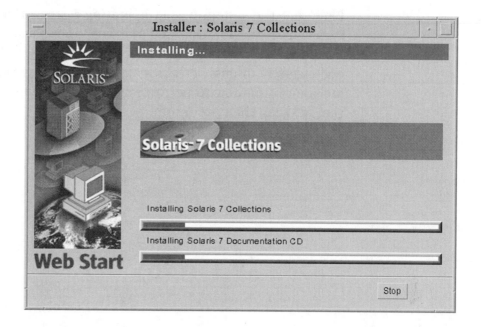

A software installation in progress.

When the installation is complete, the "Installer" will provide an Exit button, which allows the user to close the install window. The Product Registry window will remain open until the user elects to close it using the Exit button in the lower right corner of the window.

Summary

This chapter expanded on the Solaris definition of a software package. Command line and graphical interface software installation tools were examined, and examples detailing the installation, removal, and maintenance commands available for software management were provided. The chapter also covered the strengths and weaknesses of each method of software management, and how to determine which method would be most efficient for a particular package.

Now that the system has an operating system, the mass storage system is ready for use, and the end user packages have been installed, it is time to prepare the system for users. The next several chapters discuss activities aimed at enabling the system, and include tips on how to perform daily system administration functions to keep the system running smoothly

CHAPTER 13
Adding Terminals and Modems

With an understanding of the mass storage subsystem and file system, and software management, you are ready to tackle other daily administrative tasks. One of the more common tasks on any computer system is the installation, setup, and policing of network services and serial lines.

The serial ports on most systems are used for dial-up modem access, *tty* terminal access, and line printer output. A discussion of line printers on serial ports is found in Chapter 14, "Managing Printers." This chapter is focused on serial ports and how they should be configured if used for dial-up modems and *tty* terminal access. Examples show how to set up ports for particular services, and how to monitor system access through the serial ports.

In many applications, the system administrator will be asked to monitor and control access to the systems via network links. As you may recall, system security was discussed in Chapter 7, "Managing System Security." This chapter will re-examine several security topics, including how the system software should be configured to set up and manage the network services for optimum system security.

Working with Serial Devices

Before examining the methods used to add and control serial ports on Solaris systems, a few basic terms and concepts need to be defined. These follow.

- A serial port is an input/output channel between a serial device (terminals, printers, modems, and plotters, to name a few) and the operating system.

- The data is transmitted through a serial port in a bit-by-bit (serial) manner.

- The speed at which the data is transferred is referred to as the baud rate.

The baud rate is a measure of bits per second. For instance, a baud rate of 300 means that the data is being sent at 300 bits per second. There are numerous standard baud rates provided by computer systems. The most common are 300, 600, 1200, 2400, 4800, 9600, 19,200, and 38,400. Most modern terminals and printers operate at transmission speeds of 9600 baud or faster. Solaris provides the means to set the serial port speed as required by the device, or as desired by the user.

➡ *NOTE: Although the terms baud and bits-per-second do not necessarily equate to the same thing, it is common practice to refer to them as the same. To be more accurate, baud refers to the number of times per second a signal changes its value. Bits per second refers to the data rate between the computer and the device. The terms are treated as interchangeable in this chapter.*

Most of the terminals on the market offer a set of unique features in addition to those found on all similar types of units. In order for the operating system to allow the use of different terminal types, a table of terminal capabilities must be available to the operating system. This table is used to determine how to set the serial port appropriately for the connected terminal. Solaris contains such a database of terminal capabilities for many common terminals. In addition, Solaris allows the addition of new terminal definitions via this database. Under Solaris, the terminal capabilities database is located in */usr/lib/terminfo*.

Because serial ports allow access to the operating system, the administrator must be aware of the use of all such ports on the system. Unattended terminals are often used by unauthorized users attempting to access computer systems. The administrator should be aware of all serial port connections to the system, and all ports not currently in use should be disabled. The administrator should also remind authorized system users about the dangers of leaving a log-in session unattended.

Using the Service Access Facility

Under Solaris there are two methods available to the system administrator for managing serial port facilities. First is the *admintool* utility, which graphically interfaces to the Service Access Facility (SAF), and second, manual invocation of utilities provided by the SAF. Both methods of serial port administration use the same underlying set of utilities. With the SAF you can manage serial ports and network devices, including the following.

- Manage and troubleshoot TTY devices.

- Manage and troubleshoot network print service requests.

- Manage and troubleshoot the Service Access Controller.

- Add and manage listen port monitor services.

- Add and manage *ttymon* port monitor services.

The SAF is not a single program, but rather a series of background processes and commands that controls those processes. Selected components of the SAF are described in the following.

- The top-level SAF program is the service access controller (SAC). The SAC controls port monitors administered through the *sacadm* command.

- The *sacadm* command is used to administer the SAC, which controls the *ttymon* and *listen* port monitors.

- The *pmadm* command administers the *ttymon* and *listen* services associated with ports.

- The *ttymon* facility is used to monitor serial line service requests.

- The *listen* facility monitors requests for network services.

The following sections examine each component of the SAF in detail to see how these facilities interact, and what each facility does for system administrators.

Service Access Controller (SAC)

The service access controller (SAC) program manages the operation of all port monitors. These port monitors are actually programs that continuously watch for any request to access devices connected to the serial ports. The SAC process is started automatically when the system is booted to the multi-user *init* state.

When invoked, the SAC initializes the system environment by interpreting the */etc/saf/_safconfig* start-up script. The configuration files allow local customization of serial port parameters.

After the SAC has created a custom environment as directed by the start-up scripts, it reads the */etc/saf_sactab* file and starts port monitors as required. The content of a typical file is shown in the following illustration.

Content of a typical / etc/saf/_sactab file.

The SAC starts up a new process for each port monitor as directed by the */etc/saf/_sactab* file. The SAC maintains control of new processes, and can therefore be considered the "parent" process. Thus, the new processes are called "child" processes.

✓ *TIP: The process hierarchy can be viewed using the* ps -ef *command. The output column labeled PID is the process ID. The output column labeled PPID shows the parent process ID.*

Each child process started by the SAC invokes the appropriate */etc/saf/port_monitor/_config* script to customize its own environment and start the processes as determined by the configuration file. In most cases, the process started for the port is the login process. The content of a typical file is shown in the following illustration.

Content of a typical /etc/saf/zsmon/_pmtab file.

Port Monitors

Solaris implements the *ttymon* and *listen* monitors. Every time a user attempts to log in to a system through a directly connected terminal or modem, the *ttymon* monitor is invoked. Once invoked, *ttymon* monitors serial port lines for incoming data. When data is present, the *ttymon* monitor determines the proper line disciplines and baud rate by consulting the */etc/ttydefs* file, a typical content of which is shown in the following illustration.

Typical content of the /etc/ ttydefs file.

```
Console
Window  Edit  Options                                Help

# cat /etc/ttydefs
# VERSION=1
460800:460800 hupcl:460800 hupcl::307200
307200:307200 hupcl:307200 hupcl::230400
230400:230400 hupcl:230400 hupcl::153600
153600:153600 hupcl:153600 hupcl::115200
115200:115200 hupcl:115200 hupcl::76800
76800:76800 hupcl:76800 hupcl::57600
57600:57600 hupcl:57600 hupcl::38400
38400:38400 hupcl:38400 hupcl::19200
19200:19200 hupcl:19200 hupcl::9600
9600:9600 hupcl:9600 hupcl::4800
4800:4800 hupcl:4800 hupcl::2400
2400:2400 hupcl:2400 hupcl::1200
1200:1200 hupcl:1200 hupcl::300
460800E:460800 hupcl evenp:460800 evenp::307200
307200E:307200 hupcl evenp:307200 evenp::230400
230400E:230400 hupcl evenp:230400 evenp::153600
153600E:153600 hupcl evenp:153600 evenp::115200
115200E:115200 hupcl evenp:115200 evenp::76800
76800E:76800 hupcl evenp:76800 evenp::57600
57600E:57600 hupcl evenp:57600 evenp::38400
38400E:38400 hupcl evenp:38400 evenp::19200
19200E:19200 hupcl evenp:19200 evenp::9600
9600E:9600 hupcl evenp:9600 evenp::4800
4800E:4800 hupcl evenp:4800 evenp::2400
2400E:2400 hupcl evenp:2400 evenp::1200
1200E:1200 hupcl evenp:1200 evenp::300
auto:hupcl:sane hupcl:A:9600
console:9600 hupcl opost onlcr:9600::console
console1:1200 hupcl opost onlcr:1200::console2
console5:19200 hupcl opost onlcr:19200::console
contty:9600 hupcl opost onlcr:9600 sane::contty1
contty1:1200 hupcl opost onlcr:1200 sane::contty2
contty5:19200 hupcl opost onlcr:19200 sane::contty
4800H:4800:4800 sane hupcl::9600H
9600H:9600:9600 sane hupcl::19200H
19200H:19200:19200 sane hupcl::38400H
38400H:38400:38400 sane hupcl::2400H
2400H:2400:2400 sane hupcl::1200H
1200H:1200:1200 sane hupcl::300H
300H:300:300 sane hupcl::4800H
conttyH:9600 opost onlcr:9600 hupcl sane::contty1H
contty1H:1200 opost onlcr:1200 hupcl sane::contty2H
contty5H:19200 opost onlcr:19200 hupcl sane::conttyH
#
#
```

The *ttymon* facility uses the information in the */etc/ttydefs* file to determine the default line settings and proceeds to set the line accordingly. Once this is done, *ttymon* passes control to the log-in process. The log-in process is shown in the following illustration.

Steps involved in invoking the log-in process.

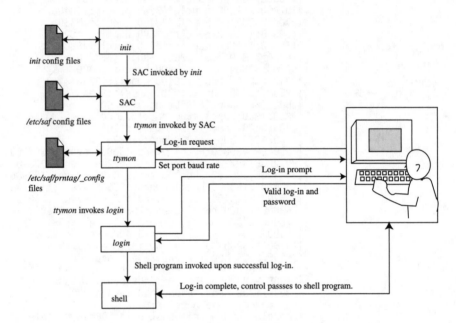

TTY Port Monitor (ttymon)

The *ttymon* facility is a streams-based port monitor. The streams interface is a standardized input/output scheme for use with serial port devices. The streams facility allows the system to communicate with serial port devices through standard system calls and facilities used to communicate with other system devices. Streams are defined by the System V Release 4 interface description (SVID).

The *ttymon* facility uses the streams interface to monitor serial ports and set serial line modes, line disciplines, and baud rates. The *ttymon* facility also invokes the log-in process.

Ttymon runs under the control of the SAC program and is configured via the *sacadm* command. Each invocation of *ttymon* can monitor multiple ports. The ports for each *ttymon* process are specified in the port monitor's administrative file. Configuration files specify the ports managed by the administrative file. The configuration files, in turn, are created by the *pmadm* and *ttymadm* commands.

The *ttymadm* command, a *ttymon*-specific administrative command, formats *ttymon* information and writes it to standard output. In brief, *ttymadm* provides the ability to present such information to the *sacadm* and *pmadm* commands. The *ttymadm* command does not administer *ttymon*, but rather is a means for the *sacadm* and *pmadm* commands to provide such administration.

Network Listener Monitor (listen)

The *listen* port monitor monitors the network for service requests. As requests arrive and are accepted, *listen* invokes the proper server to process the request. The *listen* monitor is configured via *sacadm*. Each invocation of the *listen* monitor may provide multiple services. The services provided by *listen* are specified in the port monitor administrative file, which is in turn configured via the *pmadm* and *nlsadmin* commands.

The listen monitor may be used with any connection-oriented provider that conforms to transport layer interface (TLI) specifications. Under Solaris versions prior to Solaris 2.6, *listen* monitors provided some services not managed by *inetd*, such as the network printing service. Beginning with Solaris 2.6, the print services are managed by the *inet* daemon.

The *nlsadmin* command, a *listen* monitor-specific administrative command, formats *listen* information and writes it to standard output. This command provides the ability to present such information to the *sacadm* and *pmadm* commands. The *nlsadmin* command does not administer *listen*, but rather supplies a means for the *sacadm* and *pmadm* commands to provide such administration.

The *sacadm* command provides administration of the port monitor for the SAC. With the use of command line options, the system administrator can use *sacadm* to accomplish the following tasks.

- Add or remove port monitors

- Start or stop port monitors

- Enable or disable port monitors

- Install or replace a configuration script

- Install or replace a port monitor configuration script

- Print port monitor information

Non-privileged users may request the status of port monitors, and print port and system configuration scripts. All other commands require *root* access. The *pmadm* command provides administration of services for the SAC. By using command line options, the system administrator can use *pmadm* to execute the following tasks.

- Add or remove a network service

- Enable or disable a network service

- Install or replace a service configuration script

- Print service information

The *pmadm* command may be used by non-privileged users to list the status of the service, or to print out configuration scripts. All other access to the *pmadm* command requires *root* privileges.

terminfo Database

The *terminfo* database built by the terminal information compiler (*tic*) describes the capabilities of terminals and printers. The devices are described in *terminfo* files by specifying a set of capabilities for the device. Entries in the *terminfo* database files consist of comma-separated fields. These fields contain capabilities of the device represented with Boolean, numeric, and string values. The database is located in */usr/share/lib/terminfo/*.

Information in the database is used by the *curses* library routines to define terminal cursor movement capabilities. By using the *terminfo* database, these utilities can work with a variety of terminals without the need to modify the programs for each type of terminal on the system. The information in these database files is compiled from the ASCII definition files in */usr/share/lib/termcap*. The content of a typical entry is shown in the following illustration.

```
#
# w: Wyse (although some are "y" entries instead)
#
ye|w50|wyse50|wyse-50|Wyse 50 in Wyse mode:\
        :al=\EE:am:bs:bt=\EI:cd=\EY:ce=\ET:cl=^Z:cm=\E=%+ %+ :co#80:\
        :da:db:dc=\EW:dl=\ER:ei:ei=\Er:im=\Eq:is=\E`\072\200\EC\EDF\E`\E(:\
        :kd=^J:kl=^H:kr=^L:ku=^K:li#24:nd=^L:up=^K:us=\EG8:ue=\EG0:\
        :so=\EG4:se=\EG0:sg#1:sr=\Ej:ho=^^:ug#1:\
        :if=/usr/share/lib/tabset/stdcrt:
# it is not known if the status line works with sysline
yh|w50-s|wyse50-s|Wyse 50 for sysline:\
        :hs:ts=\Ef:fs=\r:ds=\Ef\r:es:tc=w50:
yg|w50-w|w50-132|wyse50-132|Wyse 50-132:\
        :cm=\Ea%i%dR%dC:co#132:is=\E`;\200\EC\EDF\E`\E(:tc=w50:
# The following seems to be the best mode to run a Wyse-50 in.  It works
#  with Emacs using this termcap entry.  Wyse mode seems to be a loser.
#  Note that you have to manually set emulation mode to ADDSVP and turn
#  ENHANCE on. --tpm
# Tab support, better is and rs, and wyse-nk added Jan85 by sun!gnu
wv|wyse-vp|wyse|Wyse 50 in ADDS Viewpoint emulation mode with "enhance" on:\
        :am:do=^J:if=/usr/share/lib/tabset/wyse-adds:\
        :le=^H:bs:li#24:co#80:cm=\EY%+ %+ :cd=\Ek:ce=\EK:nd=^F:\
        :up=^Z:cl=^L:ho=^A:ll=^A^Z:kl=^U:kr=^F:kd=^J:ku=^Z:kh=^A:\
        :pt:so=^N:se=^O:us=^N:ue=^O:dl=\El:al=\EM:im=\Eq:ei=\Er:dc=\EW:\
        :is=\E`\072\E`9^O\Er:rs=\E`\072\E`9^O\Er:
# This is the same as above, but with no keys.  This makes vi ^F, ^B, work.
wk|wyse-vp-nk|wyse-nk|Wyse 50 in ADDS Viewpoint enhanced mode with cursor keys gone:\
        :kl@:kr@:kd@:ku@:kh@:tc=wyse-vp:
# The Wyse-50 does not correctly emulate a Televideo 925.
vw|w9|wyse925|wyse-925|Wyse-50 emulating tvi925:\
        :xn@:tc=tvi925:
# ▮
```

Content of a typical termcap *database entry.*

> ➟   *NOTE: The complexities of* terminfo *entries are beyond the scope of this chapter. For more information on the data contained in the* terminfo *database, consult the on-line* terminfo *manual page.*

Setting Up Terminals

With an understanding of the functions performed by each portion of the SAF, it is time to examine how the SAF may be used to manage terminal setup tasks. As previously mentioned, there are two interfaces to the SAF. The *admintool* program employs a graphical user interface for user interaction. The individual SAF facility commands accomplish the same tasks, but the operator must supply these commands with the proper arguments in order to accomplish the same functions.

A few simple port setup examples will show how each of these facilities operates, while also highlighting the strengths and weaknesses of each method of interaction. One of the more common tasks is to add a terminal to a system. This requires that the port be configured via the *sacadm* utility or *admintool*. The *admintool* Serial Ports manager is shown in the following illustration.

Serial Ports manager in admintool.

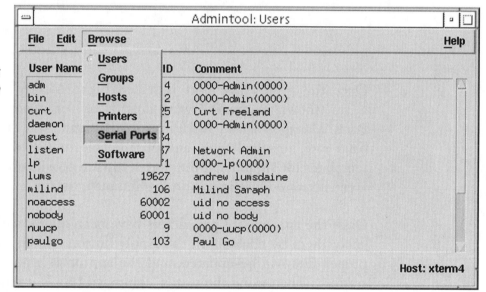

The *admintool* offers a Serial Ports manager selection, which is a screen-oriented process that leads the user through serial port setup. The *sacadm* command is a program that uses command line options to perform the same tasks. Using the *admintool*, you select the name of the port to be modified, as shown in the following illustration.

Selecting the port
to be modified.

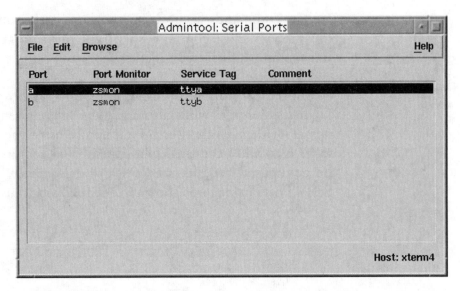

Once the appropriate port is selected, the Modify Serial Port
menu (shown in the following illustration) of *admintool's* Serial
Ports Manager allows modification of parameters such as the
baud rate, terminal type, and port enable or disable. Standard
templates for simplifying the port setup are provided in a pop-up
menu activated by clicking on the Template button.

Once the appropriate parameters have been set for a serial port,
apply them by clicking on the Apply button on this menu. The
proper files will be updated, and the appropriate port monitors
started. When using the *sacadm* command to add a terminal to a
system, you would perform the following basic steps.

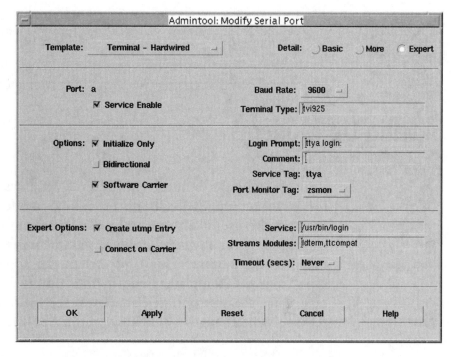

Setting port parameters with the Modify Serial Port menu.

1. Determine the version number of the port monitor by using the *ttyadm -v* command, as follows. The numeric response is needed in the next step.

   ```
   # ttyadm -v
   ```

2. Once the version of the port monitor program is known, it is possible to enable the port via the *sacadm* command. Substitute the version number from above for *version*, and substitute the port monitor tag for *pmtag,* as follows.

```
# sacadm -a -p pmtag -t ttymon -c /usr/lib/saf/ttymon -v version
```

3. When the port is enabled, start up a log-in process. Add a log-in service for the port just enabled by using the *sacadm* command. Substitute the port monitor tag for *pmtag*, the version number for *version*, the device name for *dev_path*, and the *tty* label for *ttylabel*, as follows.

```
# sacadm -a -p pmtag -s svctag -i root -fu -v version
-m "`ttyadmin -S y -d dev_path -l ttylabel -s /usr/bin/login -l ldterm, ttcompat`"
```

As illustrated by the previous example, adding a terminal is much easier to accomplish using the Serial Port Manager.

Setting Up Modems

Another common request encountered by a system administrator is to set up and administer dial-up modems. Again, this requires the use of either the *sacadm* command or the *admintool* command. Using *admintool*, follow the same steps for adding a terminal to the system. This time, select the appropriate modem type from the Template menu in the Modify Serial Port dialog box. Once the appropriate parameters for the modem have been set, apply the changes by clicking on the Apply icon.

✓   *TIP: To use UNIX-to-UNIX copy protocol (UUCP) with the newly installed modem, set the line discipline to use eight data bits, with no parity. This is accomplished by selecting the Baud Rate menu item and clicking on the Other menu item. Set the appropriate baud rate (as defined in /etc/ttydefs) that sets the parity and the correct number of data bits. One such entry would be 9600E for a 9600-baud modem.*

The steps for adding a dial-in or bidirectional modem to a system using the *sacadm* command follow.

1.   Display all port monitors.

    ```
    # sacadm -l -t ttymon
    ```

2.   If the port already has a port monitor, skip the next step.

3.   If there is no port monitor on the chosen port, use the *sacadm* command to start one.

```
# sacadm -a -p pmtab -t ttymon -c /usr/lib/saf/ttymon -v `ttyadm -V`  -y "comment"
```

4.   Determine if the chosen port already has a configured service by using the *pmadm* command.

```
# pmadm -l -s svctag
```

5.   If there is no service configured on the port, skip the next step.

6.   Use the *pmadm* command to delete the service attached to the chosen port.

```
# pmadm -r -p pmtag -s svctag
```

7.   Create the desired type of modem service. Use the port service tag listed in the */dev/term* directory for the *svctag* variable.

Modem Services

At this point the modems should be configured according to their intended use. If the modem will be used exclusively to allow remote users to log in to the system, it is referred to as a "dial-in" modem. Conversely, if the modem will be used to allow users on this system to contact other computers, the modem is referred to as a "dial-out" modem. Modems that allow incoming and outgoing connections are referred to as "bidirectional" modems.

The port setup for the three types of modems are different. The commands discussed in the following sections initialize the serial port for the appropriate type of modem.

Setting Up a Dial-in Modem

A dial-in modem requires a login session manager so that users may log in to the system on this port. The following command provides a typical port setup for a dial-in modem.

```
# pmadm -a -p pmtag -s svctag -i root -v `ttyadm -V` -fu \
-m "`ttyadm -S n -d dev_path -s /usr/bin/login -l ttylabel -m ldterm,ttcompat`" \
   -y "comment"
```

Setting Up a Dial-out Modem

A dial-out modem has no need for a login manager, but it does require a method by which the user can "open" the device for communication. To add a dial-out modem to a system, perform the following steps.

1.   Add the modem to */etc/uucp/Devices*. The format of this file is ACU *cua/svctag - speed type*. The *svctag* will be the port identifier as listed by an *ls -lsa* of the */dev/term* directory.

       ```
       # vi /etc/uucp/Devices
       ```

2.   Disable log-ins on the chosen port. Typical content of the */etc/uucp/Devices* file is shown in the following illustration.

       ```
       # pmadm -d -p pmtag -s svctag
       ```

Setting Up a Bidirectional Modem

A bidirectional modem requires both the log-in session manager and a method by which users can open it for outbound communications. To add a bidirectional modem to a system, perform the following.

1.   Determine whether services have been configured for the port.

       ```
       # sacadm -l -t ttymon
       ```

2.   If no port monitor is present, start one with the following command.

```
# sacadm -a -p pmtag -t ttymon -c /usr/lib/saf/ttymon -v `ttyadm -V` \ -y "comment"
```

```
# cat /etc/uucp/Devices

#ident  "@(#)Devices   1.6    92/07/14 SMI"  /* from SVR4 bnu:Devices 2.7 */
# Some sample entries:
# NOTE - all lines must have at least 5 fields use '-' for unused fields
# The Devices file is used in conjunction with the Dialers file.
# Types that appear in the 5th field must be either built-in functions (TCP, Sytek, DK)
#  or standard functions whose name appears in the first field in the Dialers file.
# Two escape characters may appear in this file:
# - \D which means don't translate the phone #/token
# - \T translate the phone #/token using the Dialcodes file
# Both refer to the phone number field in the Systems file (field 5)
# \D should always be used with entries in the Dialers file, since the
# Dialers file can contain a \T to expand the number if necessary.
# \T should only be used with built-in functions that require expansion
# NOTE: - if a phone number is expected and a \D or \T is not present a \T is used for
# a built-in, and \D is used for an entry referencing the Dialers file.

TCP,et - - Any TCP -

########        AT&T Transport Interface
# ---To use a STREAMS network that conforms to the AT&T Transport Interface
#       with a direct connection to login service (i.e., without
#       explicitly using the Network Listener Service dial script):
# networkx,eg devicex - - TLIS \D
#       The Systems file entry looks like:
# systemx Any networkx - addressx in:--in: nuucp word: nuucp
#       You must replace systemx, networkx, addressx, and devicex with system
#       name, network name, network address and network device, respectively.
#       For example, entries for machine "sffoo" on a STARLAN NETWORK might look like:
#             sffoo Any STARLAN - sffoo in:--in: nuucp word: nuucp
# ---To use a STREAMS network that conforms to the AT&T Transport Interface and that
#       uses the Network Listener Service dial script to negotiate for a server:
# networkx,eg devicex - - TLIS \D nls
#
# ---To use a non-STREAMS network that conforms to the AT&T Transport
#       Interface and that uses the Network Listener Service dial script
#       to negotiate for a server:
# networkx,eg devicex - - TLI \D nls
########
# NOTE: blank lines and lines that begin with a <space>, <tab>, or # are ignored.
#       protocols can be specified as a comma-subfield of the device type either in
#       the Devices file (where device type is field 1) or in the Systems file.
########
#
ACU cua/b - Any hayes
Direct cua/b - Any direct
#
```

Typical content of the /etc/uucp/Devices file.

3.   Determine whether services have been configured for the port.

```
# pmadm -l -t ttymon
```

4.   If services are associated with the port, delete them.

```
# pmadm -r -p pmtag -s svctag
```

5.   Create a bidirectional port service.

```
# pmadm -a -p pmtag -s svctag -i root -v `ttyadm -V` -fu \
    -m "`ttyadm -b -S n -d /dev/term/svctag -s /usr/bin/login -l ttylabel -m
  \ ldterm,ttcompat`" -y "comment"
```

6.   Add the modem to */etc/uucp/Devices*. For example, to use a *ttya* as a 9600-baud Telebit Trailblazer modem line, you could enter the following statements.

```
grumpy% ls -lsa /dev/term
total 16
 2 drwxrwxr-x 2 root root 512 Jul 2 1996 .
10 drwxrwsr-x 17 root sys 4608 May 2 11:10 ..
 2 lrwxrwxrwx 1 root root 38 Jul 2 1996 a -> ../../devices/sbus@1f,0/
   zs@f,1100000:a
 2 lrwxrwxrwx 1 root root 38 Jul 2 1996 b -> ../../devices/sbus@1f.0/
   zs@f,1100000:b
```

Connection via Modem

For small sites and sites with small remote or mobile offices, a common connection method is via modem. Solaris provides some extra security features for modem connections.

There are several ways in which a modem can be configured and used. From a security standpoint, a connection via a modem is a method of remote access that requires control. Dial-in services include both terminal access and network connections via the point-to-point protocol (PPP).

Dial-out service is secure in the sense that your machine is initiating the connection. Users making the dial-out connection should be familiar with the system to which they connect. Users also need to examine any downloaded software or data for potential security problems (as would be the case when downloading such items to a PC or Macintosh). Although viruses are very rare in the Solaris environment, programs that masquerade as useful tools but execute undesirable functions, such as deleting files, have been distributed.

Dial-in service is a more significant problem. Dial-in modems will respond for anyone who knows or accidentally discovers the modem's phone number. Merely making the number unlisted is not always sufficient because determined individuals will use techniques such as automated dialers or reverse phone directories to identify unlisted numbers.

The modem connection will provide a log-in prompt to the person connecting to it, just like a local terminal. At this point, the security of each account depends on its password. Consider the password aging and inactivity controls presented in Chapter 6, "Creating, Deleting, and Managing User Accounts."

Using Dial-in Passwords

Solaris provides a dial-in password, an additional aid for defending dial-in terminal access. The dial-in password is controlled by two files, */etc/dialups* and */etc/d_passwd*. As seen in the following steps, creating a dial-in password is not as easy as it should be.

1. Create */etc/dialups*. This file contains a single entry per line with the path of each dial-in modem line (e.g., */dev/term/a*).

2. Create */etc/d_passwd*. This file contains a single entry for each possible shell a dial-in modem user might use. The following is a sample *d_passwd* file.

```
/usr/lib/uucp/uucico: passwd-goes-here:
/usr/bin/csh: passwd-goes-here:
/usr/bin/ksh: passwd-goes-here:
/usr/bin/sh: passwd-goes-here:
```

➡   *NOTE: The* passwd-goes-here *components will be filled in during a
later step.*

3.   Set the owner, group, and permission bits for the files using
the following commands.

```
chown root /etc/dialups /etc/d_passwd
chgrp root /etc/dialups /etc/d_passwd
chmod 600 /etc/dialups /etc/d_passwd
```

4.   Insert an encrypted password for each shell in the place of
the *passwd-goes-here* strings shown previously. There are several ways to accomplish this. One way is to temporarily set
the *root* password to each of the shell passwords, save the
results, and then set the *root* password back to what it was
when you started. This procedure is inconvenient, but not
difficult. An example of the sequence for the *uucico* shell
follows. Note that the password is typed after each of the
password prompts but does not appear on the screen.

```
# passwd
New password:
Re-enter new password:
grep root /etc/shadow | awk -F: '{print $2}' > /tmp/ foo
# passwd
New password:
Re-enter new password:
#
```

The */tmp/foo* file now contains the encrypted password string,
which will be placed in the */etc/d_passwd* file. You can now use *vi*
or your favorite editor to insert this string between the colons
after the shell you want to protect.

Point-to-Point Protocol (PPP)

The point-to-point protocol (PPP) allows serial port devices to
become network interfaces. Solaris implements the PPP protocol via the following three optional software packages.

- SUNWpppk: PPP kernel modules.

- SUNWapppu: Link manager and login service modules.

- SUNWapppr: Configuration files for PPP.

The PPP packages consist of the following components.

- */etc/asppp.cf* configuration file: Contains setup information used by the link manager as required to manage the PPP connection.

- */usr/sbin/aspppd* link manager: Package that handles the dynamic link setup and shutdown for PPP connections.

- */usr/sbin/aspppls* log-in service: Package that manages the PPP log-in services.

- */var/adm/log/asppp.log* log file: Log file that contains information about the PPP link usage.

- */tmp/.asppp.info* FIFO file: Named pipe file used for communications between the components of the PPP package.

- */etc/init.d/asppp* file: Script run by *init* to start and stop the PPP software.

- */etc/uucp/Dialers* file: Contains information about the dialer used on this system. Used only with dial-out connections.

- */etc/uucp/Devices* file: File that contains information about the serial port devices used for PPP connections. Used only with dial-out connections.

- */etc/uucp/Systems* file: Describes the systems that can communicate with the system via PPP connections. Used only for dial-out connections.

The PPP version shipped with Solaris allows for several modes of operation. Refer to the *Solaris Administrator's Answerbook* to identify the configuration most appropriate for individual applications. The Solaris version of PPP also allows for dynamic or static IP address assignment, and inbound or outbound connections. Security options include the challenge handshake authentication protocol (CHAP), and the password authentication protocol (PAP). The use of PAP and CHAP is discussed in more detail in Chapter 18, "Network Security."

Customizing PPP

In order to use PPP on a Solaris system, the administrator must customize several configuration files. The information required prior to beginning customization follows.

- Addressing method: Will the link employ static or dynamically assigned addresses? This information is required for the */etc/asppp.cf* file.

- Network interface: Which serial port will be used? Available options are *ipdptpN* and *ipdN*, where *N* is the link number. The *ipdN* interface is used for multipoint links. This information is required for the */etc/asppp.cf* file.

- Type of connection supported: Single end point to single end point or multiple end point to single end point? This information is required for the */etc/asppp.cf* file.

- Dial-in or Dial-out: What type of service will this system provide? This information is required for the */etc/asppp.cf* and */etc/uucp/** files.

- Type of name service used: Will the link employ NIS+ (Network Information Service Plus), DNS (Domain Name Service), FNS (Federated Naming System), or other name services? The name service, routing, and authentication methods are required in the */etc/asppp.cf* file.

- Type of authentication to be used: Will the link use the challenge handshake authentication protocol (CHAP) or password authentication protocol (PAP)? This information is required for the *letc/asppp.cf* file.

- Routing requirements: Will data from this link be routed to other network destinations via dynamic or static routing?

Refer to the following illustration for the typical content of the *letc/asppp.cf* file in a single end point to single end point setup. Substitute the local net mask for the net mask value, and the host names for the "remote" and "local" references in the illustration. Refer to the "Solaris TCP/IP and Data Communications Administration Guide" in the *Answerbook* for more details about the information contained in the *letc/asppp.cf* file.

Typical content of the /etc/asppp.cf file.

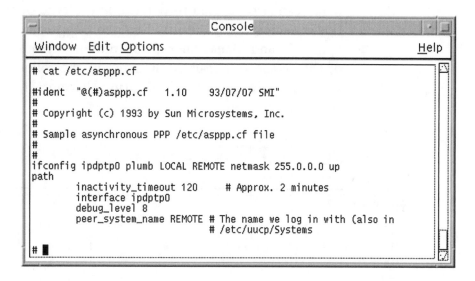

```
# cat /etc/asppp.cf

#ident  "@(#)asppp.cf   1.10    93/07/07 SMI"
#
# Copyright (c) 1993 by Sun Microsystems, Inc.
#
# Sample asynchronous PPP /etc/asppp.cf file
#
#
ifconfig ipdptp0 plumb LOCAL REMOTE netmask 255.0.0.0 up
path
        inactivity_timeout 120      # Approx. 2 minutes
        interface ipdptp0
        debug_level 8
        peer_system_name REMOTE # The name we log in with (also in
                                # /etc/uucp/Systems

#
```

Starting and Stopping PPP

Once all configuration files are customized, the PPP software can be started manually or automatically at boot time. To manually start PPP, enter the following command.

```
# /etc/init.d/asppp start
```

To manually stop PPP, enter the following.

```
# /etc/init.d/asppp stop
```

To arrange for automatic start-up of the PPP software the administrator needs to add links from the appropriate */etc/rc.d* directory to the */etc/init.d/asppp* file. Refer to Chapter 5, "Refining System Boot and Shutdown Procedures," for more information about configuring services for automatic start-up.

Checking for PPP Connections

The administrator can use the *netstat* command to determine whether a PPP connection is currently active. The *snoop* command can also be used to determine the type of traffic currently flowing across the PPP link. Refer to Chapter 17, "Network Configuration and Management," for more information about the use of the *netstat* and *snoop* commands. In the following illustration, the *ipdptpo* link is active, indicating that an active PPP connection is in progress.

Using netstat *to determine if PPP links are active.*

PPP Security

Network connections over modems using PPP also provide additional security measures in the form of the password access protocol (PAP) and the challenge-handshake authentication protocol

(CHAP). PAP adds a password check on top of any password required to log in and begin the PPP session. CHAP provides a more secure authentication method by periodically verifying the identity of the connected systems. CHAP works as follows: a challenge message is sent from one PPP node to the other, a response is returned, and the response is checked against a secret key that is not exchanged.

Configuring PPP to add one or both of these authentication schemes requires editing the PPP configuration file, */etc/asppp.cf*.

PAP and CHAP Setup

Both the server and client *asppp.cf* files must be modified to enable authentication and include the PAP and CHAP identification strings and passwords. Starting with the server, the lines added include *require_authentication*, *will_do_authentication*, *pap_id*, *pap_password*, *chap_name*, and *chap_secret*. The *require_ authentication* and *will_do_authentication* directives control the authentication protocol to be used. The system administrator can choose to use PAP, CHAP, both PAP and CHAP, or to leave these lines out and use no authentication. An example of a PPP server's *asppp.cf* file using both PAP and CHAP follows.

```
# PPP server grissom, serving client shuttle
ifconfig tpdptp0 plumb shuttle down
path
    interface ipdptp0
    peer_system_name shuttle
    require_authentication pap chap
    pap_peer_id atlantis
    pap_peer_password bigboote
    chap_peer_name abilene
    chap_peer_secret 23$56%afbf(-+=P
```

�»   *NOTE: Whereas the PAP name and password are exchanged in the
clear, the CHAP secret is never transmitted over the modem link. To
preserve the security of the CHAP secret, it should be provided to the
remote host by talking to the remote system administrator directly
rather than by using e-mail.*

The client's *asppp.cf* is almost exactly a mirror of the server's
except for the use of the *will_do_authentication* directive in place
of the *require_authentication* directive. The following is an example
for *shuttle*, the remote host listed in the server example.

```
# PPP client shuttle, connecting via server grissom
ifconfig ipdptp0 plumb shuttle grissom up
path
    interface ipdptp0
    peer_system_name grissom
    will_do_authentication pap chap
    pap_peer_id atlantis
    pap_peer_password bigboote
    chap_peer_name abilene
    chap_peer_secret 23$56%afbf(-+=P
```

Once the PPP link has been established, security between the
remote host and server depends on the same factors as for hosts
directly connected to the network.

Disabling and Removing Serial Services

There may be instances when disabling serial port access is desirable. Again, the *sacadm* or *admintool* command may be used to
achieve this goal. Using *admintool*, select the Modify Serial Port
menu, click on the Disable Port button, and then click on the

Apply icon to disable the port. This is shown in the following illustration.

Using the serial port monitor to disable a serial port.

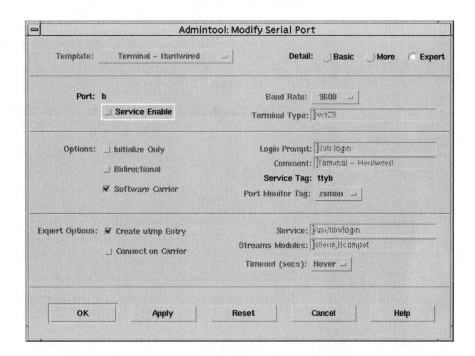

Under most circumstances, however, it is simpler to disable or enable a port by manually invoking the appropriate commands. The basic steps to manually disable a port follow.

1. Determine the monitor tag for the port to be disabled by using the *pmadm* command.

   ```
   # pmadm -l
   ```

2. Disable the port by using the *pmadm* command. Substitute the information provided in step 1 for the *pmtag* field.

   ```
   # pmadm -d -p pmtag -s svctag
   ```

3. To remove a serial port service entirely, use the Serial Port Manager or the command line interface.

4.   To remove a port with the Serial Port Manager, activate
the Edit menu and select Delete Service, as shown in the
following illustration.

Using the Serial Port Manager to remove a serial port.

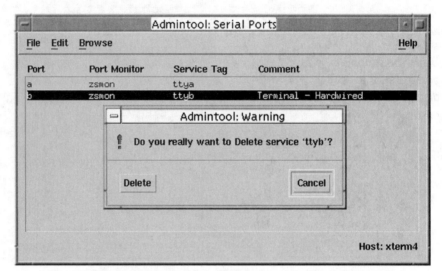

5.   To delete the port service with the command line interface, disable the port by following step 1, and then proceed
to step 3.

6.   Delete the port service using the *pmadm* command.

```
# pmadm -r -p pmtag -s svctag
```

Summary

This chapter examined how serial port and network services
requests are administered. Each utility in the service access facility (SAF) suite was examined to determine the function of the
utility, and how the utilities interact for administration of serial
port devices. Examples demonstrated the use of the SAF utilities
and the *admintool*'s serial ports monitor to manage serial ports.

This chapter also examined how to use the *sacadm* and *pmadm*
commands with command line arguments to add, modify, and

delete serial port services. One of the more common uses of serial ports is the connection of modems. This chapter also discussed the connection of modems, and configuration of serial ports for *tty*-based dial-up modems, as well as how to use the modem to provide PPP services.

CHAPTER 14
Managing Printers

The Solaris printing service provides a complete network printing environment. This environment allows sharing of printers across machines, management of special printing situations such as forms, and filtering output to match special printer types such as those that use the popular PostScript page description language.

There are two sides to managing printers. On one side is the technical aspect of connecting a printer to a system and configuring the software to work with the printer. On the other is printing policy, including who should be allowed to use a particular printer or certain forms, or whether a particular printer should be shared among workstations on the network. The printing service provides tools for making connections to printers and managing printing policy.

Printing Terminology

Printing services on Solaris use a set of special commands, daemons, filters, and directories. Files are not directly sent to the printer, but are spooled and then printed, freeing whatever application submitted the print request to move on to other tasks. The term *spooling* refers to the temporary storage of a file for later use. It comes from the early history of computing when files to be printed were saved temporarily on spools of magnetic tape.

Here spooling is defined as the process of placing a file in a special directory while it waits to be printed. Spooling puts the SunOS multi-tasking capability to good use by allowing printing to occur at the same time as other activities. An example of a printing service is shown in the following illustration.

Printing service.

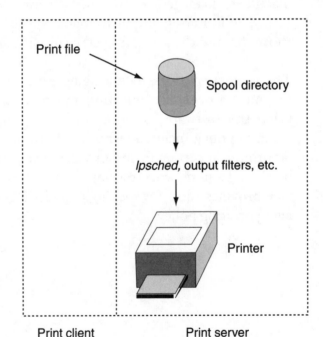

The actual work of printing is executed by a printing *daemon*. Daemon is the nickname given to all system processes running on a UNIX system. Using *ps -e* to take a look at all processes on a workstation, one would see a number of daemons. The main printing daemon is *lpsched*.

Most output bound for a printer will require filtering before it is printed. The term *filter* is used to describe a program that transforms the content of one file into another format as the file passes through the program. For instance, when an ASCII file is sent to a printer that accepts the PostScript page description language, the print service first runs the file through a filter to transform the file from ASCII into PostScript.

The resulting PostScript file contains complete instructions in a form the printer can use to print the page, and is somewhat larger than the original file due to the addition of this information. After filtering, the PostScript file is sent to the printer, which reads the description of the pages to be printed and then prints them.

Printing on the Network

Printing can be set up as a network-wide service. There are three varieties of network printing. Machines with directly connected printers that accept print requests from other machines are called *print servers*. Machines that submit print requests over the network to other machines are called *print clients*. A machine can be both a print server and a print client. Printers directly attached to the workstation are known as *local printers*, whereas printers attached to other workstations reached via the network are known as *remote printers*.

Finally, it is increasingly common for printers to be directly attached to the network. These *network printers* can either be served from a print server or act as their own print server, depending on the facilities provided in the printer by the manufacturer.

Many commands are used to set up the printing service, enforce printing policies, and handle special devices. Some of these commands, such as *lpadmin*, are used throughout the printing process to manage all aspects of the printing service. Other commands, such as *lpusers*, are used to manage a single aspect of printing. Table 14-1, which follows, summarizes printing service commands. These commands are described in detail later in the chapter as the discussion focuses on printer setup and management situations.

Table 14-1: Printing Service Commands

Command	Description
lpadmin	Printer setup and a few management aspects.
admintool	Graphical system management tool that includes printer setup.
printmgr	Graphical printer setup tool.
lpusers	Manages printer job priorities.
lpmove	Changes the destination of print jobs in the spool area.
lpforms	Manages forms and associated parameters.
lpshut	Stops the entire printing service.
lpsched	Starts the printing service.
lpset	Defines printer configuration in */etc/ printers.conf* or FNS.
lp	Submits print jobs.
printtool	OpenWindows graphical interface for submitting print jobs.
lpstat	Displays the status of the printing service and the status of individual jobs and printers.
cancel	Stops individual print jobs.
enable	Starts a printer printing.
disable	Stops printing on a particular printer.

Command	Description
accept	Starts print jobs accumulating in the print spool for a particular printer.
reject	Stops the print service from accepting jobs to be put in the printer spool area for printing on a particular printer.

Printer Setup

A printing service setup consists of four phases. First, a connection to the printer must be made. The printer is physically connected to a machine called the *print server*, or the printer must be connected to the network and configured to communicate with the other systems on the network.

Next, the print server or network printer is configured, possibly with the addition of any special communications software needed for the specific printer being installed. Then the print service is configured for other machines that will share the printer to act as *print clients*. Finally, configuration of special features such as printwheels, forms, and printer access controls may be required to complete the installation.

Setting Up a Local Printer

The most basic printing service configuration is the local printer. Remote printing and print servers are extensions of the local printing process. To configure a system for local printing, a printer is attached to a workstation or server. Because there are nearly as many types of printers as system types, the details of connecting a printer will not be covered here.

In general, the printer is connected to the workstation or server and configured as necessary to use the connection. Common connection methods include serial connections using ports available on the rear of many workstations, and parallel connections using special parallel ports or special-purpose interface cards, such as the S-Bus card used with some SPARC printers. Printers have a variety of configuration methods, from small DIP switches to keypads and display panels.

For example, a common PostScript printer such as an Apple LaserWriter uses a serial cable to connect the printer to the workstation. The printer and workstation manuals should provide information on the necessary cable. Changes to the settings on the printer to work with a serial cable may also be required. On the LaserWriter, the small DIP switches on the rear of the printer should be set as shown in the printer manual under "serial port connections."

Printers connected by a network are becoming increasingly common. Some printers can be treated as if they are print servers (see the "Setting Up Remote Printers" section later in this chapter for more on this type of printer). Some network-attached printers are treated like local printers; the "connection" is obtained over the network by installing communications software provided by the printer manufacturer.

In this case, be sure that the print server is already connected to the network and is able to communicate with other systems before beginning to install the printer software. To set up the printing service for a directly connected printer, the system administrator requires the following information on how the printer is connected.

- The port to which the printer is connected.

- If the printer is connected via a serial line, the basic serial parameters, such as baud rate, parity, data, and stop bits.

- The printer type. Solaris supports numerous printers by model name (e.g., HP LaserJet, Apple LaserWriter, NEC Spinwriter).

- The input accepted by the printer. Some printers will accept only a certain input language such as PostScript, whereas others will accept a variety of input languages.

Consider how much disk space is available for the printer spool area. This area is associated with the */var/spool/lp* directory. Depending on how Solaris was initially installed, the spool area may be in a separate disk slice or part of the *root* directory (/). Sun recommends that at least 8 MB of disk space be reserved to service a single printer for one to three users. The amount needed on a specific system depends on the number and size of the print jobs likely to be run through a particular printer.

Factors to consider include the number of users who will submit print jobs, the type of printer, and the speed of the printer. Remember that files with graphical images are larger than text files of the same number of pages. If you are setting up a server that will be servicing several printers, consider placing the */var* directory on a separate disk slice, with sufficient space for all printers to be serviced.

✓ *TIP: Allocate more disk space for PostScript printers than for other printer types. PostScript print jobs, especially those containing graphics, can be quite large, and are typically much larger than ASCII print jobs. A good rule of thumb is to allow for 1.5 times as much space for a Post-Script printer that will be printing mostly text, two to three times as much space for PostScript printers handling grayscale graphics, and five times as much space for printers handling color PostScript images.*

Setting Up an Apple LaserWriter IINT

With the previous details in mind, you can walk through an example of the process of printer setup. In the example, the printer is an Apple LaserWriter IINT directly connected via a serial line on port *b* of the workstation. The LaserWriter accepts only PostScript data, and expects the serial connection to be 9600 baud, 7 data bits, 1 stop bit, and no parity.

To begin the printer setup process, the system administrator should be logged in as the *root* user. From the CDE Workspace menu, select Printer Administration from the Tools sub-menu. If you are using OpenWindows, start the *print manager* by typing the following.

```
# /usr/sadm/admin/bin/printmgr &
```

➡ *NOTE: The* printmgr *command is new in Solaris 8. If you are using an earlier version of Solaris, use the* admintool *command and select Printers from the Browse menu. To add a local printer, select Add Local Printer from the Edit menu. The data entry screens are the same for* admintool *and* printmgr.

The opening screen of the *print manager* is shown in the following illustration.

Opening printmgr
screen.

The first time you start *printmgr*, it will ask which naming service
you use. In this example, use local files to hold the printing service information by selecting "files" from the Select Naming Service dialog box. The *printmgr* can be used with other naming
services such as NIS+, and the naming service can be changed by
recalling the dialog by selecting Select Naming Service from the
Print Manager menu, as shown in the following illustration.

Once the naming service is selected, a new printer can be added. The *printmgr* supports the addition of three classes of printers: directly attached, network, and printers attached to remote servers or workstations. In this example, the printer is directly attached to the workstation. Select New Attached Printer from the Printer menu. The resulting window is a form used to specify parameters associated with the printer to be added. A blank form is shown in the facing illustration.

The first item in the form, Printer Name, can be up to 14 characters long. Making the name descriptive of printer usage, location, and ownership aids users in selecting the correct printer for their job. Locations with a large number of printers distributed across a network often adopt a uniform naming scheme that encodes the location and type of printer in the name. For example, *Lillyps* might be used to name the PostScript printer in the Lilly building.

The next field in the form is automatically filled in, and lists the name of the local machine as the print server. For the printing service of a local printer, the local machine is both client and server.

The third field is for a description of the printer that can be displayed by various printing commands. Additional information about the printer, such as the room in which it is located and model number or name, are entered here.

*Attached
printer form.*

The fourth field asks for the port to which the printer is connected. This is one of the items the system administrator must remember when a printer is connected to a workstation. Select from the menu that appears under the small box to the left of the field. Selecting Other produces a small entry window for entering the SunOS device name for the port the printer is connected to if the printer is not connected to one of the listed ports.

The fifth field for Printer Type refers to the particular manufacturer and model, such as HP Laser or Diablo. Most laser printers that accept the PostScript language can be set to either Post-Script or PostScript Reverse. PostScript Reverse causes the printing service to reverse the order of the pages printed. The reversed output order is handy for printers that stack output face up, as this eliminates the need to collate the printed pages by hand after they are printed.

If the printer being installed is not listed in the pop-up menu near the fifth field, check the *terminfo* directories located in */usr/share/lib/terminfo*. A series of directories with single-letter and single-number names is located in the *terminfo* directory. The *terminfo* entries are grouped by the first letter or number of the entry in the directories. Each entry describes a terminal or printer, and the name of an entry can be used to fill in the printer type field.

If no entry is available for the printer being installed, use an entry for a similar printer. For example, if a dot matrix printer is being installed, try one of the *epson* entries. If this tactic does not meet with acceptable results, try copying the entry for a similar printer and editing it to describe the features of the printer being installed. The manual page for *terminfo* lists the format and keywords used in *terminfo* entries.

The sixth field, File Contents, which looks like a duplicate of the fifth field, refers to the type of information that can be sent to the printer rather than the type of printer. Some printers can automatically detect PostScript and ASCII files, whereas others can accept only one type of input or the other.

Selecting the type of information the printer accepts causes the *print manager* to configure the printing service for filtering the print job as necessary to create the appropriate input type for the printer. For example, if an ASCII text file is sent to a printer whose File Contents field is set to PostScript, the printing service will automatically filter the file to be printed with a program that converts the ASCII file to PostScript and then sends the converted file to the printer.

The seventh field, Fault Notification, indicates where problem messages from the printing service are to be sent. "Write to superuser" sends the problem messages directly to the terminal of a user logged in as *root*. Alternatively, problem messages can be mailed to the *root* user, or simply not sent, by using the None option. The importance of printer problems and who needs to be informed depend on the printer, its location, and local policies.

The eighth field offers two check boxes. If the Default box is checked, the printer described in this form is where output printed on a particular workstation will go when a print command is run without listing an explicit printer name (e.g., *lp file*). The Banner box asks if a separator page should be printed at the start of each job. Because the banner will include the name of the user who printed the job, it is handy at locations with many print jobs being requested by different users.

The final field is an access list for the printer. By default, any user can print to any local printer. You may want to limit printer access to the user who owns the printer, or to members of the local work group. A more detailed discussion of printer access is covered in the "Access Control" section that follows, and the setup of a remote or network accessible printer appears in "Setting Up Remote Printers."

When the form is complete, click on the OK button. The *admintool* will install the new printer and make it ready for use.

Printers can also be set up via command lines. The *lpadmin* command covers a super-set of the functions available with the Print

Manager. To perform the same printer setup shown in the completed local printer form, use the following *lpadmin* commands.

```
lpadmin -p 1st-floor -D "Apple LaserWriter" -v /dev/term/b -T ps -I postscript \
    -A write -o banner
accept 1st-floor
```

The first command creates the printer *1st-floor*. The *-p* option specifies the printer name. The *-D* option includes a comment, and the *-v* option specifies the device to which the printer is attached. The *-T* and *-I* options specify the printer type and file content, respectively. The *-A* option sets the alert type for announcing printer problems. The *-o* option disables the printing of a banner page unless requested by the user. The second command, *accept*, tells the printing system to begin accepting print jobs for the new printer.

Some printers, although connected to the network, are treated as local printers. Typically these printers require special communications software to transmit print jobs to them, rather than using the standard LPD protocol. Printers of this type will need to be installed according to their manufacturer's directions using the software supplied with them. One such example is the HP JetDirect software, which includes its own installation and printer setup utility.

Changing printer parameters is a matter of changing the options in the *lpadmin* command or selecting the printer from the *printmgr* window by double clicking on it. A form similar to that filled in to create a printer will appear, allowing you to edit the fields to be changed. To delete a printer, select a printer from the list by clicking on the printer's entry line and selecting Delete from the Edit menu.

The Modify Printer form, shown in the following illustration, contains a different set of fields than the Local Printer form. In particular, if the printer port needs to be changed, the printer entry must be deleted and re-created with the new type. The new fields that appear in the Modify Printer form refer to the control of print file printing and queuing, and are duplicated by the *enable/disable* and *accept/reject* commands, respectively.

Modifying a printer's configuration using the Modify Printer form.

Setting Up Remote Printers

A local printer setup configures the local machine as print server and print client. The default access list created allows other workstations to send jobs to the printer spool to be printed. The only task remaining for remote printing is to provide information about the printer to each print client. Three ways of accomplishing this task follow.

- If the systems sharing the printer are running Solaris 2.6 or later and are using NIS+ (as described in Chapter 21, "Network Name Services"), sharing the information is easy. The *printmgr* or *admintool* will automatically add the printer information to the NIS+ database. If *lpadmin* was used to set up the printer, the NIS+ databases must be manually updated.

- If the systems sharing the printer are running Solaris 2.6 or later, but not using NIS+, the */etc/printers.conf* file can be distributed to each print client. A painless way of accomplishing the distribution is to use the *rdist* command, as described in Chapter 16, "Automating Administrative Tasks."

- If the print server is not a Solaris 2.6 system, access to the printer via the common Berkeley *lpd* printer protocol can be set up by selecting Access to Printer from the pull-right Add item in the Edit menu of *admintool*, or by selecting Add Access to Printer from the Printer menu in *printmgr*. The following illustration shows an example of the ensuing form.

*Setting up remote
printer access.*

The printer name refers to the printer on the print server, and the printer server is the name of the machine the printer is physically connected to. A description can be included as well, and a remote printer can be set as the default printer.

Network-connected printers are handled differently from print servers and attached printers, although they have similar attributes. The *printmgr* has a separate form, shown in the following illustration, for network printers that handles the differences.

Network printer form.

As before, setting up a network printer with the *printmgr* involves filling in the form. The major difference here is represented by the seventh and eight fields, which specify the destination and protocol to be used in communicating with the printer. The destination is the name of the print queue offered by the network-attached printer. The printer's manual should contain this information.

The protocol is the method of communication used to transmit the print job over the network to the printer. Many printers support the Berkeley LPD protocol specified by the BSD option in field eight. The TCP option uses a raw TCP connection to the printer.

➥   *NOTE: Some popular network-attached printers support their own communications protocol, such as Hewlett Packard's JetDirect protocol. For these printers, additional communications software will need to be installed to use the printer's own communications protocol, or the printer will need to be configured to use the Berkeley LPD protocol if it is supported by the printer. When using* lpadmin *to set up remote printer access, the -s option is used to specify the remote system. On* grissom, *the following* lpadmin *command line would be used to specify access to the basement printer on* glenn.

```
# lpadmin -p basement -s glenn
```

Access Control

The printing service provides access control for both local and remote printers. Access control is implemented through *allow* and *deny* lists.

You can select a printer through the Printer Manager window, and edit the access list at the bottom of the Modify Printer form. Take a look at the form in the illustration appearing in the previous section. The list managed by the form is an *allow* list, meaning that users on this list and only users on this list can print on the printer. This technique is handy for situations where the printer is in an individual's office or is used by a small work group.

However, sometimes you need the ability to prevent a certain user or two from using a printer. This is done with a *deny* list. These lists can be created only by using the *lpadmin* command. The following command line denies access to the *basement* printer to user *mary*.

```
# lpadmin -p basement -u deny:mary
```

Printing

· ·

Solaris provides many tools for submitting print jobs to the printing service. There are three basic commands: *lp* for submitting jobs, *cancel* for halting print jobs, and *lpstat* for checking on the progress of a submitted job and the general status of the printing system. There is also an OpenWindows interface named *printtool*, which combines aspects of both *lp* and *lpstat*, as does the CDE printer icon found on the right side of the CDE "dashboard."

lp and cancel Commands

The *lp* command allows a user to submit print jobs with a large number of options, and to change a print job's options while it is in the print spool waiting to be printed. An example of the simplest form of the *lp* command follows.

```
% lp my-stuff
```

This command will print the file named *my- stuff* to the default printer. If no file is listed, *lp* reads from the standard input, making it suitable for sending the output of other commands via a pipe. To better control how *my- stuff* or other print jobs are printed, *lp* has a variety of option flags. Some of the most frequently used options are listed in table 14-2, which follows.

Table 14-2: Frequently Used lp Command Options

Option	Description
-c	Copy before printing. Normally the printing service makes a link to the file to be printed and reads it at the time printing takes place. However, if you plan to remove or overwrite the file before it is printed, using this option forces the printing service to make a copy of the file. The printing service removes this temporary copy at the successful conclusion of the print job.
-d destination	To print on a specific printer or class of printers, provide the name of the printer or class as the *destination*. A destination of *any* will submit the job to any printer that can satisfy other options that may have been set.
-f form	This option specifies the form or paper to be used when printing the job. With a destination of *any* (see previous option), the *form* option will route the print job to any printer that has the specified form mounted on it. See the discussion of forms that follows.
-H handling	Specifies job handling. By using *hold* as the *handling* request, a job can be placed in the print spool but not printed. Using *resume* releases a held job to be printed. Specifying *immediate* allows users in the *lpadmin* group to get a print job processed ahead of jobs waiting in the printer spool.
-m	This option causes the print service to send the user electronic mail when a print job concludes.
-n copies	Specifies the number of copies of the file to be printed.
-o option	This flag allows a number of printer-specific options. These include options that modify the line and character spacing, and even those that send special control sequences to certain printers. A common usage is to specify the *nobanner* option to prevent a banner page from being printed. Such an option saves paper for printers that handle few print jobs, and where individual jobs are easily sorted out.
-q priority	Sets the print job *priority* in the print queue. The default print priority and the priority limits can be set and controlled on a per-user basis, as described later in this chapter.
-t title	Puts the *title* of the print job on the banner page.

Option	Description
-T content	Specifies the *content* type of the print job, such as *PostScript*. Print jobs are routed to printers that can either directly handle the content type, or printers for which a filter is available to convert the content type into a form the printer can handle.
-w	This option causes the print service to write a message on the terminal, *commandtool*, or *shelltool* window when a submitted print job finishes printing.

With such a wide variety of options, a job is occasionally submitted that specifies incorrect options. When this happens, the printing service cannot satisfy the request and the job will be rejected. If the job is accepted, but the options should be corrected, you need to know the *request id* of the print job. This is the magic code *lp* provides when a print job is submitted. The following illustration shows an example of a print job that could not be satisfied, followed by one in which the options were wrong and then corrected by using *lp*. Note how the *request id* is used.

Correcting an incorrect print job using request id.

```
                              Terminal
 Window  Edit  Options                                          Help
glenn% lp -d 1st-floor -f checks /etc/hosts
UX:lp: ERROR: Form or character set not available.
       TO FIX: Check the spelling of the form/character
               set name. Use the lpstat -f -S command
               to list all known forms and character
               sets.
glenn% lp -H hold -n 1000 -d 1st-floor /etc/hosts
request id is 1st-floor-3 (1 file(s))
glenn% lp -i 1st-floor-3 -n 1
glenn% lp -i 1st-floor-3 -H resume
glenn% cancel 1st-floor-3
1st-floor-3: cancelled
glenn% █
```

cancel Command

The *cancel* command accepts either a *request id*, as shown in the previous example, or the name of a printer. Given a *request id, cancel* removes the print job from the printer spool. Given a printer name, cancel stops the currently printing job. The *cancel* command can be used only to stop a user's own print jobs. However, the *root* user can cancel any job submitted by any user. But what if the *request id* for a print job has been forgotten?

lpstat Command

The *lpstat* command provides information on the status of print jobs, printers, and the printing service daemons. Similar to *lp*, *lpstat* has numerous options. By default, keying in *lpstat* will list all print jobs submitted to the printing service. Every job is listed along with the *request id* of the job so that a user or the system administrator can cancel or modify the request as needed. An example of the use of this command is shown in the following illustration.

Checking on print jobs with lpstat.

More information is available upon using one of the commonly used options listed in the table 14-3, which follows.

Table 14-3: Frequently Used lpstat Command Options

Option	Description
-a list	Check to see if destinations or classes in the list are accepting print jobs. If no list is present, all destinations and classes are checked.
-c list	List the members of a class. If no list is given, all classes and respective members are listed.
-d	Report the default print destination.
-f list -l	Checks to determine if the form in the list is available. Adding the -l flag will list the description of the form.
-o list	Report on print requests in the list.

CDE Printer Icon

The CDE printer icon provides drag-and-drop access to the default printer. It is located in the CDE dashboard to the right of the center set of screen switch buttons. Dragging a file from the CDE File Manager onto the printer icon submits a print job. Clicking on the icon brings up a small window that graphically displays printer status and lists jobs in the print queue. Options are described in the built-in help system found in the menu at the upper right corner of the window. The CDE print screen is shown in the following illustration.

CDE print screen.

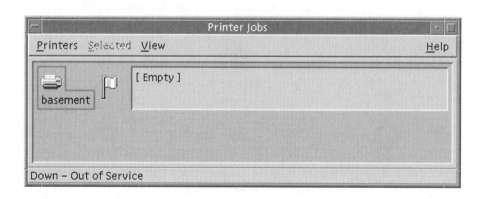

Printer and Print Job Management

Sometimes printing problems develop that require printer queues to be managed in more sophisticated ways. In larger organizations with many printers and users, there is a constant need to shift printing resources around, and to route print jobs in response to printing demands, hardware failures, and the movement of personnel and equipment. Several commands allow the system administrator to deal with these situations.

Disable and Enable, Reject and Accept

Printers can be stopped by using the *disable* command. By default, *disable* will stop the currently printing job and save it to be reprinted when the printer is returned to service. The system administrator can cause the printing system to wait for the current job to finish printing and then stop the printer by using the -*W* option, or cancel the current job and not reprint it later by using the -*c* option. She can also leave a message to be displayed when the *lpstat* command is used to explain why the printer is disabled, by using the -*r* option flag. Restarting the printer is a simple matter of using the *enable* command. An example appears in the following illustration.

```
 ┌─                                 Terminal                              ·  □
  Window  Edit  Options                                                   Help
 ┌──────────────────────────────────────────────────────────────────────────┐
 │glenn% disable -r "Replacing toner cartridge" 2nd-floor                     │
 │UX:disable: ERROR: You aren't allowed to do that.                           │
 │             TO FIX: You must be logged in as "lp" or "root".               │
 │glenn% su                                                                   │
 │Password:                                                                   │
 │# disable -r "Replacing toner cartridge" 2nd-floor                          │
 │printer "2nd-floor" now disabled                                            │
 │# enable 2nd-floor                                                          │
 │printer "2nd-floor" now enabled                                             │
 │# ▉                                                                         │
 └──────────────────────────────────────────────────────────────────────────┘
```

Disabling and enabling a printer.

As the example shows, the *disable* and *enable* commands are reserved for use by the *root* or *lp* users. While a printer is disabled, print jobs can still be submitted for printing on that printer. Such jobs are held in the printer spool area until printing resumes. The spooling of jobs to be printed is controlled by the *accept* and *reject* commands. These two commands start and stop the acceptance of jobs to be spooled. The only option flag, *-r*, provides a means of presenting a reason when a printer spool rejects jobs. An example appears in the following illustration.

```
 ┌─                                 Terminal                              ·  □
  Window  Edit  Options                                                   Help
 ┌──────────────────────────────────────────────────────────────────────────┐
 │# reject -r "Paper feeder broken" 1st-floor                                 │
 │destination "1st-floor" will no longer accept requests                      │
 │# lp -d 1st-floor /etc/motd                                                 │
 │1st-floor: queue is disabled                                                │
 │# lpstat -a                                                                 │
 │foobar accepting requests since Jul 09 21:01 1997                           │
 │1st-floor not accepting requests since Wed Jul  9 21:01:10 EST 1997 -       │
 │        Paper feeder broken                                                 │
 │basement accepting requests since Jul 09 21:01 1997                         │
 │2nd-floor accepting requests since Fri Jun 20 15:09:36 EST 1997             │
 │# accept 1st-floor                                                          │
 │destination "1st-floor" now accepting requests                              │
 │# ▉                                                                         │
 └──────────────────────────────────────────────────────────────────────────┘
```

Rejecting and accepting print jobs.

Once a printer spool has been set to *reject*, print jobs cannot be sent using *lp*, and the reason indicated by the *-r* option of the *reject* command is printed when the *lpstat* command is used to check on the printer.

Moving Print Jobs

The printer and printer spool control commands are used in concert with other printing system commands to handle various situations. Suppose a printer is out of commission or is otherwise unusable for a period, and print jobs are pending. Fortunately, the printing service allows print jobs to be moved from one printer spool to another.

The approach to handling this situation follows: (1) reject print jobs for the unavailable printer to prevent more jobs from accumulating, and (2) move the print jobs to an alternative printer. An example appears in the following illustration.

Moving print jobs to another printer.

Canceled jobs cannot be moved, but any other jobs in a print spool can be directed to another printer.

Removing a Printer

Removing a printer from the print service follows the same general strategy as moving print jobs. First, the printer and spool are stopped and then the printer entry is removed. There are two ways to remove the printer. One is to use the Solaris Print Manager, select the printer from the listing, and select Delete Printer from the Printer menu. You can also use the *lpadmin* command, the command line equivalent to the Print Manager. The command *lpadmin -x printer-name* removes the printer named *printer-name* from the system.

➡   *NOTE: Neither the Print Manager nor* lpadmin *will remove a printer with jobs waiting in its printer spool. Before removing a printer, move or cancel all print jobs in the spool.*

Starting and Stopping the Printing Service

The entire printing service can be stopped and started as necessary. With the print service shut off, no jobs are printed or accepted for spooling for any local or remote printer accessible by the workstation. Stopping and starting are accomplished using the *lpshut* and *lpsched* commands. An example appears in the following illustration.

Stopping and starting the print service.

The first command shows -*r*, a special option to the *lpstat* command that indicates whether the print service scheduler is running. The *lpshut* command stops the scheduler. Starting the scheduler is a matter of running the scheduler by typing the full path to it, as shown in the previous illustration.

Forms, Printwheels, and Fonts

Some older printers support features such as downloadable fonts, interchangeable printwheels, and font cartridges. In addition, many printers support printing on preprinted forms or paper sources of differing sizes.

Printers with downloadable fonts allow you to change typeface by changing a printwheel or font cartridge, and printing on preprinted forms or different paper sizes can be handled by the printing service. These special situations are handled by the *lpadmin* command and the special forms handling command, *lpforms*.

The general strategy for forms and different fonts or printwheels depends on where the font or form is located. Fonts or forms resident in the printer, such as fonts loaded in the printer's read-only memory and forms located in additional paper feeders, are handled by informing the print service of their presence. Once informed and with a proper *terminfo* entry for the printer, the print service can automatically switch between printer resident fonts and forms based on the selections made for each print request.

Changing Fonts

Modern PostScript printers handle font changes in the page description sent to them; the printing service requires no special processing to handle this. However, older printers that use hardware such as font cartridges or printwheels require special handling. The *lpadmin* command's -S option allows the system administrator to specify which printwheels are available. If you are working with older printers, consult the manual page for *lpadmin*.

Changing Paper

Changing forms on many printers requires changing and aligning paper or switching paper trays. For these situations, the system administrator must load the definitions for the alternative forms into the print service and specify how to announce a change of form. The print service will automatically hold print jobs that require a special form until the system administrator informs the print service that the form has been mounted on the printer. In addition, a user access list can be created to control the users with permission to request and print on a given form, such as preprinted checks.

Configuring the print service to use a form requires a couple of steps. First, the *lpforms* command is used to define the form. This can be executed interactively, or by feeding the answers to *lpforms* from a file. The following illustration shows two form definitions used to specify letter- and legal-size paper for a laser printer.

Defining forms using the lpforms command.

The first form pertaining to the *lpforms* command almost completely describes the form by explicitly providing each option. The only option not used here is *comment*, which allows a description of the form to be added to its definition, and an alignment pattern, which can be printed to aid in aligning pre-printed forms.

In the second form definition, only options that differ from the defaults are used. Results can be viewed by using the *-l* option. Use the *-x* option to remove an incorrect form from the print service. Alerts and user lists can also be attached to forms, as indicated in the following illustration.

Assigning user lists and alerts to forms.

```
 ┌─────────────────────────── Terminal ────────────────────────────┐
 │ Window  Edit  Options                                    Help    │
 ├─────────────────────────────────────────────────────────────────┤
 │ # lpforms -f checks -                                            │
 │ page length: 30                                                  │
 │ page width: 80                                                   │
 │ number of pages: 1                                               │
 │ line pitch: 12                                                   │
 │ character pitch: 10                                              │
 │ character set choice: any                                        │
 │ ribbon color: any                                                │
 │ # lpforms -f checks -A mail -Q 2 -W 5 -u allow:accounting!bob    │
 │ # ▉                                                              │
 └─────────────────────────────────────────────────────────────────┘
```

In the second example, a form is defined for pre-printed checks. The second *lpforms* command assigns an alert message to the form with the *-A*, *-W*, and *-Q* options. The *-A* option can be used to send the alert message via mail, writing it on the user's screen (*write*), running a specified command, or doing nothing (*none*). The *-W* option, which specifies the *alert* message timing, is in the example set to send the message every two minutes. The *-Q* option specifies how many print jobs must be in the printer spool waiting to be printed on this form before the alert is sent.

The *-u* option specifies a user access list. In this case, an *allow* list is used. As discussed in the previous section on setting up a printer, an *allow* list specifies the user or users who can print on this form. The notations *machine!user* and *machine!all* are used to specify a user on a specific machine or all users of a machine, respectively.

Printing Problems and Solutions

The two basic printing problems are not producing output or producing incorrect output. In both cases, a useful approach toward identifying the problem is to walk backward along the path the print job takes. Start with the printer itself and work toward the command used to submit the print request.

At the printer, start with the obvious. Is the printer on? Is the machine loaded with paper, ribbon, toner, and so forth? Laser printers in particular are intelligent devices and will halt due to a variety of problems, including paper jams and other internal faults. Check the printer to be sure it is ready to accept a print job.

Next, check the connection from the printer to the print server. For intelligent printers, such as most laser printers, begin by checking the printer's configuration. Some printers have communications settings that are easily changed by mistake. Verify that the printer is using the correct port and that the parameters for the port are accurate, such as baud rate and parity. Determine whether the cable between the printer and computer is plugged in.

Network parameters should be checked for printers directly connected to the network. Checking may be required for the IP address number, subnet mask, and broadcast address. See Chapter 17, "Network Configuration and Management," for a discussion of these parameters.

At this point, check the print server. A common source of problems is the incorrect definition of the printer when it was set up. Check each field of the local printer form, viewed using the Printer Manager screen accessed from the *admintool*. Determine whether the communications parameters match the settings on the printer. If incorrect output is the problem, check to determine whether the printer type and file content fields are correct for the printer in question. In some cases, the system administrator may need to experiment with these settings to find a combination that works.

If the printing problem occurs on a print client, try sending a print job to the printer directly from the print server. This procedure may help determine the location of the problem (client or server). If the problem is on the client, check to determine whether the printer configuration file, */etc/printers.conf*, is up to date.

Once convinced that the printer is properly connected, the printer server is operating, and the print client is correctly configured, move on to check the network. Use a utility such as *ping* to verify whether communications are working between the print client and server. See Chapter 17, "Network Configuration and Management," for more information on setting up, monitoring, and troubleshooting network problems. Finally, check the network log file, */var/lp/logs/lpNet*. This file lists problems found by the printing service when it sends or receives a print request over the network.

Problems Printing from Windows NT

A common problem when printing from a system running Windows NT to a Solaris printer is the inability to print anything but text files. Documents printed from Microsoft Word and other applications print as a jumble of characters. This is corrected by changing the printer configuration on the Windows NT system as follows. (The menu selections appearing in the following pertain to Windows NT 4.0.)

1. Select Settings from the Start menu.

2. Select Printers.

3. If the Solaris printer does not exist, add the printer according to the instructions that ship with Windows NT.

4. Double click on the printer icon.

5. Select Document Properties from the File menu.

6. Select PostScript Options, and set the following options to No: Send CTRL-D Before Each Job; Send CTRL-D After Each Job; and Generate Job Control Code.

Summary

The Solaris printing service consists of a collection of commands that work together to handle a variety of printing situations. When setting up printing, pay careful attention to the printer's connection to the workstation or server, and to the allocation of sufficient printer spool space. Next, employ the printing service's access control mechanisms to implement printer and form access policies.

# CHAPTER 15

System Backups and Disaster Recovery

Previous chapters have explored Solaris installation, managing users, peripherals, security, and network services. One very important area yet to be explored is recovering from a disaster. Whether the disaster is caused by a user removing a critical file, or loss of data due to a disk drive failure, the system administrator needs to be prepared. One of the simplest methods of protecting data is to perform file system backups at regular intervals.

Importance of System Backups

A system administrator's most important task is to ensure the availability of user data. Every file, every database, every byte of information stored on the system must be available. This requirement dictates the need for backup copies of all data in the event there is a need to retrieve such information at a later date.

However, making a copy of the data and storing it in another office in the same building may not be a solution to the problem.

For instance, in the event of a natural disaster the corporation may need to have copies of the data in remote locations to ensure survivability and accessibility of the backup media.

In the event of a hardware failure or some other form of disaster, the system administrator should be able to reload the bulk of user information from backup media. There will almost always be some loss of information. It is impractical to back up every key-stroke as it occurs. Therefore, it is possible to lose information entered into the system between backups. The goal of a backup procedure is to minimize data loss and allow the corporation to reload quickly and continue with business.

Which Files Should Be Backed Up?

Backing up all files on the system at regular intervals is good prac-tice. Some files, however, rarely change. The administrator needs to examine the content of system disks to identify the files that should be a part of file system backups, as well as files to be elimi-nated from regular backups.

Some programs are part of the operating system. Administrators often decide not to back up the operating system binaries, opting to reload them from distribution media in the event of a loss. This practice requires making a backup of any operating system-related files that are created or changed. A few examples might be the NIS+ database, the password file, shadow file, rc files, and customizations that affect these files.

Another set of files that may be eliminated from regular backups are vendor-supplied binaries. If the distribution media for com-mercially purchased software is available, regular backups of these files are not necessary. Again, the administrator should ensure that any customized files or start-up scripts required for those packages are on the backup list.

Files that should be backed up regularly include user files, corporate databases or other important data, and files that have been changed since the time of the last backup. On large systems the amount of data to be backed up may exceed several gigabytes a day. Consequently, the administrator should decide how and when to perform backups in order to minimize impact on the system and loss of important data.

✓ TIP: *One method of determining which files change frequently is to use the* tripwire *program as described in Chapter 7, "Managing System Security."* Tripwire *can produce a list of files that have changed subsequent to the creation of the* tripwire *database. This list should be parsed to ensure that files that change are backed up to tape.*

How Often Should Backups Be Performed?

Frequency is another very important factor in successful file system backup strategies. The administrator must determine how much data can be lost before seriously affecting the organization's ability to do business. In many cases losing the data from a single day is deemed an acceptable loss. This is generally the case for activities such as university teaching laboratories, customer service databases, and Internet service providers.

At the other end of the spectrum are situations that do not allow the loss of any data. Corporations that use computers for financial transactions, stock and commodities trading, and insurance activities would be a few examples. In these cases the loss of a single transaction is not acceptable.

A reasonable backup schedule for a typical corporation would call for some form of backup to be performed every day. In addition, the backups must be performed at the same time every day so that the operator knows which tape to use for reload, and to force the administrator into a schedule. Backups are too important to leave to haphazard scheduling. The fact that backups must be maintained on a rigid schedule lends support to the automated backup methods discussed later in this chapter and in Chapter 16, "Automating Administrative Tasks."

> ◆ *NOTE: The examples in this chapter assume computer system operation seven days a week, and thus the need for seven backups per week. In some instances, corporate computer systems are idle on weekends, enabling the administrator to use a five day per week backup schedule.*

Backup Strategies and Scheduling

Once the administrator has determined which files to back up, and how often to perform backups, several additional decisions must be made. These decisions include identifying the type of backups to perform, and the devices to use to perform the backups. Such decisions will serve to delineate the parameters of backup scheduling.

Before scheduling the backups, the administrator needs to determine the type of backup strategy to use, the type of backup system to use, and how much media will be required for the chosen methods. The following sections examine several backup schedules, the media requirements for each schedule, and the strengths and weaknesses of each backup method.

Types of Backups

Several types of backups may be performed, and the simplest type is a full dump. A full dump is also referred to as a level 0 (zero) backup. A full dump copies every disk file to the backup media. On systems with large disk storage capacity, it may not be possible to make full dumps very often due to the cost of the tapes, tape storage space requirements, and the amount of time required to copy files to the tapes.

Because full dumps are expensive in terms of time and resources, most programs written to perform backups allow for some form of incremental dump. In an incremental dump, files created or modified since the last dump are copied to the backup media. For instance, if a full dump is performed on a Sunday, the Monday incremental dump would contain only the files created or changed since the Sunday dump.

The type of dump to be performed will typically rely on the chosen backup strategy. The strategy includes factors such as how long to keep copies of the dump media on hand, what type of dump to perform on which day of the week, and how easily information can be restored from the backup media.

Backup Strategies

Because performing dumps at regular intervals is vital, developing a dump schedule is recommended. The dump schedule notes when the operator should perform full and incremental dumps on the systems. Full dumps allow you to establish a snapshot of the system at a given point in time. All corporate data should be contained on a full dump.

Incremental dumps enable the administrator to recover active files with a minimum of media and time. Incremental dumps capture data that have changed since the last incremental dump of a lower level. For example, a level 1 dump would back up all files created or changed since the last level 0 dump. A level 2 dump would back up all files since the last level 1 dump, and so on. In general, a level N dump will back up all files altered or created since the last dump at a lower level.

In order to ensure that the backups capture as much of the system data as possible, the interval between backups should be kept to a minimum. With these constraints in mind, it is time to examine a few of the more popular backup strategies.

➡️   *NOTE: In order to ensure successful dumps, it is recommended that the system be shut down to the single-user* init *state.*

Volume/Calendar Backup

The volume/calendar backup strategy calls for a full system backup once a month. An incremental backup is performed once a week for files that change often. Daily incremental backups catch the files that have changed since the last daily backup.

A typical schedule would be to perform the full (level 0) backup one Sunday a month, and weekly level 3 backups every Sunday of the month. Daily level 5 backups would be performed Monday through Saturday. This would require eight complete sets of media (one monthly tape, one weekly tape, and six daily tapes), as detailed in table 15-1.

Recovering from complete data loss with the volume/calendar scheme requires restoring from the most recent full backup, then restoring from the most recent weekly backup, and finally, restoring from each daily backup tape written since the weekly backup.

An advantage to this backup scheme is that it requires a minimum of media. One problem with this backup scheme is that the tapes are immediately reused. For example, every Monday overwrites last Monday's backup information. Consider what would happen if one of the disk drives failed during the second Monday backup. It would not be possible to recover all the data, because the system was in the process of overwriting the backup tape when the drive failed.

Table 15-1: Typical Scheduling for Volume/Calendar Backup

Sun	Mon	Tue	Wed	Thu	Fri	Sat
Date: 1	Date: 2	Date: 3	Date: 4	Date: 5	Date: 6	Date: 7
Level: 0	Level: 5	Level: 5	Level: 5	Level: 5	Level: 5	Level: 5
Tape: A	Tape: B	Tape: C	Tape: D	Tape: E	Tape: F	Tape: G
Date: 8	Date: 9	Date: 10	Date: 11	Date: 12	Date: 13	Date: 14
Level: 3	Level: 5	Level: 5	Level: 5	Level: 5	Level: 5	Level: 5
Tape: H	Tape: B	Tape: C	Tape: D	Tape: E	Tape: F	Tape: G
Date: 15	Date: 16	Date: 17	Date: 18	Date: 19	Date: 20	Date: 21
Level: 3	Level: 5	Level: 5	Level: 5	Level: 5	Level: 5	Level: 5
Tape: H	Tape: B	Tape: C	Tape: D	Tape: E	Tape: F	Tape: G
Date: 22	Date: 23	Date: 24	Date: 25	Date: 26	Date: 27	Date: 28
Level: 3	Level: 5	Level: 5	Level: 5	Level: 5	Level: 5	Level: 5
Tape: H	Tape: B	Tape: C	Tape: D	Tape: E	Tape: F	Tape: G
Date: 29	Date: 30	Date: 31	–	–	–	–
Level: 3	Level: 5	Level: 5	–	–	–	–
Tape: H	Tape: B	Tape: C	–	–	–	–

Grandfather/Father/Son Backup

The grandfather/father/son backup strategy is similar to the volume/calendar strategy. The major difference between the two schemes is that the grandfather/father/son method incorporates a one-month archive in the backup scheme. This eliminates the problem of overwriting a tape before completing a more recent backup of the file system.

Implementing the grandfather/father/son strategy requires performing a full (level 0) dump once a month to new media. Once a week, an incremental (level 3) backup must be performed that captures all files changed since the last weekly backup. This weekly backup should also be saved on new media. Each day an incremental level 5 backup must be performed to capture files that have changed since the last daily backup. The daily backups reuse the tapes written one week earlier, as shown in table 15-2.

Table 15-2: Typical Scheduling for Grandfather/Father/Son Backup

Sun	Mon	Tue	Wed	Thu	Fri	Sat
Date: 1	Date: 2	Date: 3	Date: 4	Date: 5	Date: 6	Date: 7
Level: 0	Level: 5	Level: 5	Level: 5	Level: 5	Level: 5	Level: 3
Tape: A	Tape: B	Tape: C	Tape: D	Tape: E	Tape: F	Tape: G
Date: 8	Date: 9	Date: 10	Date: 11	Date: 12	Date: 13	Date: 14
Level: 5	Level: 5	Level: 5	Level: 5	Level: 5	Level: 5	Level: 3
Tape: H	Tape: B	Tape: C	Tape: D	Tape: E	Tape: F	Tape: I
Date: 15	Date: 16	Date: 17	Date: 18	Date: 19	Date: 20	Date: 21
Level: 5	Level: 5	Level: 5	Level: 5	Level: 4	Level: 5	Level: 3
Tape: H	Tape: B	Tape: C	Tape: D	Tape: E	Tape: F	Tape: J
Date: 22	Date: 23	Date: 24	Date: 25	Date: 26	Date: 27	Date: 28
Level: 5	Level: 5	Level: 5	Level: 5	Level: 5	Level: 5	Level: 3
Tape: H	Tape: B	Tape: C	Tape: D	Tape: E	Tape: F	Tape: K
Date: 29	Date: 30	Date: 31	--	--	--	--
Level: 0	Level: 5	Level: 5	–	–	–	–
Tape: L	Tape: B	Tape: C	–	–	–	–
--	--	--	Date: 1	Date: 2	Date: 3	Date: 4
–	–	–	Level: 5	Level: 5	Level: 5	Level: 3
–	–	–	Tape: D	Tape: E	Tape: F	Tape: M
Date: 5	Date: 6	Date: 7	Date: 8	Date: 9	Date: 10	Date: 11
Level: 5	Level: 5	Level: 5	Level: 5	Level: 5	Level: 5	Level: 3
Tape: H	Tape: B	Tape: C	Tape: D	Tape: E	Tape: F	Tape: G

Sun	Mon	Tue	Wed	Thu	Fri	Sat
Date: 12	Date: 13	Date: 14	Date: 15	Date: 16	Date: 17	Date: 18
Level: 5	Level: 5	Level: 5	Level: 5	Level: 5	Level: 5	Level: 3
Tape: H	Tape: B	Tape: C	Tape: D	Tape: E	Tape: F	Tape: I
Date: 19	Date: 20	Date: 21	Date: 22	Date: 23	Date: 24	Date: 25
Level: 5	Level: 5	Level: 5	Level: 5	Level: 5	Level: 5	Level: 3
Tape: H	Tape: B	Tape: C	Tape: D	Tape: E	Tape: F	Tape: J
Date: 26	Date: 27	Date: 28	--	--	--	--
Level: 0	Level: 5	Level: 5	–	–	–	–
Tape: A	Tape: A	Tape: C	–	–	–	–
--	--	--	Date: 1	Date: 2	Date: 3	Date: 4
–	–	–	Level: 5	Level: 5	Level: 5	Level: 3
–	–	–	Tape: D	Tape: E	Tape: F	Tape: K
Date: 5	Date: 6	Date: 7	Date: 8	Date: 9	Date: 10	Date: 11
Level: 5	Level: 5	Level: 5	Level: 5	Level: 5	Level: 5	Level: 3
Tape: H	Tape: B	Tape: C	Tape: D	Tape: E	Tape: F	Tape: M
Date: 12	Date: 13	Date: 14	Date: 15	Date: 16	Date: 17	Date: 18
Level: 5	Level: 5	Level: 5	Level: 5	Level: 5	Level: 5	Level: 3
Tape: H	Tape: B	Tape: C	Tape: D	Tape: E	Tape: F	Tape: G
Date: 19	Date: 20	Date: 21	Date: 22	Date: 23	Date: 24	Date: 25
Level: 5	Level: 5	Level: 5	Level: 5	Level: 5	Level: 5	Level: 3
Tape: H	Tape: B	Tape: C	Tape: D	Tape: E	Tape: F	Tape: I
Date: 26	Date: 27	Date: 28	Date: 29	Date: 30	Date: 31	--
Level: 0	Level: 5	Level: 5	Level: 5	Level: 5	Level: 5	–
Tape: L	Tape: B	Tape: C	Tape: D	Tape: E	Tape: F	–

In order to maintain a one-month archive, the monthly full backup tape should be placed in storage. Each weekly full backup should also be placed in storage. When the time comes to perform the second monthly full backup, new media should be used. When the third monthly backup is due, the first month's full backup media should be reused. The weekly backups are archived in a similar manner.

This scheme requires two sets of monthly backup media (one in storage, one active), five sets of weekly backup media, and six sets of daily backup media. A total of 13 sets of media are required to implement this strategy with a one-month archive of information.

In order to recover from complete data loss, first restore the most recent level 0 backup tape. Next, restore from the most recent of the level 3 backups, if that backup was written after the level 0 backup. When the level 3 backup has been restored, the operator would restore from each of the level 5 backups that were written after the level 3 backup.

This backup strategy requires much more media than the simple volume/calendar strategy. Although media cost is increased with this plan, data survivability also increases.

Tower of Hanoi Backup

The Tower of Hanoi backup strategy is a variation of "exponential backup." Both strategies rely on mathematical functions of powers of 2. For example, the use of five backup tapes provides for a 32-day schedule. The use of six tapes would provide for a 64-day schedule.

The Tower of Hanoi backup schedule provides outstanding data survivability and a minimum of media. Unfortunately, on a seven-day backup system, the scheduling of full backups as opposed to partial backups can become a problem for the operator.

One way to avoid operator confusion is to perform a special level 0 backup on the first day of each month. This tape would not be one of the five tapes used in the backup cycle. Total media requirements in this scheme would be seven sets of media.

To recover from complete data loss, first restore from the most recent level 0 backup, and then restore from the level 1 backup if that backup was written after the level 0 backup. Next, restore consecutively from the most recent level 3 and 4 backups if both were written after the level 0 backup. Finally, restore each of the level 5 backups that were written after the level 0 backup. The tower of Hanoi backup is detailed in table 15-3.

A Reasonable Alternative

The following four-week schedule offers a reasonable backup schedule for most sites. Performing a full dump on the first Sunday of the month provides a monthly snapshot of the system data. Using two sets of dump media allows the operator to store information for two full months.

Note that in the example the Tuesday through Friday incremental backups contain extra copies of files from Monday. This schedule ensures that any file modified during the week can be recovered from the previous day's incremental dump.

To recover from complete data loss, restore the most recent full (level 0) backup tape. Next, restore from the most recent of the weekly (level 3) backups. Once the weekly backups are restored, restore from each of the daily (level 5) backups. Table 15-4 shows the specifics of one reasonable alternative backup strategy.

Table 15-3: Typical Scheduling for Tower of Hanoi Backup

Sun	Mon	Tue	Wed	Thu	Fri	Sat
Date: 1	Date: 2	Date: 3	Date: 4	Date: 5	Date: 6	Date: 7
Level: 0	Level: 5	Level: 4	Level: 5	Level: 3	Level: 5	Level: 4
Tape: E	Tape: A	Tape: B	Tape: A	Tape: C	Tape: A	Tape: B
Date: 8	Date: 9	Date: 10	Date: 11	Date: 12	Date: 13	Date: 14
Level: 5	Level: 1	Level: 5	Level: 4	Level: 5	Level: 3	Level: 5
Tape: A	Tape: D	Tape: A	Tape: B	Tape: A	Tape: C	Tape: A
Date: 15	Date: 16	Date: 17	Date: 18	Date: 19	Date: 20	Date: 21
Level: 4	Level: 5	Level: 0	Level: 5	Level: 4	Level: 5	Level: 3
Tape: B	Tape: A	Tape: E	Tape: A	Tape: B	Tape: A	Tape: C
Date: 22	Date: 23	Date: 24	Date: 25	Date: 26	Date: 27	Date: 28
Level: 5	Level: 4	Level: 5	Level: 1	Level: 5	Level: 4	Level: 5
Tape: A	Tape: B	Tape: A	Tape: D	Tape: A	Tape: B	Tape: A
Date: 29	Date: 30	Date: 31	--	--	--	--
Level: 3	Level: 5	Level: 4	–	–	–	–
Tape: C	Tape: A	Tape: B	–	–	–	–
--	--	--	Date: 1	Date: 2	Date: 3	Date: 4
–	–	–	Level: 5	Level: 0	Level: 5	Level: 4
–	–	–	Tape: A	Tape: E	Tape: A	Tape: B
Date: 5	Date: 6	Date: 7	Date: 8	Date: 9	Date: 10	Date: 11
Level: 5	Level: 3	Level: 5	Level: 4	Level: 5	Level: 1	Level: 5
Tape: A	Tape: C	Tape: A	Tape: B	Tape: A	Tape: D	Tape: A
Date: 12	Date: 13	Date: 14	Date: 15	Date: 16	Date: 17	Date: 18
Level: 4	Level: 5	Level: 3	Level: 5	Level: 4	Level: 5	Level: 0
Tape: B	Tape: A	Tape: C	Tape: A	Tape: B	Tape: A	Tape: E

Sun	Mon	Tue	Wed	Thu	Fri	Sat
Date: 19	Date: 20	Date: 21	Date: 22	Date: 23	Date: 24	Date: 25
Level: 5	Level: 4	Level: 5	Level: 3	Level: 5	Level: 4	Level: 5
Tape: A	Tape: B	Tape: A	Tape: C	Tape: A	Tape: B	Tape: A
Date: 26	Date: 27	Date: 28	--	--	--	--
Level: 1	Level: 5	Level: 4	–	–	–	–
Tape: D	Tape: A	Tape: B	–	–	–	–

Table 15-4: Typical Scheduling for Reasonable Backup

Sun	Mon	Tue	Wed	Thu	Fri	Sat
Date: 1	Date: 2	Date: 3	Date: 4	Date: 5	Date: 6	Date: 7
–	Level: 0	Level: 5	Level: 5	Level: 5	Level: 3	–
–	Tape: A	Tape: B	Tape: C	Tape: D	Tape: E	–
Date: 8	Date: 9	Date: 10	Date: 11	Date: 12	Date: 13	Date: 14
–	Level: 5	Level: 5	Level: 5	Level: 5	Level: 3	–
–	Tape: F	Tape: B	Tape: C	Tape: D	Tape: E	–
Date: 15	Date: 16	Date: 17	Date: 18	Date: 19	Date: 20	Date: 21
–	Level: 5	Level: 5	Level: 5	Level: 5	Level: 3	–
–	Tape: F	Tape: B	Tape: D	Tape: D	Tape: E	–
Date: 22	Date: 23	Date: 24	Date: 25	Date: 26	Date: 27	Date: 28
–	Level: 5	Level: 5	Level: 5	Level: 5	Level: 3	–
–	Tape: F	Tape: B	Tape: D	Tape: D	Tape: E	–
Date: 29	Date: 30	Date: 31	--	--	--	--
–	Level: 5	Level: 5	–	–	–	–
–	Tape: F	Tape: B	–	–	–	–

Types of Backup Devices

Once a backup strategy has been chosen, it is time to devote some attention to backup devices. Principal backup device requirements follow.

- User ability to write data to the device

- Media capable of storing the data for long periods

- Supports standard system interconnects

- Supports reasonable input/output throughput

Most devices used for system backups consist of a magnetic media (tape or disk) and the drive mechanism. The drive mechanism provides the electronics and transport functions to allow the device to store digital signals on the magnetic media. Some backup devices use optical rather than magnetic media to store information. Still other backup devices use a combination of magnetic and optical storage methods. The following sections examine the characteristics of a few of these devices.

Tape Backup Devices

Tape backup devices are probably the most common backup media in use. The media is relatively inexpensive, the performance is reasonable, the data formats are standardized, and tape drives are easy to use. These factors combined make magnetic tape backups an attractive option. One complication of tape backups is deciding which type of tape system to use. Several of the more popular options are explored in the following sections.

1/2-inch 9-Track Tape Drive

Half-inch tape drives were once the mainstay of backup media. The early 1/2-inch tape drives allowed the operator to write information on the tape at 800 bits per inch of tape. Later drives allowed 1600 bits per inch, and the most recent drives allow 6,250 bits per inch. At 6,250 bits per inch, a 1/2-inch tape can hold up to 100 Mb of information.

A typical 1/2-inch tape consists of 2,400 feet of 1/2-inch-wide tape on a 10-inch plastic reel. The 1/2-inch tape drives are typically large and expensive, but the tape itself is relatively inexpensive. The 1/2-inch format is still a standard media for many industrial operations.

Older 1/2-inch tape drives required special bus interfaces to connect them to a system, whereas the SCSI interface is common today. Due to the vast quantity of information stored on 1/2-inch tape, the 1/2-inch tape drive will be required for many years to come simply to access the stored information. However, with systems containing literally hundreds of gigabytes of disk storage, 1/2-inch tape becomes impractical for data backup.

Cartridge Tape Drive

Cartridge tape drives store between 10 Mb and several Gb of data on a small tape cartridge. Cartridge tape drives are usually smaller and less expensive than 1/2- inch tape drives. Tape cartridges are typically more expensive than a reel of 1/2-inch tape. Because the data cartridge stores more information, a cartridge tape must be manufactured to tighter tolerances than 1/2-inch tapes.

Because the media is smaller and the drives less expensive, cartridge tape systems have become a standard distribution media for many software companies. Moreover, cartridge tapes are particularly attractive for disaster recovery situations because many systems still include boot PROM code to boot cartridge distribution tapes.

A cartridge tape distribution of the operating system is easily stored in a remote location. In the event of a failure or disaster at the primary computer location, these distribution tapes could be used to reload/rebuild a system to replace the damaged host.

Most cartridge tape systems use SCSI interconnections to the host system. These devices support data transfer rates up to 5 Mb per second. This transfer rate may be a little misleading, however, as the information is typically buffered in memory on the tape drive. The actual transfer rate from the tape drive memory to the tape media is typically about 500 Kb per second.

8-mm Tape Drive

These tape drives are also small and fast, and use relatively inexpensive tape media. The 8-mm media can hold between 2 and 40 GB of data. Because of high density storage, 8-mm drives have become a standard backup device on many systems. Several companies also use 8-mm tape as a distribution media for software.

The 8-mm drives use the SCSI bus as the system interconnection. Low-density 8-mm drives can store 2.2 Gb of information on tape. These units transfer data to the tape at a rate of 250 Kb per second. High-density 8-mm drives can store between 5 and 40 GB of information on a tape.

At the "low" end, the 8-mm drives do not use data compression techniques to store the information on tape. At the "high" end, the drives incorporate data compression hardware used to increase the amount of information that can be stored on the tape. Regardless of the use of data compression, high-density 8-mm drives transfer data to tape at a rate of 500 Kb per second. High-density drives also allow the user to read and write data at the lower densities supported by the low-density drives. When using the high-density drives in low-density mode, storage capacities and throughput numbers are identical to low-density drives.

Digital Audio Tape Drive

Digital audio tape (DAT) drives are small, fast, and use relatively inexpensive tape media. Typical DAT media can hold between 2 and 40 GB of data. The DAT media is a relative newcomer to the digital data backup market. The tape drive electronics and media are basically the same as the DAT tapes used in home audio systems.

The various densities available on DAT drives are due to data compression. A standard DAT drive can write 2 Gb of data to a tape. By using various data compression algorithms, manufacturers have produced drives that can store between 2 and 40 GB of data on a tape. DAT drives use SCSI bus interconnections to the host system. Because DAT technology is relatively new, it offers performance and features not available with other tape system technologies. For instance, the DAT drive offers superior file search capabilities as compared to the 8-mm helical scan drives on the market.

Digital Linear Tape (DLT)

Digital linear tape backup devices are among the newest devices on the backup market. These tape devices offer huge data storage capabilities, high transfer rates, and small (but somewhat costly) media. Digital linear tape drives can store up to 70 Gb of data on a single tape cartridge. Transfer rates of 5 Mb/second are possible on high-end DLT drives, making them very attractive at sites with large on-line storage systems.

Where 8-mm and DAT tapes cost (roughly) $15 per tape, the DLT tapes can run as much as $60 each. However, when the tape capacity is factored into the equation, the costs of DLT tapes become much more reasonable. (Consider an 8-mm tape that holds 14 Gb on average versus a DLT cartridge, which can hold 70 Gb of data.)

> ➼   *NOTE: Many operators elect not to enable the compression hardware on tape systems, opting instead for software compression before the data are sent to the tape drive. In the event of a hardware failure in the tape drive's compression circuitry, it is possible that the data written to tape would be scrambled. By using the software compression techniques, the operator can bypass such potential problems.*

Jukebox System

Jukebox systems combine a "jukebox" mechanism with one or more tape drives to provide a tape system capable of storing several hundred Gb of data. The tape drives in the jukebox systems are typically DAT, 8-mm, or DLT devices. Many of these systems employ multiple tape drives, and special "robotic" hardware to load and unload the tapes.

Jukebox systems require special software to control the robotics. The software keeps track of the content of each tape and builds an index to allow the user to quickly load the correct tape on demand. Many commercially available backup software packages allow the use of jukebox systems to permit backup automation.

Optical Backup Devices

Magnetic tape has been used as a backup media for many years. Until recently, magnetic tape was the most economical way to back up mass storage devices. Recently, optical storage devices have become another economical means to back up mass storage systems. A few of the more popular optical backup devices are described in the following sections.

Compact Disk

Compact disk read-only-memory devices (CD-ROM) are useful for long-term archival of information. Although the name implies that these are read-only devices, recent technology has made it possible to mass market the devices that create the encoded CD-ROM media. These CD-ROM writers (also called CD-recordables) make it possible to consider CD-ROM as a backup device.

One of the major decisions in choosing a backup device is the ability of the media to store information for long periods. CD-ROM media offers excellent data survivability. Another advantage to the CD-ROM is the availability of reliable data transportability between systems. This reliability is possible due to the CD-ROM's adherence to industry standardized data formats.

Along with these advantages, the CD-ROM offers a few unique disadvantages. The foremost disadvantage to the CD-ROM as a backup device is the setup cost to create a CD. Not only is the CD-ROM writer an expensive device, setting up and creating a CD is a time-intensive operation. Other disadvantages include the (relatively) high priced media, and the slow speed at which the CD-ROM writers operate compared to tape drives.

Some small sites may decide to back up to CD-ROM Rewritable media. (CDR). The CDR format allows the reuse of the optical media, thereby reducing the cost of backing up to optical devices. Unfortunately, the CDR media is still a low-density solution, providing a mere 650 Mb of storage per disk. The setup and record time for CDR media is comparable to CD media, making CDR less interesting for backups at large sites.

WORM Disk

Write once read many (WORM) storage systems have been around longer than CD-ROM backup systems, and ways have been found to overcome some of the throughput problems associated with CD-ROM writers. These devices use standard SCSI system interconnects to interface with the system. Most WORM devices provide data rates similar to SCSI tape drives (500 kilobits per second).

The media used by WORM devices can often hold several Gb of data and is less expensive than the CD-ROM media. Unfortunately, there are no clear industry standards for recording methods on WORM devices. This has led to a plethora of manufacturers who produce drives with proprietary recording standards. Consequently, the data written on one brand of WORM drive (quite often) cannot be read on another brand of WORM drive.

Magneto-optical Backup Devices

Optical storage systems and associated media are typically expensive. They are also relatively slow devices. Consequently, optical storage systems are rarely used as backup devices. In contrast, magnetic tape (or disk) storage systems are inexpensive and fast. Unfortunately, the media is bulky and susceptible to damage and data loss. By combining the two storage systems into a single system, manufacturers have been able to provide fast, inexpensive, and reliable backup systems.

Many of the magneto-optical systems are hierarchical, meaning that they keep track of how long a file has been in storage since the last modification. Files that are not accessed or modified are often eligible to be stored on the slower optical storage section of the system. Frequently accessed files are maintained on the magnetic storage section of these systems, which allows for faster access to files.

Most magneto-optical storage systems use standard SCSI bus system interconnections. These systems can typically provide the same (or better) data transfer rates as SCSI tape and disk systems.

Disk Systems as Backup Devices

One problem involved in using tape devices for backups is the (relatively) low data throughput rate. If the operator had to back up several gigabytes or terabytes of data daily it would not take long to realize that tape drives are not the best backup method. Although optical backup devices offer high storage capacity, the optical devices are often much slower than tape devices. So how can an administrator back up large-scale systems?

One popular method of backing up large-scale systems is to make backup copies of the data on several disk drives. Disk drives are orders of magnitude faster than tape devices, and therefore offer a solution to one of the backup problems on large-scale systems. However, disk drives are much more expensive than tapes. Disk backups also consume large amounts of system resources. For example, you would need 100 2-Gb disks to back up a hundred 2-Gb disks. Fortunately, there are software applications and hardware systems available to transparently perform this function.

RAID Disk Arrays

One operating mode of redundant arrays of inexpensive disks (RAID) enables the system to make mirror image copies of all data on backup disk drives. RAID disk arrays also allow data striping for high-speed data access. Yet another mode stores the original data, as well as parity information on the RAID disks. If a drive should fail, the parity information may be used to re-create the data from the failed drive. See Chapter 11, "Disk and File System Maintenance," for an in-depth discussion of RAID disk arrays.

Problems with Disks as Backup Devices

Although backing up to disk devices is much faster than backing up to other devices, it should be noted that disk devices present a potentially serious problem. One of the important considerations of backup planning is the availability of the data to users. In the event of a natural disaster, it may be necessary to keep a copy of the corporate data off-site.

When tape devices are employed as the backup platform, it is a simple matter to keep a copy of the backups off-site. When disk drives are employed as a backup media, the process of keeping a copy of the backup media off-site becomes a bit more complicated (not to mention much more expensive). In the case of a RAID disk array, the primary copy of the data is stored on one disk, and the backup copy of the data is stored on another disk. However, both disks are housed in a single box. This makes the task of moving one drive off-site much more complicated.

RAID disk arrays have recently been equipped with fiber channel interfaces. The fiber channel is a high-speed interconnect that allows devices to be located several kilometers from the computer. By linking RAID disk arrays to systems via optical fibers, it is possible to have an exact copy of the data at a great distance from the primary computing site at all times.

In applications and businesses where data accessibility is of the utmost importance, the use of RAID disk arrays and fiber-channel interconnections could solve most, if not all, backup and survivability problems. A RAID fiber-channel implementation is shown in the following illustration.

Providing high data accessibility, automatic backups, and survivable systems via RAID and fiber channel.

Floppy Disk Backups

Another type of system that offers a challenge for backups is the nomadic computer. These are the laptops and other portable computers that are carried by salespersons, executives, and others in the course of their travels. Because space is at a premium on these systems, they often do not contain tape drives for backup purposes. They do, however, usually contain a floppy disk drive.

The floppy disk drive is also a useful backup device on larger desktop systems. Many users may want to make backup copies of their own files. Some users will not have access to system tape drives, optical drives, or other more conventional backup devices.

However, they do have access to floppy disk drives in their desktop workstations.

Floppy disks offer an ideal, low-cost method of backing up files in such instances. Although the floppy diskette does not allow storage of very much information, it is an inexpensive and convenient way for users to perform their own backups. In addition, due to the industry standard floppy diskette formats, it also allows users to carry files between systems of different architectures and operating systems.

High-Density Removable Media Backups

A relatively recent addition to the backup market is the high-density removable media drive. These drives have become very popular in the PC and Macintosh markets, with many vendors offering them as factory-installed equipment. Examples of these devices include the Iomega ZIP and JAZ drives, and the Imation Superdisk drive. These devices are capable of recording 100 Mb to 2 Gb of data on removable media that resemble floppy disk–ettes.

Until recently, Solaris could not make use of these high-density removable media devices. Many of these devices employ a parallel port interface. A few of them offer SCSI interfaces, allowing them to be connected to the external SCSI port on a SPARC workstation. Beginning with Solaris 8, the operating system contains the device drivers to enable these devices to be used on a Solaris system. The drivers allow the media to be mounted as a *ufs* file system, or for use with the *pcfs* file system.

Backup Commands

As noted in previous sections, several devices are available to use as backup media. There are also several methods available for copying data onto these backup devices. Solaris provides several commands that may be used to perform this task. The most versatile of these commands is *ufsdump* and its companion, *ufsrestore*. The *ufsdump* command is a system-level backup utility that uses a special format when writing information to the tape.

The *ufsrestore* command understands the special format used by *ufsdump*, and provides the ability to read *ufsdump* tapes. The *ufsdump/ufsrestore* utilities are geared toward providing entire file system backups. These utilities are not particularly useful for building distribution media, nor for performing personal file backups.

In contrast, the *tar* utility is more efficient at building distribution tapes or personal archives. The *tar* command is not very efficient for use as a general backup utility. The *tar* command is standard across UNIX platforms; therefore, data written under one UNIX variant are usually accessible under another variant. The *tar* package uses command line flags to control whether it is writing, or reading from, the tape.

The *cpio* command is a longstanding utility in the UNIX arena. It was available in the earliest releases of UNIX software, and therefore has a large base of users who know and trust it. When coupled with commands that can locate files by attributes such as the modify date, *cpio* can be used as a backup command.

The *dd* command allows the user to perform high-speed copies from one device to another. One caveat must be understood: when *dd* is used to copy files to tape for backup purposes, the files must be restored to a device identical to the original device that was dumped. The *dd* command (in essence) creates a binary image of the data on the new device. Several filters are built into *dd* to allow some control of this behavior.

ufsdump Command

This *ufsdump* application was developed to allow the backup of entire systems one at a time. The *ufsdump* program allows the operator to specify the files to be "dumped" (or backed up to tape), and options to use during the dump. In addition, *ufsdump* enables scheduling of different levels of dumps on different days. The *ufsdump* command also allows for dumps that occupy multiple tape reels. Probably the easiest way to learn about the *ufsdump* command is to examine a few typical instances of how it is used.

In order to perform a full dump of a system, the operator must first know which disk devices are present on the system. To obtain this information, use the *df* command, as shown in the illustration that follows.

✗   *WARNING: Several releases of Solaris have included versions of ufsdump and ufsrestore that contain a buffer that is not bounds checked. This has allowed the creation of an exploit that gives a hacker root access to your system! Many security experts recommend that the file permissions of ufsdump and ufsrestore be changed to 0755. This disables the exploit, as the program no longer operates with special permissions, but it also means that the account used to run backups must be given read permission on the system disks.*

Using df *to identify the file systems present on a system.*

```
┌─┬─────────────────────────── Console ───────────────────────┬─┬─┐
│ ─ │                                                          │ · │ ▫ │
├───┴──────────────────────────────────────────────────────────┴───┤
│ Window   Edit   Options                                     Help │
├──────────────────────────────────────────────────────────────────┤
│ # df                                                            ▲ │
│ /               (/dev/dsk/c0t3d0s0 ):    17026 blocks   15617 files │
│ /usr            (/dev/dsk/c0t3d0s6 ):   193848 blocks  130585 files │
│ /proc           (/proc            ):        0 blocks     768 files │
│ /dev/fd         (fd               ):        0 blocks       0 files │
│ /var            (/dev/dsk/c0t3d0s3 ):    52768 blocks   15408 files │
│ /export/home    (/dev/dsk/c0t3d0s7 ):  1065914 blocks  461951 files │
│ /opt            (/dev/dsk/c0t3d0s5 ):   412120 blocks  145083 files │
│ /usr/openwin    (/dev/dsk/c0t3d0s4 ):    70324 blocks  112389 files │
│ /tmp            (swap             ):   290744 blocks    8569 files ▼ │
│ # █                                                               │
└──────────────────────────────────────────────────────────────────┘
```

➥ *NOTE: The* ufsdump *command requires that the user have read access privileges on the system disks.*

Once the operator has determined which file systems are present on the system, it is easy to determine the command line options to use for backing up the file systems to tape. The syntax of the *ufsdump* command follows.

```
/usr/sbin/ufsdump [options] [arguments] files_to_dump
```

➥ *NOTE: Consult the on-line manual page for more information about the* ufsdump *command, and the* AnswerBook *for more complete descriptions of* ufsdump *and available options.*

Some of the commonly used command line options for *ufsdump* follow.

- *0-9*: These numeric values specify the dump level. All files listed in the *files_to_dump* list that have been modified since the last *ufsdump* at a lower dump level are copied to the *dump_file* destination.

- *b*: Signifies the blocking factor to be used. The default is 20 blocks per write for tape densities of 6,250 BPI (bytes per inch) or less. The blocking factor of 64 is used for tapes with 6,250 BPI or greater density. The default blocking factor for cartridge tapes is 126.

➡ *NOTE: The blocking factor is specified in 512-byte blocks.*

- *c*: Signifies that the backup device is a cartridge tape drive. The option sets the density to 1,000 BPI and the blocking factor to 126.

- *d*: Signifies the density of the backup media in BPIs. The default density is 6,250 BPI except when the *c* option is used. When the *c* option is used, the density is set to 1,000 BPI per track. Typical values for selected tape devices follow.

  - 1/2" tape (6,250 BPI)
  - 1/4" cartridge (1,000 BPI)
  - 2.3-Gb or 5.0-Gb 8-mm tape (54,000 BPI)

- *D*: Signifies that the dump device is a floppy diskette.

- *f*: Signifies the *dump_file*. This option causes *ufsdump* to use *dump_file* as the file to dump to, instead of */dev/rmt/0*.

- *s*: Signifies the size of the backup volume. This option is not normally required because *ufsdump* can detect end-of-media. When the specified size is reached, *ufsdump* waits for the operator to change the volume. The size parameter is interpreted as the length in feet for tapes and cartridges, and as the number of 1,024-byte blocks for diskettes. Typical values used with the *s* option follow.

  - 1/2" tape (2,300 feet)
  - 60-Mbyte 1/4" cartridge (425 feet)
  - 150-Mbyte 1/4" cartridge (700 feet)
  - 2.3-Gb 8-mm (6,000 feet)
  - 5.0-Gb 8-mm (13,000 feet)
  - diskette (1,422 blocks)

- *u*: This option causes *ufsdump* to annotate which file systems were dumped, the dump level, and the date in the */etc/dumpdates* file.

- *v*: This letter signifies that *ufsdump* should verify the content of the backup media after each tape or diskette is written.

For illustration purposes, assume that an 8-mm tape system is connected to a system as */dev/rmt/o*. To make a full backup of the entire disk *cot3do*, the operator could issue the following command.

```
# ufsdump 0fu /dev/rmt/0 /dev/rdsk/c0t3d0s2
```

As shown in the illustration, *ufsdump* sends several messages to the controlling terminal. The messages provide several pieces of information about the status of the dump. Most of the dump information messages are self-explanatory (date of the dump, date of last dump, dump is done). Some of the dump information messages merely echo the command line option settings (files to dump, dump device, block size). Messages that actually provide new information about the status of the current dump include the following items.

- *Estimated 47952 blocks (23.41MB)*: Estimated size of the dump.

- *XX.XX% done, finished in H:MM*: What proportion of the dump is completed, and how long the remainder will take to complete.

- *47948 blocks (23.41MB) on 1 volume at 1038 KB/sec*: This message tells how many blocks were dumped, and how many tapes (volumes) were required.

Using ufsdump *to perform a full dump of the system disk.*

```
┌──────────────────────────── Console ──────────────────────────┐
│ Window   Edit   Options                               Help     │
├────────────────────────────────────────────────────────────────┤
│ # ufsdump 0bdfu 50 54000 /dev/rmt/0bn /dev/rdsk/c0t3d0s0        │
│   DUMP: Writing 25 Kilobyte records                            │
│   DUMP: Date of this level 0 dump: Wed Apr 16 19:53:40 1997    │
│   DUMP: Date of last level 0 dump: the epoch                  │
│   DUMP: Dumping /dev/rdsk/c0t3d0s0 (grumpy:/) to /dev/rmt/0bn  │
│   DUMP: Mapping (Pass I) [regular files]                      │
│   DUMP: Mapping (Pass II) [directories]                       │
│   DUMP: Estimated 47952 blocks (23.41MB).                     │
│   DUMP: Dumping (Pass III) [directories]                     │
│   DUMP: Dumping (Pass IV) [regular files]                     │
│   DUMP: 47948 blocks (23.41MB) on 1 volume at 1038 KB/sec      │
│   DUMP: DUMP IS DONE                                          │
│   DUMP: Level 0 dump on Wed Apr 16 19:53:40 1997              │
│ #                                                              │
└────────────────────────────────────────────────────────────────┘
```

ufsrestore Command

Now that file systems have been copied onto a tape, how is this information retrieved? Solaris provides an application to restore data from the backup media to the system mass storage devices. This application is called *ufsrestore*. The syntax of the *ufsrestore* command follows.

```
/usr/sbin/ufsrestore options [ arguments ] [ filename ... ]
```

Some of the most useful options to the *ufsrestore* commands follow.

- *i*: Places *ufsrestore* in the interactive mode. Commands available in this mode follow.

  - *add [filename]*: Adds the named file or directory to the list of files to be extracted.

  - *cd directory*: Changes to *directory* on the dump media.

  - *delete [filename]*: Deletes the current directory or file from the list of files to be extracted.

- *extract*: Extracts all files on the extraction list from the dump media.
- *help*: Displays a summary of available commands.
- *ls [directory]*: Lists files in *directory* (dump media) or the current directory, which is represented by a period (.).
- *pwd*: Prints the full path name of the current working directory.
- *quit*: Exits immediately.
- *verbose*: Toggles the verbose flag (the program prints a line for every action it takes).

- *r*: Restores the entire content of the media into the current directory.

- *x*: Extracts the named files from the media.

- *b*: Sets the *ufsrestore* blocking factor.

- *f [dump_file]*: Tells *ufsrestore* to use *dump_file* instead of */dev/rmt/o* as the file to restore from.

- *R*: Resumes restore after volume is changed.

- *t*: Prints table of contents for dump file.

- *n*: Skips to the *nth* file when multiple dump files exist on the same tape.

- *v*: Displays the name and *inode* number of each file restored.

Note that the *i*, *r*, *R*, *t*, and *x* arguments are mutually exclusive. Only one of these arguments may be used at a time.

Example ufsrestore

This section provides a few examples of *ufsrestore* usage, such as that shown in the following illustration. In the examples, assume that the system has an 8-mm tape drive connected as */dev/rmt/0*.

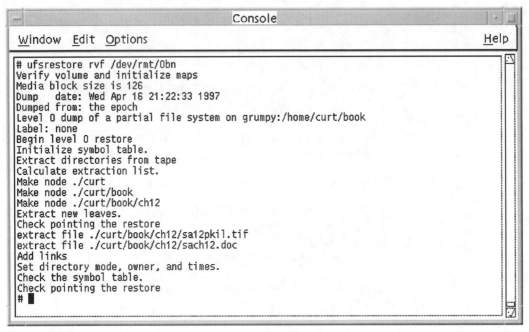

```
                          Console
 Window  Edit  Options                                      Help

 # ufsrestore rvf /dev/rmt/0bn
 Verify volume and initialize maps
 Media block size is 126
 Dump    date: Wed Apr 16 21:22:33 1997
 Dumped from: the epoch
 Level 0 dump of a partial file system on grumpy:/home/curt/book
 Label: none
 Begin level 0 restore
 Initialize symbol table.
 Extract directories from tape
 Calculate extraction list.
 Make node ./curt
 Make node ./curt/book
 Make node ./curt/book/ch12
 Extract new leaves.
 Check pointing the restore
 extract file ./curt/book/ch12/sa12pkil.tif
 extract file ./curt/book/ch12/sach12.doc
 Add links
 Set directory mode, owner, and times.
 Check the symbol table.
 Check pointing the restore
 #
```

Using ufsrestore to perform a restore to the system disk.

To completely restore an entire disk (*c0t2dos2*), change directory to the file system to be restored (e.g., */home*), and then issue a command similar to the following, substituting the appropriate disk partition information.

```
# ufsrestore rbf 50 /dev/rmt/0 /dev/rdsk/c0t2d0s2
```

To perform a partial restore to a file system, use the interactive mode of *ufsrestore*, as shown in the following illustration. First, change directory to a temporary storage space (e.g., */tmp/restore*).

Next, issue a command similar to the following string, substituting the appropriate disk partition information.

```
# ufsrestore ibf 50 /dev/rmt/0
```

```
                            Console
 Window  Edit  Options                                      Help

# ufsrestore if /dev/rmt/0bn
ufsrestore > ls
.:
 .Xauthority        TT_DB/          kernel/         sbin/
 .desksetdefaults   bin             lib             swap/
 .dt/               dev/            lost+found/     tmp/
 .dtprofile         devices/        mnt/            usr/
 .hotjava/          dodump          net/            var/
 .mailcap           etc/            opt/            vol/
 .solregis/         export/         platform/       xfn/
 26dump             floppy/         proc/
 AccessX            home

ufsrestore > add etc
ufsrestore > add 26dump
ufsrestore > extract
You have not read any volumes yet.
Unless you know which volume your file(s) are on you should start
with the last volume and work towards the first.
Specify next volume #: 1
set owner/mode for '.'? [yn] y
ufsrestore > quit
#
```

Using ufsrestore *in interactive mode.*

> ✓   *TIP: A side effect of using* ufsrestore *is that it creates a "symbol table" file in the file system where the files are restored. This file is named* restoresymtable. *It is wise to remove this file because it is usually quite large. Once the restore is complete, remove the file by issuing the command* rm ./restoresymtable.

Remote Backup and Restore

How can a file system dump be performed on a system without a backup device? The *ufsdump* and *ufsrestore* commands allow input and output to be sent to the standard input and output streams. This flexibility allows the *ufsdump/ufsrestore* output to be sent through the network to a system with a backup device. The *-f* option for *ufsdump* and *ufsrestore* specifies the dump device.

- *f dump_file*: This option tells *ufsdump* to use *dump_file* as the file to dump to. If *dump_file* is specified as a hyphen (-), *ufsdump* will dump to standard output. If the name of the file is of the form *machine:device*, the dump is carried out from the specified machine over the network using the *rmt* facility.

- ➡ NOTE: *Because* ufsdump *is normally run by* root, *the name of the local machine must appear in the* /.rhosts *file of the remote machine.*

To perform a remote restore using the *ufsrestore* command, the operator can make use of the *-f* option again.

- *f dump_file*: This option tells *ufsrestore* to use *dump_file* as the file to restore from. If *dump_file* is specified as a dash (-), *ufsrestore* reads from the standard input.

- ➡ NOTE: *Because* ufsrestore *is also normally run by* root, *the name of the local machine must appear in the* /.rhosts *file of the remote machine.*

To copy a home partition to another machine with a remote *ufsrestore*, consider the following command, substituting the appropriate directory and device names.

```
# ufsdump 0f - /dev/rdsk/c0t2d0s2 | (rsh machine ; cd /home;ufsrestore xf -)
```

tar Command

What happens if the operator does not want to dump and restore complete file systems? For example, what if a user simply wants to make a tape of the data associated with one project? Is there a simpler method than using *ufsdump* and *ufsrestore*? Most UNIX derivatives provide a standardized utility called *tar*. The *tar* command creates tape archives and provides the ability to add and extract files from these archives.

The syntax for using the *tar* command was covered in Chapter 4, "Fundamental System Administrator Tools." The following examples outline some typical uses of *tar* as a backup utility. To create an archive of */home/project* and */project/data* on the default *tar* device, the user would issue the following command.

```
% tar c -C /home/project include -C /data/project
```

A user may want to create an archive of his/her home directory on a tape mounted on drive */dev/rmt/0*. To do so, use the following command.

```
% cd ; tar cvf /dev/rmt/0
```

To read the table of contents of the *tar* file, issue the following command.

```
% tar tvf /dev/rmt/0
```

To extract files from the *tar* file, issue the following command.

```
% tar xvf /dev/rmt/0
```

It is also possible to get *tar* to transfer files across the network. To create a *tar* file on a remote host, use the following command.

```
% tar cvfb 20 filenames | rsh host dd of = /dev/rmt/0 obs = 20b
```

To extract files from a remote *tar* file into the current local directory, issue the following command.

```
% rsh -n host dd if = /dev/rmt/0 bs = 20b | tar xvBfb 20 filenames
```

cpio Command

Another command that may be used to back up and restore files is *cpio*. The *cpio* command copies file archives "in and out." The syntax for the *cpio* command follows.

```
cpio -key [options] filename
```

The key option to *cpio* determines the actions to be performed. The following flags are mutually exclusive.

- *cpio -i* (copy in) extracts files from the standard input, which is assumed to be the product of a previous *cpio -o*.

- *cpio -o* (copy out) reads the standard input to obtain a list of path names. The *cpio* command then copies those files to the standard output with path name and status information. Output is padded to a 512-byte boundary by default.

- *cpio -p* (pass) reads the standard input to obtain a list of path names of files that are conditionally created and copied into the destination directory tree.

- *NOTE: Consult the on-line manual pages for more complete information on the* cpio *command.*

To make a *cpio* copy of the files in the current directory in a file named *../newfile*, issue the following command.

```
% ls | cpio -oc >> ../newfile
```

- *NOTE: The* find, echo, *and* cat *commands can also be used as substitutes for the* ls *command to produce a list of files to be included in the* ./newfile.

To extract the files from *../newfile*, issue the following command.

```
% cat newfile | cpio -icd
```

To copy the content of the current directory to a new directory named *newdir*, issue the following command.

```
% find . -depth -print | cpio -pdlmv newdir
```

➡ *NOTE: When using* cpio *in conjunction with* find, *use the* L *option with* cpio *and the* -follow *option with* find.

dd Command

Yet another command available to use as a backup and restore utility is the *dd* utility. The *dd* command copies the input file to output, applying any desired conversions in the process. When complete, *dd* reports the number of whole and partial input and output blocks. The syntax of the *dd* command follows.

```
dd [ option = value ] ....
```

Valid options to the *dd* command follow.

- *if=filename*: Use *filename* as the input file. *stdin* is the default value.

- *of=filename*: Use *filename* as the output file. *stdout* is the default value.

- *ibs=n*: Use *n* as the input block size. *512* is the default value.

- *obs=n*: Use *n* as the output block size. *512* is the default value.

- *bs=n*: Use *n* as the input and output block size This supersedes the *ibn* and *obn* arguments. If no conversion is specified, preserve input block size.

- *files=n*: Copy and concatenate *n* input files before terminating.

- *skip=n*: Skip *n* input blocks before performing copy.

- *iseek=n*: Seek *n* blocks from the beginning of the input file before copying.

- *oseek=n*: Seek *n* blocks from the beginning of the output file before copying.

- *count=n*: Copy *n* input blocks.

- *swab*: Swap every pair of bytes.

- *sync*: Pad every input block to *ibs*.

The *dd* command has a few noteworthy restrictions, as follows.

- Do not use *dd* to copy files between file systems with different block sizes.

- Using a blocked device to copy a file will result in extra nulls being added to the file, in order to pad the final block to the block boundary.

- When *dd* reads from a pipe, using the *ibs* = *X* and *obs* = *Y* operands, the output will always be blocked in chunks of size Y. When *bs* = *Z* is used, the output block size will be whatever could be read from the pipe at the time.

To copy the files on device */dev/dsk/c1t3d0s2* to tape drive */dev/rmt/2* with a 20-block record size, issue the following command.

```
# dd if=/dev/rdsk/c1t3d0s2 of=/dev/rmt/2 bs=20b
```

To extract this information from the tape to a disk named */dev/dsk/c2t1d0s2*, issue the following command.

```
# dd if=/dev/rmt/2 of=/dev/rdsk/c2t1d0s2 bs=20b
```

Dealing with Specific Backup Issues

Certain aspects of successful backup and restore strategies require special attention. For instance, how could the operator restore the *root* file system if the *root* disk had crashed and there was no way to boot the system? Many administrators are also concerned with how to automate backups to minimize time investment while ensuring successful backups. Next, what happens if a backup requires 2 Gb of backup media, but the backup device can write only 1 Gb to the media?

These are just a few of the file system backup problems the administrator should be prepared to encounter. Fortunately, they are also topics that have been encountered and solved by the creators of Solaris. With an understanding of the basics of performing backups, it is possible to examine these specific topics in more detail.

Restoring the Root File System

It would appear that one of the most difficult problems faced when using *ufsrestore* is restoring the *root* file system. If the *root* file system is missing, it is not possible to boot the damaged system. Further, it is unlikely that there would be a file system tree to restore to.

One way to accomplish a *root* file system reload is by booting the system to the single-user state from the CD-ROM distribution media. Once the system is running, the operator could issue the proper commands to reload the *root* device from backup media.

As discussed in previous chapters, the Solaris operating system can be booted from the distribution media. Table 15-5, which follows, provides the system-type-to-boot-device mapping for a local CD-ROM drive.

Table 15-5: System Type to Solaris Boot Device Correspondence

System Type	Boot Command
Sun4	b sd(0,30,1)
SparcEngine	b sd(0,6,5)
Sun4c	boot cdrom
Sun4d	boot cdrom
Sun4m	boot cdrom
Sun4u	boot cdrom

For example, the boot command for a SPARCServer 2000 system would be as follows.

```
ok boot cdrom
```

This command will load the SunInstall program from the CD-ROM. At this point, follow the instructions in Chapter 3, "Preparing for a Solaris Installation," to load a new copy of the operating system to the new *root* drive. Alternatively, select the menu item that allows SunInstall to bring up a single-user shell on the system. Once the shell prompt appears, reload the *root* drive from the backup media instead of loading from the distribution media.

Another way to reload the *root* file system would be to boot the system to the single-user state as a client of another system on the network. At that point, the proper commands could be issued to reload the *root* device from backup media.

Automated Backups

All commands mentioned in this chapter may be adapted to provide for automated backups. Backup automation requires the creation of a shell script file containing the desired backup commands. When the script files have been created, submit a job to *cron* to execute the command files at the desired time. For example, place the following commands in a file named */dodump*.

```
#!/bin/sh
# dump /, /usr, /var, /home, /opt, /usr/openwin
#
mt -f /dev/rmt/0b rew
/usr/lib/fs/ufs/ufsdump 0bdfu 50 54000 /dev/rmt/0bn /dev/rdsk/c0t3d0s0
/usr/lib/fs/ufs/ufsdump 0bdfu 50 54000 /dev/rmt/0bn /dev/rdsk/c0t3d0s6
/usr/lib/fs/ufs/ufsdump 0bdfu 50 54000 /dev/rmt/0bn /dev/rdsk/c0t3d0s3
/usr/lib/fs/ufs/ufsdump 0bdfu 50 54000 /dev/rmt/0bn /dev/rdsk/c0t3d0s7
/usr/lib/fs/ufs/ufsdump 0bdfu 50 54000 /dev/rmt/0bn /dev/rdsk/c0t3d0s5
/usr/lib/fs/ufs/ufsdump 0bdfu 50 54000 /dev/rmt/0bn /dev/rdsk/c0t3d0s4
mt -f /dev/rmt/0b rew
mt -f /dev/rmt/0b offl
```

Place a tape in the */dev/rmt/0* tape drive, log in as *root* on the system, and type # *sh -x /dodump*. The system will respond by performing a full backup of all file systems listed in the *dodump* file. Although this is a simplistic method of automating file system backups, it serves to illustrate the point that it is possible to automate backups. Chapter 16, "Automating Administrative Tasks," discusses more elegant solutions to the backup automation problem using the *cron* command.

Multi-volume Dumps

Two of the backup commands mentioned in this chapter also allow for multi-volume backups. The *ufsdump* command and the *cpio* command allow a backup to be stored over multiple media sets. The other commands (*tar* and *dd*) will allow the user to split the backups onto several sets of media, but these commands require that the user perform much of the work manually.

The *cpio* command watches for the "end of medium" event. When *cpio* detects this event, it stops and prints the following message on the terminal screen.

```
If you want to go on, type device/file name when ready.
```

To continue the dump, the operator must replace the medium and type the character special device name (e.g., */dev/rdiskette*) and press <Enter>. At this point, the operator may choose to have *cpio* continue the backup to another device by typing the device's name at the prompt. An example of multi-volume dumps using *cpio* is shown in the following illustration.

✗   WARNING: *Simply pressing <Enter> at the prompt will cause* cpio *to exit.*

Multi-volume dumps using cpio.

```
# find . -depth -print |cpio -ocv >/dev/rdiskette
status_note
minutes_31397
description
to_do
progress
End of medium on "output".
To continue, type device/file name when ready.
/dev/rdiskette
milestones
specifications
.
3603 blocks
#
# mkdir /tmp/restore
# cd /tmp/restore

# cpio -icdv </dev/rdiskette
status_note
minutes_31397
description
to_do
progress
End of medium on "input".
To continue, type device/file name when ready.
/dev/rdiskette
specifications
.
3602 blocks
#
# ls -las
total 3288
    8 drwxr-xr-x    2 root     other         372 Apr 16 20:43 .
    8 drwxrwxrwt    8 sys      sys           645 Apr 16 20:41 ..
  200 -rw------T    1 root     other      102400 Apr 16 20:38 description
  272 -rw-------    1 root     other      139264 Apr 16 20:39 milestones
  600 -rw------T    1 root     other      307200 Apr 16 20:37 minutes_31397
  400 -rw------T    1 root     other      204800 Apr 16 20:38 progress
  400 -rw------T    1 root     other      204800 Apr 16 20:43 specification
s
  400 -rw------T    1 root     other      204800 Apr 16 20:37 status_note
 1000 -rw------T    1 root     other      512000 Apr 16 20:38 to_do
#
```

Using *ufsdump* as the backup command is somewhat simpler. Like *cpio*, *ufsdump* will detect the end of medium event and stop operation. The *ufsdump* command will then wait for the operator to change the media before it continues. Unlike *cpio*, *ufsdump* does not need the name of the backup device to continue. The operator simply needs to confirm that the media has been changed and that everything is ready for *ufsdump* to continue operation. An example of multi-volume dumps using *ufsdump* is shown in the following illustration.

```
┌─────────────────────────────── Console ───────────────────────────┬─┐
│ Window  Edit  Options                                        Help  │ │
├────────────────────────────────────────────────────────────────┬──┤
│ # ufsdump 0oDfu /dev/rdiskette ./demo                          │▲││
│   DUMP: Writing 10 Kilobyte records                            │ ││
│   DUMP: Date of this level 0 dump: Wed Apr 16 21:07:10 1997    │ ││
│   DUMP: Date of last level 0 dump: the epoch                   │ ││
│   DUMP: Dumping /dev/rdsk/c0t3d0s6 (grumpy:/usr) to /dev/rdiskette. │
│   DUMP: Mapping (Pass I) [regular files]                       │ ││
│   DUMP: Mapping (Pass II) [directories]                        │ ││
│   DUMP: Estimated 5516 blocks (2.69MB).                        │ ││
│   DUMP: Dumping (Pass III) [directories]                       │ ││
│   DUMP: Dumping (Pass IV) [regular files]                      │ ││
│   DUMP: End-of-file detected                                   │ ││
│   DUMP: 51.78% done, finished in 0:01                          │ ││
│   DUMP: Change Volumes: Mount volume `#2´ on `grumpy:/dev/rdiskette´ │
│   DUMP: NEEDS ATTENTION: Is the new volume (#2) mounted on `grumpy:/dev/rdiskette´ │
│                                    and ready to go?: ("yes" or "no") yes │
│   DUMP: Volume 2 begins with blocks from inode 85493          │ ││
│   DUMP: 5498 blocks (2.68MB) on 2 volumes at 19 KB/sec        │ ││
│   DUMP: DUMP IS DONE                                           │ ││
│ # █                                                            │▼││
└────────────────────────────────────────────────────────────────┴──┘
```

Multi-volume dumps using ufsdump.

Summary

This chapter explored the commands that can be used to make backup copies of system data, why it is important to make such backup copies of the data, and selected methods of avoiding data loss due to natural or other disasters. The authors hope that readers never have to use any of these backup copies to restore the operation of their systems, but such restorations are inevitable. Good backups require a lot of time and attention, but having a reliable copy of the data is much more acceptable than the time and expense of rebuilding a system without such backup copies.

CHAPTER 16
Automating Administrative Tasks

As demonstrated in previous chapters, administering Solaris systems can be a time-consuming process. Simply keeping track of system files could be a full-time job, let alone managing security and user accounts. This chapter explores methods of automating certain mundane system administration chores. The goal of this chapter is to demonstrate ways the administrator can "free up" time for more urgent tasks, such as helping users.

Updating System Files with the rdist Command

One of the most frequent housekeeping jobs an administrator faces is keeping system files current. For example, an administrator at a small site might manage 25 to 30 machines. Each of the machines has a private copy of the */usr/lib/sendmail* software installed on the local disk. How could the administrator efficiently update the *sendmail* software on 30 machines?

The */bin/rdist* command is a remote file distribution utility. The *rdist* utility allows the administrator to set up a central repository of files that must be updated on multiple machines within the corporate network. To update all machines on the network, the administrator makes changes to the files in the central repository, and then uses the *rdist* utility to distribute these files to the other hosts on the network.

The *rdist* utility requires setup before it is used the first time. When the setup is complete, the task of updating files on machines becomes a trivial task.

rdist Setup

The setup for *rdist* consists of editing two files. First is the *.rhosts* file, which grants permission for users to access the system through the network, and second is the *Distfile*, an ASCII file that lists the commands and files *rdist* will use to update the other machines on the network.

/.rhosts File

The */.rhosts* file on each of the machines on the network should contain an entry allowing the distribution server access. For example, if the distribution server were *mercury.astro.com*, and the *rdist* command required *root* access, the *rdist* client machines should have the following line in their respective */.rhosts* files.

```
mercury.astro.com root
```

➡   *NOTE: The* rcp, rsh, *and* rlogin *commands make use of the* /.rhosts *file to determine access. Access is granted if the machine and user being contacted have a* /.rhosts *file, and if the file contains a line listing the machine from which the connection derives. Exercise great care upon allowing access to the root account via entries in the* /.rhosts *file. See*

Chapter 18, "Network Security," for more information on root account security.

Distfile

The *rdist* command searches for a file named *Distfile* in the current directory. If the file exists, *rdist* executes the commands in the file to update the other hosts on the network. The *Distfile* allows the administrator to list the hosts and respective files to be updated. The syntax of the *Distfile* entries is very similar to the UNIX regular expression syntax. The user may define groups of hosts that require special handling, or macro commands that define a series of commands to be executed on a host.

✓   *TIP: It is also possible to use* rdist *in a limited mode without all of the setup. To distribute a file from the current machine to another, invoke* rdist *with the following format:* rdist -c *filename host. For example, the following command would distribute the* /etc/inet/hosts *and* /etc/mail/aliases *files from the current host to host mercury:* rdist -c /etc/inet/hosts /etc/mail/aliases mercury.

Instead of attempting to list the details of how to use the *rdist* command, the following section will describe typical usage.

rdist Example

Assume that the SpaceCadets Corporation system administrator is attempting to stay current with file maintenance on the corporate hosts. One of the tools the administrator decides to use is the */bin/rdist* command. The following example details the steps the administrator must follow in order to get *rdist* running on the corporate Solaris systems.

Challenge

Corporate management has requested that the administrator automate the distribution of several files on the corporate machines. However, not all machines will be updated with the same information. For example, the machines in the engineering department will not allow employees from the marketing department to log in. Conversely, the machines in the marketing department will not allow employees of the engineering department to log in.

Management has also decided that a single e-mail server will suffice for all machines within the corporation. Because the e-mail server is to handle sending and receiving e-mail for all corporate computer systems, the *sendmail* configuration on the server should allow the server to accept mail bound for any system within the corporate domain. Client machines should be configured to send their mail to the server for delivery, and to load the users' e-mail from the mail spool area on the mail server system. This network configuration is shown in the facing illustration.

Identifying Files to rdist

One of the first implementation issues to be dealt with is identifying *which* files to *rdist*. The administrator reviews the list of tasks set forth by corporate management, and develops the following file list.

- */.rhosts*: Specifies which hosts have privileged network access to the system. All machines must have this file.

- */etc/inet/hosts*: Contains the mapping of IP addresses to host names for the hosts within the corporation. All hosts must have this file.

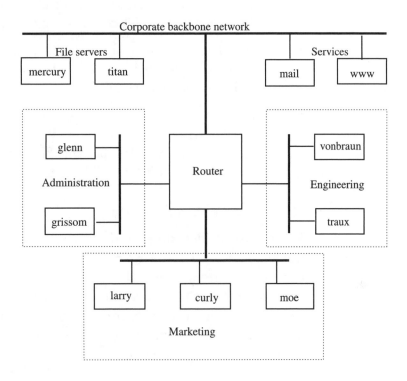

SpaceCadets Corporation network configuration.

- *   */etc/passwd*: System password file. The administrator decides that two password files are necessary: one for engineering and one for marketing. The administrator also decides that others may be required in the future.

- *   */etc/mail/sendmail.cf*: Configuration file for the *sendmail* program. Separate *sendmail.cf* files include one for the e-mail server and another for all other hosts on the network.

- *   */etc/mail/sendmail.cw*: Informs the e-mail server of the hosts for which to accept e-mail. The file resides only on the e-mail server system, and lists all corporate host names.

- *   */usr/lib/sendmail*: The *sendmail* program. All hosts on the network will have the same *sendmail* binary.

- */etc/mail/aliases*: Lists the e-mail aliases used by the corporation. This file resides on all corporate hosts.

- */etc/groups*: System group file. The administrator decides to make the group files on all machines identical in order to avoid problems in the future.

- */etc/shells*: Lists the shell programs allowed on the systems. In order to keep users from loading unsupported shells, the administrator decides that this file should be centrally maintained. All hosts will have this file.

Creating a Central Repository

With a list of files in hand, the administrator starts planning the implementation of the *rdist* server. The *rdist* server contains a central repository that holds the master copy of the files to be distributed to other corporate hosts. The administrator decides to name the directory that contains this central repository of files */common*. Subdirectories in the */common* directory with familiar names include the following.

- */common/etc*: Contains copies of files that will end up in the */etc* directory of the *rdist* client machines.

- */common/usr*: Contains copies of files that will end up in the */usr* directory of the *rdist* client machines.

This scheme allows the administrator to tailor the central repository in an easy-to-decipher hierarchical storage area. The administrator copies the appropriate files into the central repository and then edits each file to tailor it to the appropriate task.

Identifying Machine Groupings

The next challenge the administrator has to meet is grouping machines into easy to maintain sets. To achieve this objective, the administrator creates several groups of machines, as follows.

- SERVERS: List of file servers to be managed with *rdist*. This group includes the systems that provide the file systems containing applications and user file space for the corporation.

- SERVICES: List of service machines to be maintained by *rdist*. This group includes the e-mail server and a web server.

- CLIENTS: List of all non-file server machines to be managed under *rdist*. These machines are desktop workstations. They obtain applications and user file storage space from the corporate file servers, and services (e-mail and web) from the corporate services machines.

- ENGR: Subset of the CLIENTS group. These machines are located in the engineering department.

- MARKET: Another subset of the CLIENTS group. These machines are located in the marketing department.

- ADMIN: Another subset of the CLIENTS group. These machines are the administrative systems.

Implementation

When planning is complete, the administrator creates a *Distfile* within the */common* directory. The following is an example of what the administrator's *Distfile* might contain.

```
# SpaceCadets Corporation rdist Distfile
# First, define the hosts to be managed.
ALL=(${SERVERS} ${CLIENTS} ${SERVICES} )
SERVERS=(mercury.astro.com titan.astro.com)
CLIENTS=( ${ADMIN} ${ENGR} ${MARKET} )
ADMIN=( glenn.astro.com grissom.astro.com)
ENGR=( vonbraun.astro.com truax.astro.com )
MARKET=( larry.astro.com curly.astro.com moe.astro.com)
SERVICES=( mail.astro.com www.astro.com )
```

```
# Next, define the files to be installed on ALL hosts.
hosts:
/common/etc/hosts -> ${ALL}
        install /etc/inet/hosts;
group:
/common/etc/group -> ${ALL}
        install /etc/group;
shells:
/common/etc/shells -> ${ALL}
        install /etc/shells;
sendmail:
/common/usr/lib/sendmail -> ${ALL}
          install /usr/lib/sendmail;
      special "/usr/local/bin/killj sendmail";
      special "chown bin /usr/bin/sendmail";
      special "chmod 4755 /usr/bin/sendmail";
      special "/usr/lib/sendmail -bd -q1h";
aliases:
/common/etc/aliases -> ${ALL}
        install /etc/aliases;
        special "/usr/ucb/newaliases";
# Define the files to be installed on the ENGineeRing hosts.
prtconf:
/tmp/trash -> ${ENGR}
        install /tmp/trash;
        special "rm /tmp/trash";
        special "/etc/prtconf";
passwd.engr:
/common/etc/passwd.engr -> ( ${ENGR} )
        install /etc/passwd.engr;
        special "mv /etc/passwd.engr /etc/passwd"
        special "/bin/pwconv";
rhosts.engr:
/common/.rhosts.engr -> ${ENGR}
        install /.rhosts.engr;
        special "mv /.rhosts.engr /.rhosts";
```

```
        special "chmod 600 /.rhosts";
# Define the files to be maintained on the MARKETing hosts.
passwd.market:
/common/etc/passwd.market -> ( ${MARKET} )
        install /etc/passwd.market;
        special "mv /etc/passwd.market /etc/passwd"
        special "/bin/pwconv";
rhosts.market:
/common/.rhosts.market -> ${MARKET}
        install /.rhosts.market;
        special "mv /.rhosts.market /.rhosts";
        special "chmod 600 /.rhosts";
# Define the files to be maintained on other subsets of hosts.
sendmail.cf.clients:
/common/etc/sendmail.cf.clients -> ${CLIENTS}
        install /etc/mail/sendmail.cf.clients;
        special "mv /etc/mail/sendmail.cf.clients /etc/mail/sendmail.cf";
        special "ln -s /etc/mail/sendmail.cf /etc/sendmail.cf";
sendmail.cf.server:
/common/etc/sendmail.cf.server -> mail.astro.com
        install /etc/mail/sendmail.cf.server;
        install /etc/mail/sendmail.cw;
        special "mv /etc/mail/sendmail.cf.server /etc/mail/sendmail.cf";
        special "ln -s /etc/mail/sendmail.cf /etc/sendmail.cf";
```

Running rdist

When the setup is complete, it is time to run the *rdist* command. Using the debugging mode of *rdist* is recommended when a new *Distfile* is used for the first time. To do so, you invoke *rdist* as follows. (See also the following illustration.)

```
# rdist -n -f Distfile
```

Typical output of /bin/rdist –n for testing rdist rules.

```
┌─────────────────────────────────────────────────────────────┐
│ ─                        Console                      · □     │
├─────────────────────────────────────────────────────────────┤
│  Window  Edit  Options                              Help      │
├─────────────────────────────────────────────────────────────┤
│ grumpy# /bin/rdist -n -f /common/Distfile                    │
│ updating host mercury.astro.com                               │
│ install /common/etc/inet/hosts /etc/inet/hosts               │
│ updating host titan.astro.com                                 │
│ install /common/etc/inet/hosts /etc/inet/hosts               │
│ updating host glenn.astro.com                                 │
│ install /common/etc/inet/hosts /etc/inet/hosts               │
│ updating host grissom.astro.com                               │
│ install /common/etc/inet/hosts /etc/inet/hosts               │
│ updating host vonbraun.astro.com                              │
│ install /common/etc/inet/hosts /etc/inet/hosts               │
│ updating host truax.astro.com                                 │
│ install /common/etc/inet/hosts /etc/inet/hosts               │
│ updating host larry.astro.com                                 │
│ install /common/etc/inet/hosts /etc/inet/hosts               │
│ updating host curly.astro.com                                 │
│ install /common/etc/inet/hosts /etc/inet/hosts               │
│ updating host moe.astro.com                                   │
│ install /common/etc/inet/hosts /etc/inet/hosts               │
│ updating host mail.astro.com                                  │
│ install /common/etc/inet/hosts /etc/inet/hosts               │
│ updating host www.astro.com                                   │
│ install /common/etc/inet/hosts /etc/inet/hosts               │
│ updating host mercury.astro.com                               │
│ install /common/etc/group /etc/group                          │
│ updating host titan.astro.com                                 │
│ install /common/etc/group /etc/group                          │
│ updating host glenn.astro.com                                 │
│ install /common/etc/group /etc/group                          │
└─────────────────────────────────────────────────────────────┘
```

The administrator should study the resulting output to determine if *rdist* will perform the desired steps. Several changes to the *Distfile* may be necessary before the process works perfectly. Once the *Distfile* is correct, run *rdist* manually with the following command to ensure that everything works the first time.

```
# rdist -f Distfile
```

Subsequent sections of this chapter will discuss methods of making the system automatically run the *rdist* (and other) commands at specified times using the *cron* facility.

✓   *TIP: The* rdist *command may be used to install local packages and configuration files on new machines. Once the administrator installs the op-*

erating system and has a new host running on the network, rdist *may be used to automatically install all of the "localisms" on the new system.*

�10   *NOTE: The* rdist *command may also be used to run commands on a remote machine. However, because* rdist *was not intended to perform remote execution of commands, a few extra steps are required. A simple way of accomplishing this is to distribute an empty file to the target machine, then use the "special" directive to cause the machine to remove the empty file, and then execute the desired commands. For instance, in the example Distfile, the ENGR machines are commanded to run the* prtconf *command every time* rdist *runs. The output of the command may be e-mailed to the administrator along with the rest of the* rdist *report. This allows the administrator to keep track of the hardware/ software configuration of the systems within the engineering department.*

Finding and Removing Unused Files

An old system administration adage states that "Disk usage will expand to fill all available disk space." What steps are necessary to ensure that enough disk space is available for users? One way of policing free disk space might be to locate and remove all temporary files that had not been accessed (used) during a predetermined period. Although a valid management trick, this becomes a very time-consuming process if it is done manually. Fortunately, UNIX provides a utility called *find* that permits automation of the process.

find Command

The *find* utility allows the user to search for specific information on system disks. The find utility may be used to search for files, directories, or strings of characters on the disks. When an item is found, the find command provides many additional options.

You could have find execute other commands such as using *rm* to remove files, or printing the location of the item on the screen. You could cascade several functions together and use find to perform more complex operations. The syntax for using the *find* command follows.

```
find pathlist expression
```

The *pathlist* argument is a list of files to be searched by *find*. The file list may consist of file names, directory names, or file system names. If the *root* (/) directory is specified as the *pathlist*, *find* will traverse all directories on the system.

The *expression* argument includes the string of characters that *find* will search for and what to do when items are found. Commonly used expressions for the *find* utility follow.

- *-name*: Search for files called "name."

- *-atime N*: Search for files accessed within the past *N* days.

- *-ls*: Print current path name and related information.

- *-mtime N*: Search for files modified *N* days ago.

- *-exec command*: Execute "command" if file is found.

- *-print*: If file is found, print path name to file.

- *-type C*: Search for files of type *C*.

Valid values for C include the following.

- *b*: Search for block special files.

- *c*: Search for character special files.

- *d*: Search for directories.

- *l*: Search for symbolic links.

- *p*: Search for FIFO files.

- *s*: Search for sockets.

- *f*: Search for plain files.

- *-fstype type*: Search for file systems of *type*.

The following sections provide a few examples of how the *find* utility might be used in daily system administration activities.

Eliminating Old Files

In a previous chapter the *find* command was used to locate and remove all *a.out* and *.o* files in a user's home directory that were more than seven days old. The find statement to perform this task follows.

```
% find $HOME \(-name a.out -o -name '*.o'\) -atime +7 -exec rm {} \;
```

> ➥   *NOTE: The use of parentheses, quotation marks, backslashes, and curly braces is explained in detail later in this chapter. These symbols are required to cause the shell to interpret the command in a particular manner.*

In the previous example, the parentheses are used to inform find to search for multiple file names. The curly braces following the *rm* command tell *find* to execute the *rm* command, and remove any files that match the search specifications. The backslashes are used so that the shell does not interpret the parentheses and semicolon characters, but rather passes them to the *find* command for interpretation.

How could you modify this command to find the same files in all system directories? Simply change the *pathlist* argument from *$HOME* to *root* (*/*) as follows.

```
# find / \(-name a.out -o -name '*.o'\) -atime +7 -exec rm {} \;
```

> ✸   *NOTE: The previous example should be run as* root *because* root *is the only user that can access all files on the system. Note the* root *prompt in the example.*

Other Uses for find

Because *find* can locate so many different things on disks, there are other uses for this tool. Experienced system administrators use *find* on a daily basis. The following sections present a sampling of tasks that can be automated with the *find* command.

Locating Files

Assume you want to locate all instances of a particular file so that information in the file(s) can be updated. With the *find* utility, this process becomes a simple operation, as shown in the following illustration.

Using find *to locate all instance files named* .login.

Locating Text Strings

The *find* command can also be used to locate a file that contains a particular string of characters. Why is this useful? One popular method of breaking into systems is to replace system binaries with new versions that contain backdoor entry points for use by people who want to break in. Quite often these programs contain a unique string of characters (such as a compiled-in password) that make them easy to search for. The *find* command makes it easy to search the disk for a specific string of characters, as shown in the following illustration.

Using find *to locate all instances of a particular string.*

```
┌─────────────────────── Console ────────────────────────┐
│ Window  Edit  Options                              Help │
├─────────────────────────────────────────────────────────┤
│ # find / -type f -exec grep -l "floobydust" {} \;       │
│ /opt/tmp/.xxsh                                           │
│ /opt/tmp/login                                          │
│ /usr/bin/ .login                                        │
│ /usr/tmp/.xxsh                                          │
│ /var/mail/.xxsh                                         │
│ /tmp/()                                                 │
│ /var/tmp/.xxsh                                          │
│ # █                                                     │
└─────────────────────────────────────────────────────────┘
```

Checking File Permissions

Another common use of the *find* command is to locate files with particular access permissions. For instance, files that can be written by users other than the file owner could be considered security risks. Changing the access permission of these files so that only the owner has write permission is recommended. Rather than spending hours looking for such files manually, *find* can be used to automate this operation, as shown in the following illustration.

Using find *to locate and change permissions on files with mode 777.*

```
┌─────────────────────── Console ────────────────────────┐
│ Window  Edit  Options                              Help │
├─────────────────────────────────────────────────────────┤
│ # find /home -perm 777 -print -exec chmod 750 {} \;     │
│ /home/curt/.login                                       │
│ /home/guest/.login                                      │
│ /home/users/milind/.login                               │
│ /home/users/paulgo/.login                               │
│ /home/users/rich/.login                                 │
│ /home/users/terry/.login                                │
│ /home/users/lums/.login                                 │
│ /home/sally/.login                                      │
│ /home/erin/.login                                       │
│ /home/shaun/.login                                      │
│ /home/varmint/.login                                    │
│ /home/ontherange/.login                                 │
│ # █                                                     │
└─────────────────────────────────────────────────────────┘
```

In certain situations it may be desirable to change the ownership of files owned by a particular user. For instance, you may want to change a user's log-in name or user ID for some reason. In order for the user to be able to access his/her files after the *uid/username* change, the files would have to be "chowned" and "chmoded" to the new *uid/username*. The *find* command used to locate all files owned by the user and change ownership to the new user name is shown in the following illustration.

Using find *to locate and change the ownership of files owned by a particular user.*

```
┌──────────────────────────── Console ──────────────────────────────┐
│  Window  Edit  Options                                    Help      │
├────────────────────────────────────────────────────────────────────┤
│ # find /home -user varmint -print                                   │
│ /home/varmint                                                       │
│ /home/varmint/.login                                                │
│ /home/varmint/project                                               │
│ /home/varmint/project/secret                                        │
│ /home/varmint/project/data                                          │
│ /home/varmint/project/bugs                                          │
│ /home/varmint/project/src                                           │
│ # find /home -user varmint -exec chown erin {} \;                   │
│ # find /home -user varmint -print                                   │
│ # find /home -user erin -print                                      │
│ /home/varmint                                                       │
│ /home/varmint/.login                                                │
│ /home/varmint/project                                               │
│ /home/varmint/project/secret                                        │
│ /home/varmint/project/data                                          │
│ /home/varmint/project/bugs                                          │
│ /home/varmint/project/src                                           │
│ # █                                                                 │
└────────────────────────────────────────────────────────────────────┘
```

You may want to build script files that automatically run the *find* commands necessary to perform normal housekeeping, such as cleaning temporary file systems. The following section examines one way to automate the execution of these scripts.

Automating Commands with cron

The *cron* command is a chronometer (clock) based function that executes commands at specified dates and times. One of the more common uses of *cron* is to run batch jobs at off-peak hours. Another common use of *cron* is to run housecleaning and administrative functions at regular intervals. Because *cron* runs under the control of a system daemon program, understanding the types of processes available under Solaris is crucial.

Process Types

The Solaris operating system is a multi-user/multi-tasking operating system. This means that many users can be active on the system simultaneously, and many jobs (processes) can be active simultaneously. Although only one job is active (per processor) at any given time, many jobs can be in the run queue.

The run queue is a list of jobs awaiting respective slices of run time from the scheduler. Because the slice of run time allotted to each process is so short, many jobs appear to be active simultaneously. In many cases, these jobs are interactive processes started by users sitting at a terminal or using a window on a workstation. These processes are generically referred to as foreground processes.

Foreground Processes

When users sit at the terminal typing commands to the shell, they are typically executing foreground processes. Foreground processes maintain control of the terminal (window) until the execution is complete, and are therefore referred to as "interactive" processes. Foreground processes typically read input from the terminal and write output to the terminal. Some typical foreground processes include system commands such as *ls, cat, vi,* and *sh.*

Background Processes

Some of the processes running on the system are not interactive processes. A process that is not run interactively is referred to as a background process. When a process is run in the background, it is disassociated from the terminal. This means that the terminal is free for other uses while the background process executes. Because the background process is not associated with a terminal, the user must arrange for the background process to get input from a file or other source. Likewise, the user must arrange for the background process output to be sent to a file (or another process).

cron *Processes*

Processes executed by *cron* are background jobs. Many *cron* jobs are a special flavor of background processes known as *batch jobs*. A batch job reads its input from one place, processes the input, and then writes its output (if it produces any output) to another place.

In order to run a *cron* job, the user must create a *crontab* entry. A *crontab* entry tells the *cron* program which command to run, and more importantly, when to run it. The *crontab* (*cron* tables) command allows users to create and modify *crontab* entries. Once the user has created a *crontab* entry, the *crontab* program stores that entry in the */var/spool/cron/crontabs* directory. The *cron* software scans this directory to identify the commands to run at particular times.

By default, the *crontab* command is not limited to *root* user usage. Other users can create their own *cron* jobs. System administrators using the *root* user account can control access to *cron* by editing the */etc/cron.d/cron.allow* and */etc/cron.d/cron.deny* files. As the file names imply, *cron.allow* contains a list of all users permitted *cron* access, whereas *cron.deny* contains a list of all users denied *cron* access. By default, both files are empty.

➡️   *NOTE: Many system administrators limit cron access to system staff members. Otherwise, system users could use cron to hide (i.e., move) illegal files prior to security sweeps designed to detect such files.*

crontab Entries

Whereas the *crontab* program is used to create *cron* table entries, it is the *cron* daemon (a background process) that examines the entries to determine when *cron* jobs should be executed. In order for *cron* to understand the *crontab* entries, they must conform to a specific template. Every entry in a *crontab* file consists of the following fields.

- Minute of execution (0-59, asterisk, or list)

- Hour of execution (0-23, asterisk, or list)

- Day of the month (1-31, asterisk, or list)

- Month of the year (1-12, asterisk, or list)

- Day of the week (0-6 [Sunday = 0], asterisk, or list)

- Command string to be executed

As previously noted, the first five fields may contain a numeric argument. For example, to execute a command on the first day of the month, place a one (1) in the third field of the *crontab* file. This command would be executed on the first day of the month at the time set by the hour, minute, month, and day fields. An example of a *root crontab* file is shown in the following illustration.

```
┌─                              Console                              · □
 Window  Edit  Options                                          Help
┌─────────────────────────────────────────────────────────────────┐
│# cat /var/spool/cron/crontabs/root                                │
│#ident  "@(#)root    1.12      94/03/24 SMI"   /* SVr4.0 1.1.3.1   */│
│#                                                                  │
│# The root crontab should be used to perform accounting data collection.│
│#                                                                  │
│# The rtc command is run to adjust the real time clock if and when │
│# daylight savings time changes.                                   │
│#                                                                  │
│0 2 * * 0,4 /etc/cron.d/logchecker                                 │
│5 4 * * 6   /usr/lib/newsyslog                                     │
│15 3 * * * /usr/lib/fs/nfs/nfsfind                                 │
│1 2 * * * [ -x /usr/sbin/rtc ] && /usr/sbin/rtc -c > /dev/null 2>&1│
│# ▮                                                                │
└─────────────────────────────────────────────────────────────────┘
```

Example root crontab *file.*

Alternately, the first five fields may contain a list of numeric values separated by commas. This allows scheduling commands to be run multiple times a day with a single *crontab* entry. For example, to execute a command on the hour (and 15, 30, and 45 minutes after the hour), place the *0,15,30,45* string in the first field of the *crontab* file. The command would be executed four times an hour as specified by the other four fields.

The asterisk acts as a wild-card entry in the *crontab* file. When *cron* finds an asterisk in a field, it automatically substitutes every valid value for that field. For example, a *crontab* entry with an asterisk in the first field would instruct *cron* to execute the command once a minute as long as the conditions set in the other fields were true.

crontab *Options*

The *crontab* application allows the following options.

- *crontab -e* edits a copy of the current user's *crontab* file, or creates an empty file to edit if *crontab* does not exist.

- *crontab -l* lists the user's current *crontab* file.

- *crontab -r* removes the user's *crontab* file from the *crontab* directory.

Example crontab *Entries*

The use of the *find* command to locate and remove unused files from the system was covered earlier in this chapter. You can automate system housecleaning tasks by cascading the find and *cron* utilities. The *crontab* entries that follow execute *find* commands to clear out specific files. These files may be system-generated temporary files, or files that are otherwise easily recreated.

➥ *NOTE: Exercise caution when creating such* crontab *entries to avoid accidentally removing critical files. Manual development and testing of crontab entries is recommended prior to having the system perform these tasks automatically.*

The *crontab* entry that follows will run at 02:15AM every day of every month. The *find* command will search every file system for *.o* files (object files created by the compilers). If such files are found and are over seven days old, *find* will remove them. Note that this command will not remove such files from NFS-mounted file systems because of the *-fstype nfs -prune* directives.

```
15 02 * * * find / -name *.o -mtime +7 -exec rm {} \; -o \
-fstype nfs -prune >>/dev/null 2>>&1
```

The *crontab* entry below will run at 03:05AM every day of every month. The */bin/rdist* command will update the files on every machine listed in the */common/Distfile* control file. Because the output of this command will be mailed to the *root* account, the administrator can check file distribution results.

```
05 03 * * *   /bin/rdist -f /common/Distfile
```

Finally, the following *crontab* entry will run at 03:10AM every day of every month. The *find* command will search through the */tmp* file system for directories over one (1) day old. If such directories are found, and they are not the */tmp* directory itself, or the *lost+found* directory, they will be removed.

```
10 03 * * *   cd /tmp ; find . ! -name . ! -name lost+found -type d -mtime
    +1 -exec rmdir {} \; /dev/null 2&1
```

✓   *TIP: The* cron *command may be used to automate many functions. A few simple examples include automated distribution of new system software, file system backups, collection of network usage statistics, and creating system accounting summaries. If you can develop a shell script file to perform these tasks, it is also possible to develop a* crontab *entry that will allow the system to automatically perform the task.*

System Automation with Shell Scripts

The UNIX shell is an interactive programming language, as well as a command interpreter. The shell executes commands received directly from the user sitting at the terminal. Alternately, the shell can execute commands received from a file. A file containing shell programming commands is called a shell file, or *shell script*.

The examples in the following sections are all written as Bourne shell scripts. The Bourne shell is included with all versions of Solaris. Until recently, Sun did not include the Perl language with the operating system distribution. Beginning with Solaris 8, the Perl language has become a part of the operating system distribution.

Many of the operations performed by a system administrator are accomplished by typing commands at the terminal. Due to file access permissions, these commands must often be issued by *root*. Under Solaris, the standard shell for *root* is */sbin/sh*, a statically linked version of the Bourne shell. On the other hand, the program in */usr/bin/sh* is a dynamically linked version of the Bourne shell. What is the difference between these two versions of the *sh* program, and why are both needed?

➡️ *NOTE: Although shell scripts are powerful system administration tools, they do not fulfill every need.* Perl, awk, *and* Tcl/Tk, *among other utilities, are often found in administrator's toolboxes. For more information on these utilities, consult the on-line manual pages or the Answerbook. There are also several good reference books available on these utilities.*

Review of Dynamic Versus Static Linking

A statically linked program is one that links (or includes) all library functions at compile time. If changes are made to a library routine due to compiler or other system software upgrades, the program must be recompiled to take advantage of those changes. A statically linked program will usually be quite large because it includes all library functions.

A dynamically linked program loads the library functions at run time, which allows the program to include the latest version of all library functions upon invocation. No recompilation is required to enjoy the advantages of updated library routines. Dynamically linked programs are usually smaller than statically linked counterparts because the library functions are not included in the on-disk binary image.

✗ *WARNING: Many of the programs in the* /sbin *directory are statically linked. Although it may be tempting to replace these programs with dynamically linked versions to save disk space, this practice is not recommended. The programs in* /sbin *are statically linked for a very good reason. In the event the dynamic libraries (typically stored in the* /usr *file system) are not available due to a partial system failure, the programs in* /sbin *(such as the* root *log-in shell) will still function. Consequently, the system administrator can still log in and attempt to recover from the failure.*

The Shell Game: Selecting the Right Shell for the Job

The Bourne shell implements a language commonly referred to as shell programming language, or shell programming. However, the Bourne shell is just one of many shells available for Solaris. Another popular shell program is */usr/bin/csh* (the C shell). The C shell offers user features that are not available in the Bourne shell, such as advanced job control.

Other available shells include the Bourne Again shell (*bash*), Korn shell (*ksh*), *tcsh* (a variant of *csh*), and the *zsh* shell. Most of these variations share a common command structure with the Bourne and C shells.

Due to the fact that the shell is the primary user interface to the system, its selection is typically intensely personal. Users select the shell they are most comfortable with. Every shell has strengths and weaknesses the administrator can exploit. One of the secrets of system administration is knowing which tool will make a particular task easier to perform. In some cases, a system administrator may choose one tool over another based on familiarity alone.

Although the Bourne shell and C shell differ, they share a few concepts. Both shells allow the user to declare local variables, provide and understand variables that customize the execution environment for users, and interact with the operating system to provide error handling and other basic features.

In order to create shell scripts to aid in system administration, you must understand the language used by each shell. Unfortunately, this topic is beyond the scope of a single chapter in a systems administration book. Entire books have been dedicated to the description of the Bourne, C, and Korn shells. The following section examines a few of the most frequently used capabilities of the Bourne shell program. The reader is encouraged to seek out more comprehensive coverage of these programs in other publications.

Basic Shell Features

Most shells offer basically the same capabilities to the user, and can successfully interpret standard Solaris commands. However, the shell programming language recognized by the shell may differ from one program to another. The programming language of most shells conforms to either the Bourne or C shell language. For purposes of illustration, the examples in the following sections use the Bourne shell syntax, because the Bourne shell is the default *root* shell.

Environment and Local Variables

One of the similarities between the Bourne and C shell programs is the use of variables. Shell programs support local and environment categories of variables.

Local variables consist of variable names chosen by the user, and the variables contain data defined by the user. System binaries have no access to the variables. These variables are much like the local variables that a user would employ when writing a program in a high-level programming language.

Environment variables can be viewed as a combination of reserved words and global variables in a high-level programming language. The names of several of these variables cannot be altered by the user. The user can set many of these variables to a value that can be checked by system binaries. Users can modify some of these variables in order to customize the behavior of the shell to suit their preferences and needs. Some environment variables are reserved, and therefore cannot be modified by users.

Users can create their own environment variables using the *setenv* command. This command allows software developers to create environment variables used by the software packages they distribute. The discussion of environment variables in the following sections is limited to well-known system environment variables used in Solaris.

Environment Variables

Environment variables allow users to customize the operating environment. Some of the more familiar Bourne shell environment variables are described in the table 16-1, which follows.

Table 16-1: Commonly Used Bourne Shell Environment Variables

Variable Name	Set By	Description
CDPATH	Not set by default	Directory search path
COLUMNS	Set by shell	Width of window
EDITOR	Not set by default	Command editor
ENV	Set by user	Setup script invoked for each command
ERRNO	Set by shell	Last error code from a system call
HOME	Set by log-in	Default argument for *cd*
LOGNAME	Set by log-in	User's log-in name

Variable Name	Set By	Description
LPDEST	Not set by default	Sets the default printer destination
MAIL	Set by log-in	File that contains user's incoming mail
OLDPWD	Set by shell	Previous working directory
OPTARG	Set by shell	Current argument after *getopts*
OPTIND	Set by shell	Number of arguments remaining after *getopts*
PATH	Set by log-in	Search path for commands
PPID	Set by shell	Process ID of parent process
PS1	Set by shell	Primary shell prompt
PS2	Set by shell	Secondary shell prompt
PS3	Set by shell	Select shell prompt
PS4	Set by shell	Debugging shell prompt
PWD	Set by shell	Current working directory
RANDOM	Set by shell	A random number between 0 and 32,767
SECONDS	Set by shell	Number of seconds since login
SHELL	Set by login	Path to desired shell
TERM	Not set by default	Sets the terminal type
TMOUT	Not set by default	Idle time-out delay
VISUAL	Not set by default	Command exit mode for editors

The environment variables in the previous table are available for every log-in session under the Bourne shell. For instance, if the user opens two window sessions on a single system, the same list of environment variables is available in each window. If the user changes the value of an environment variable in one window, however, the new value affects only the shell in that window, as shown in the following illustration.

➥ *NOTE: Environment variable settings propagate from the shell where they are set (the parent shell) to any commands or shells invoked from the parent shell. Local variables do not propagate to commands or shells invoked by the parent shell.*

Setting an environment variable in one window does not set it for all windows.

Local Variables

A local variable is defined for the user's personal use. Users may find local variables such as loop counters and character strings to be very handy. For example, a shell variable that contains the operating system version could be helpful. The following code segment sets a local version variable.

```
% set version = `uname -s -r`
```

To determine the value of the version variable, type *echo $version.* Similar to the environment variable, a local variable is available only under the log-in shell in which it was created (see the following illustration).

A local variable defined in one log-in session is not available in another.

To set a local variable for all log-in sessions, place the shell commands that perform this operation in the *.login* file in the *$HOME* directory. This file is executed for each log-in session initiated.

✓   *TIP: Local variables may also be used in conjunction with other shell features.*

Setting Variables from .login

In order to set an environment variable for all login sessions, users must place the shell commands that perform this operation in the *.login* file in their respective home directories. This file is executed for each log-in session initiated. The following paragraphs present an example of using code in the *.login* file to set an environment variable.

A system administrator at a large site may need to log in on several systems during the workday. The standard Bourne shell dollar sign ($) prompt complicates identifying the system the shell is operating on. It would be useful for the shell's command prompt to provide the name of the system that the user is logged in on.

Setting an environment variable makes possible customization of the login prompt such that it informs the user of the system she is logged in on. Under the Bourne shell, this variable is called *PS1* (*Prompt String 1*). To see what the prompt is currently set to, issue the command *echo $PS1*.

➥   *NOTE: To cause the shell to print the value of a variable, preface the variable name with a dollar sign. Otherwise, the shell will echo the variable name string to the terminal. For example, typing* echo PS1 *to the Bourne shell would result in the system echoing "PS1" to the terminal.*

Solaris provides a utility called *uname*, which prints information about the system. If called with the *-n* flag, *uname* will print the host name of the system. The following command will capture the output of *uname -n* and assign it to the *PS1* variable.

```
% PS1=`uname -n`
```

Some systems use fully qualified domain names instead of a simple name. For example, the response to *uname -n* on a system the users call *mercury* might be *mercury.astro.com*. In order to set the prompt to *mercury* alone, issue a command such as the following.

```
PS1=`uname -n| awk -F\. '{print $1}'`
```

The *awk* command is a pattern matching utility that contains its own programming language. In the example, the *awk* command scans the output of the *uname* command. The *-F\.* directive to *awk* tells it to use the period character as a word separator. The directive *{print $1}* tells *awk* to print out the first word of the input string (everything up to the first period in this case). This output is used as the input for the set directive. Therefore, the prompt variable gets set to the value returned by the *awk* command.

Using Quotation Marks in Scripts

Several of the previous examples contained backslashes and quotation marks. What do these characters mean to the shell? Quotes are required because the shell interprets some characters differently than might be expected. The following *metacharacters* have a special meaning to shell programs: $ (dollar sign), ^ (caret), ; (semicolon), : (colon), # (pound sign), ~ (tilde), ? (question mark), & (ampersand), { } (curly brackets), [ ] (square brackets), ' (single quote), ` (back quote),* (asterisk), ( ) (parentheses), | (pipe), < (left angle bracket), > (right angle bracket), newline space, and <Tab>. In order to pass these characters on to commands, they must be quoted or escaped when used as part of a command.

Quoting and Escaping Metacharacters

Quoting or escaping a metacharacter causes the shell to alter its interpretation of the character. There are several ways to quote or escape a metacharacter. The easiest way to tell the shell not to use special interpretations of characters is to precede the character by a backslash (\ ). This is referred to as *escaping* the character, as shown in the following illustration.

Using a backslash to escape a character.

The single quote mark (') will protect text from *any* substitution. Backslashes, wildcards, double quotes, variables, and command substitution are ignored when enclosed in single quotes (see the following illustration). The following are rules for using single quote marks.

- Only a single quote can end a single quoted string.

- Single quoted strings can span multiple lines.

- It is not possible to write a single quote in single-quoted text.

Using single quotes in the shell.

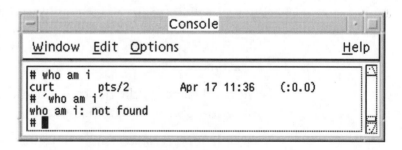

When the shell encounters commands surrounded by back quotes (` ), it performs command substitution. The commands inside the back quotes are executed, and the result is substituted in place of the backquoted command. For example, *set prompt=`uname -n`* causes the shell to execute the *uname -n* command. The result of this command is a string that provides the system name. This string is used as the input for the set command, resulting in the prompt variable being set to the string that gives the name of the machine. An example of the use of back quotes is shown in the following illustration.

Using back quotes in the shell.

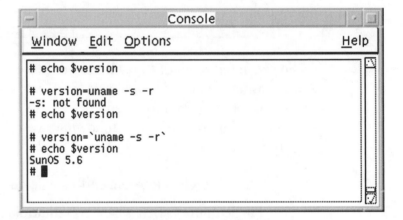

Yet another method of quoting in shell scripts is to use double quotes ("). The following rules apply to the use of double quotes, an example of which is shown in the following illustration.

- Use double quotes to protect text from wildcard substitution.

- The shell removes the quotation marks and passes the string to subsequent commands.

- Variable replacement, command substitution, and backslashes are effective inside the double quotes.

- To write a double quote (") inside double quotes, the double quote is escaped by preceding it with a backslash (e.g., "this \"escapes\" the double quotes").

- Single quotes will be treated as text inside double quotes.

- Quoted text can span multiple lines.

Using double quotes in the shell.

```
# grep this is a test /etc/passwd
grep: can't open is
grep: can't open a
grep: can't open test
/etc/passwd:guest:x:1234:60001:this is a test account:/home/guest:
/bin/csh
# grep "this is a test" /etc/passwd
guest:x:1234:60001:this is a test account:/home/guest:/bin/csh
# ▮
```

Redirection Metacharacters

The shell programs allow the user to redirect the input and output streams to programs. The metacharacters that follow are used to invoke this input/output redirection. Known as *input redirection*, the left angle bracket (<) character causes the shell to take the following word as the name of the input file. The following rules apply when using input redirection.

- *<file* causes the shell to read its input from the file.

- *<&n* causes the shell to read its input from the file descriptor *n*.

- *<&-* causes the shell to close the standard input file.

- *<Tab* tag causes the shell to read up to the line starting with tag.

- *<-* tag causes the shell to read up to the line starting with tag, while discarding leading white space.

Known as output redirection, the right angle bracket (>) causes the shell to take the following word as the name of the output file. The following rules apply when using output redirection.

The illustration that follows shows the use of input and output redirection in the shell.

- *file* causes the shell to write the output to file.

- >&*n* causes the shell to write the output to the file descriptor *n*.

- >&- causes the shell to close the standard output file.

- >> file causes the shell to append the output to file.

- The pipe character (|) informs the shell to take the output of one command and pipe it into the input of the next command.

Using input and output redirection in the shell.

```
                              Console
Window  Edit  Options                                      Help
# ls -lsa /tmp/curt
total 16
    8 drwxr-xr-x   2 root     other         69 Apr 24 12:03 .
    8 drwxrwxrwt   8 sys      sys          851 Apr 24 12:03 ..
# fmt -s <SACH16C.TXT |spell >/tmp/curt/spout
# ls -lsa /tmp/curt
total 24
    8 drwxr-xr-x   2 root     other        107 Apr 24 12:05 .
    8 drwxrwxrwt   8 sys      sys          851 Apr 24 12:03 ..
    8 -rw-r--r--   1 root     other       2714 Apr 24 12:06 spout
# █
```

Other Metacharacters

- The other metacharacters do not fit nicely into a particular group. Nonetheless, they are important characters and should be examined. These follow. The space character is considered white space. White space is used as a delimiter for commands and strings in shell programming.

- The <Tab> character is another white space character.

- The newline character signifies the end of a line of input. It is typically used to signify the end of a command.

- The tilde (~) is a C shell metacharacter, and refers to the user's home directory. For example, *ls ~shaun* would list the contents of Shaun's home directory.

- The caret (^) is used as the delimiter in C shell command substitutions.

- The dollar sign ($) is used as a prefix for local and environment variables under the shells.

- The ampersand character (&) tells the shell to run this job in the background. Commands are normally run in the foreground. This means that the terminal (or window) is not available for other uses until the command completes its execution. When a job is running in the background, the terminal is free for other activity.

- The asterisk (*) is a wildcard character. The asterisk matches any string.

- The question mark (?) is another wildcard character. It matches any single character.

The [...] construct is another form of wildcard substitution. The pattern will match one character in the input stream, but the matched character can be any of the characters enclosed in the brackets. For instance, *rm file.[ot]* will remove any *file.o* and *file.t* files in the current directory. An example of wildcard use is shown in the illustration that follows.

*Using wildcard
substitution in the
shell.*

```
┌─────────────────────────── Console ──────────────────────┐
│  Window  Edit  Options                              Help  │
├───────────────────────────────────────────────────────────┤
│ # ls                                                      │
│ SACH16C.TXT   ch13         sa16bqot.tif   sa16nogo.tif    │
│ address.blt   ch15         sa16chmd.tif   sa16rdct.tif    │
│ address.rbt   ch9          sa16chon.tif   sa16sqot.tif    │
│ address.ttt   contact.blt  sa16ctab.tif   sa16stng.tif    │
│ bad.save      done.txt     sa16dqot.tif   splat           │
│ badfsck       idea.txt     sa16envi.tif   splat.c         │
│ ch11          lostnfnd     sa16escp.tif                   │
│ ch12          order.blt    sa16fall.tif                   │
│ # ls address.[a-z]lt                                      │
│ address.blt                                               │
│ # ls address.*t                                           │
│ address.blt  address.rbt   address.ttt                    │
│ #                                                         │
└───────────────────────────────────────────────────────────┘
```

- The curly bracket characters ({ }) are operators when appearing by themselves. When embedded in a word, or when used as a command argument, they are treated as text. When used to enclose other text, the brackets are used to preserve operator priority rules.

- The parentheses characters are used to group commands much like the braces. The following are the differences between the curly brackets and parentheses. The illustration that follows shows the use of grouped metacharacters in the shell.

  - Parentheses do not require leading white space.
  - Commands enclosed in parentheses are executed by a subshell, and hence cannot alter the execution environment of the current shell.

*Using grouping
metacharacters
in the shell.*

```
┌─────────────────────────── Console ──────────────────────┐
│  Window  Edit  Options                              Help  │
├───────────────────────────────────────────────────────────┤
│ # find /home -name ch11 -o -name ch13 -print             │
│ /home/curt/ch13                                          │
│ # find /home \( -name ch11 -o -name ch13 \) -print       │
│ /home/curt/ch11                                          │
│ /home/curt/ch13                                          │
│ #                                                        │
└───────────────────────────────────────────────────────────┘
```

- The semicolon (;) is a command separator in the shell.

- The colon character (:) is a null command to the shell.

- The pound sign (#) is recognized as a comment when found at the beginning of a word.

Shell Programming

In addition to its function as the user interface, the shell allows the user to write shell programs. These programs can make use of variables, metacharacters, built-in shell features, and other UNIX system functions. The following sections touch briefly on some of the more useful shell programming constructs available in the Bourne shell. In-depth coverage of these constructs may be found in several commercially available reference books, the manual pages, and the *Answerbook*.

which *Statement*

The *which* statement determines which of multiple programs with the same name will be executed when a user types the name of the program. For instance, if a user has a program named *hello*, and a system program is also called *hello*, which program would be executed when the user types the following?

```
% hello
```

The *which* command consults the user's *.profile* file and searches the path environment variable to determine which binary occurs first in the search path. Due to the shell's traversal of the search path, the command that is executed will always be the first occurrence of that command in the search path. If a user is experiencing problems with a particular utility from his account, it is often useful to have the user log in and type the following to determine the path to the binary.

```
% which utilityname
```

In many cases, the problem is caused by the user aliasing the utility to another function, or possessing a personal program with the same name as the utility. Use of the *which* statement is shown in the following illustration.

Using the which statement to identify the version of sh that is running.

Loop Control Statements

Shells provides several statements that implement program loop capabilities. Two of the most frequently used constructs are the *for* and *while* statements. Subsequent sections explore statement usage.

for *Statement*

The *for* statement, a loop control construct in the shell, tells the shell to perform a list of instructions for each element in the argument list. The syntax of the *for* statement follows.

```
for variable
in argument_list
do
     list of functions to perform
done
```

How could the *for* statement be used to simplify system administration tasks? Assume that an administrator would like to identify the operational status of several machines on the network. One way would be to make a list of all machine names, and then log in to each machine to ensure that it is up and running. A simpler method to accomplish the same goal would be to use a *for* loop to check the machines automatically, as shown in the following illustration.

Using a for *loop to determine machine status.*

```
┌─────────────────────────────────────────────────────────────────────┐
│ ─                              Console                           ◄ ▢  │
├─────────────────────────────────────────────────────────────────────┤
│ Window  Edit  Options                                          Help   │
├─────────────────────────────────────────────────────────────────────┤
│ # for i                                                          ▲    │
│ > in grumpy wizard lsc xterm4                                         │
│ > do                                                                  │
│ > rup $i                                                              │
│ > done                                                                │
│       grumpy    up 23 days,  4:18,    load average: 1.00, 1.00, 1.00  │
│       wizard    up 89 days,  7:29,    load average: 3.50, 2.09, 1.34  │
│       lsc       up  3 days, 10:12,    load average: 0.03, 0.01, 0.00  │
│       xterm4    up 41 days, 21:46,    load average: 0.20, 0.18, 0.18  │
│ # █                                                              ▼    │
└─────────────────────────────────────────────────────────────────────┘
```

while *Statement*

The *while* statement is another loop control construct offered by the shell. The syntax of the while statement follows.

```
# while ( condition )
do
      list of actions to be performed
done
```

The *while* statement tests the *condition*, and if the condition evaluates as "true," the actions are performed until the condition is no longer true. The *while* function can be very useful when constructing shell scripts that require the occurrence of a condition prior to performance of a given action.

For example, some organizations limit outside connectivity to the use of a dial-up link in order to provide better protection from intruders. These sites typically have a single dial-out modem connected to a machine. When several users need access to this modem, competition problems may arise.

Sometimes the administrator requires access to the modem to perform functions such as downloading security patches or new versions of software. In order to automate the process of determining whether the modem is free, a simple script could be developed to check the status of the modem port. The following illustration provides an example of such a script.

Using a while loop to detect when a dial-up modem is in use.

```
Console                                              Help
 Window  Edit  Options

# cat -v modemcheck
#!/bin/sh
#
# check if modem is busy at one minute intervals
#

while [ -f /var/spool/locks/LK* ]
do
    dte=`date`
    echo "Modem busy at $dte"
    sleep 60
done

#
# Modem is free!  Inform operator!
#

echo "^G^G^G^G^G"
echo "Modem is free at $dte"
# ./modemcheck
Modem busy at Thu Apr 24 13:08:57 EST 1997
Modem busy at Thu Apr 24 13:09:57 EST 1997

Modem is free at Thu Apr 24 13:09:57 EST 1997
#
```

This script examines */var/spool/locks* to determine whether a lock file exists. If the lock file exists, the script prints the date, and goes to sleep for 60 seconds. As soon as the modem is free (the lock file is gone), the script echoes several *<Ctrl>+G*s to the terminal. The *<Ctrl>+G* character causes the terminal bell to sound, alerting the administrator that the modem is available.

Conditional Statements

The shell programs also allow the use of conditional execution statements. These statements allow modification of the shell script actions based on the results of a test on an argument or variable. Conditional statements test the value of an operation, and if the result of the test is true, a particular action is taken. If the result of the test is not true, another action may be taken.

if Conditional

One of the simple conditional statements available in the shell is the *if* statement. As seen below, this statement operates in the shell similarly to the *if* statement in high level languages such as C or Fortran.

```
if [ expression ]
then
    list of actions to perform ;
else
    list of actions to perform ;
fi
```

The expression may use flags or operators to direct the test of the condition. Some of the more useful flags appear in the table 16-2, which follows.

Table 16-2: Frequently Used Expression Flags

Flag	Result
-o	Logical OR operation (if [a or b]).
-a	Logical AND operation (if [a and b]).
!	Logical negation operation (if [a is not 0]).
=	Equality test (if [a+1 = b]).
!=	Inequality test (if [a != b]).
-r filename	True if user has read permission on *filename*.

Flag	Result
-w filename	True if user has write permission on filename.
-x filename	True if user has execute permission on filename.
-e filename	True if file exists.
-o filename	True if user owns file.
-z filename	True if file is empty.
-f filename	True if file is a plain file.

The easiest way to understand the use of these expressions is to examine sample shell code. Because UNIX systems use different directories to store the system binaries, users may become confused when they try to set up *.login* and *.profile* files on their accounts. For example, the *uname* utility resides in */bin/uname* on Solaris, but other UNIX dialects may place it in */usr/5bin/uname* or */usr/bin/uname*. Consequently, the user must search for the utility in order to use it.

The following code segment tests to determine where an executable version of the *uname* utility is installed. Once the utility is located, the variable *version* is set to the operating system version name as returned by *uname*.

```
if [ -x /usr/5bin/uname ] then
     set version = `/usr/5bin/uname -s -r` ;
else if [ -x /usr/bin/uname] then
     set version = `/usr/bin/uname -s -r` ;
else
     set version = `/bin/uname -s -r` ;
fi
```

case *Conditional*

Another useful conditional statement available in the shell is the *case* statement. The *case* statement can be used to replace an *if-*

then-else conditional in a shell script. The *case* statement allows more flexibility in the value matching process than is afforded by the *if* statement. Therefore, the *case* statement is a very popular conditional statement. The syntax of the *case* statement follows.

```
case value
in
value_1)
      list of actions to perform if value is equal to value_1
;;
value_2)
      list of actions to perform if value is equal to value_2
;;
. . .
esac
```

The shell checks the value of *value* and compares it to *value_1*. If the values are equal, the list of actions between *value_1)* and *;;* are performed. Once those statements are completed, the shell exits the switch and resumes execution at the statement following *esac*.

If the two values are not equivalent, the value of the variable is compared for equivalence with *value_2)* and so on. The *esac* statement tells the shell that there are no more values to compare against, which causes the shell to exit the *case* statement and continue execution at the statement following *esac*.

The following code segment examines the value of the *version* variable set in the previous section. Depending on the value of the version variable, shell statements in the *.login* file customize the environment for the operating system.

```
case $version
in
SunOS 5.?)
# Place code after this comment to customize the Solaris environment.
```

```
        path=( $path:/usr/sbin:$HOME/bin/solaris:/usr/bin:/sbin )
        PS1=`uname -n`%
        MANPATH=( /usr/man:/usr/local/man )
        LPDEST=myprinter
        ;;
SunOS 4.1.?)
# Place code after this comment to customize the SunOS 4.1.X environment.
        set path=( $path:/bin:/etc:/usr/bin:/usr/etc:$HOME/bin/ sunos )
        PS1=`uname -n`%
        umask 0077
        set notify
        ;;
HP-UX B.10.2)
# Place code after this comment to customize the HP-UX environment.
        set path=( $path:/bin:/etc:/usr/bin:/usr/etc:$HOME/bin/ hpux )
        set cpu = `hostname | awk '{FS = .; print $1}'`
        PS1=$cpu>>
        PAGER=/usr/bin/more
        ;;
esac
```

Summary

This chapter briefly examined methods of automating selected tasks performed on a daily basis by system administrators. Solaris provides many tools to assist with the automation process, including the use of utilities such as */bin/, rdist, cron*, and *find*, and selected built-in functions of the shell programs. The most recent Solaris distributions also include the Perl scripting language. Examples showing how to cascade these constructs to develop useful administration tools were presented.

PART III

Network Services

CHAPTER 17
Network Configuration and Management

Previous chapters covered how to manage peripherals and resources directly connected to local systems. It is now time to examine how computer networks allow the utilization of resources indirectly connected to local systems. This chapter explores a few of the more common network technologies used today, as well as a few that promise to be the network technologies of the future. In addition, network configuration, important network administration files, and basic network management and troubleshooting are discussed.

The information presented in this chapter could easily fill several chapters (or entire books). Hopefully, the overview information provided in this chapter will prepare you for managing network applications such as network file systems (NFS) and network information services (NIS) presented in subsequent chapters.

Overview of the Internet

The Internet consists of many dissimilar network technologies around the world that are interconnected. It originated in the mid-1970s as the ARPANET. Primarily funded by the Defense Advanced Research Projects Agency (DARPA), ARPANET pioneered the use of packet-switched networks using Ethernet, and a network protocol called Transmission Control Protocol/Internet Protocol (TCP/IP). The beauty of TCP/IP is that it "hides" network hardware issues from end users, making it appear as though all connected computers are using the same network hardware.

The Internet currently consists of thousands of interconnected computer networks and millions of host computers. The Internet spans Earth, and includes several links into outer space (e.g., the Mir space station, amateur packet radio relays, and the NASA Advanced Communications Technology Satellite).

Connecting to the Internet

In the international telephone system, every phone line is assigned a unique phone number. The hierarchical addressing scheme of country code, area code, exchange, and line number ensures that any phone in the system can establish a connection with any other phone in the system. Telephone numbers are assigned by the telephone companies.

If hosts on a network want to communicate, they need an addressing system that identifies the location of each host on the network. In the case of hosts on the Internet, the governing body that grants Internet addresses is the Network Information Center (NIC), affectionately known as "The Nick."

Technically, sites that do not want to connect to the Internet need not apply to the NIC for a network address. The network/ system administrator may assign network addresses at will. However, if the site decides to connect to the Internet at some point in the future, it will need to re-address all hosts to a network address assigned by the NIC. Although reassigning network addresses is not difficult, it is tedious and time consuming, especially on networks of more than a few dozen hosts. It is therefore recommended that networked sites apply for an Internet address as part of the initial system setup.

Beginning in 1995, management of the Internet became a commercial operation. The commercialization of Internet management led to several changes in the way addresses are assigned. Prior to 1995, sites had to contact the NIC to obtain an address. The new management determined that sites should contact respective network service providers (NSPs) to obtain IP addresses. Alternatively, sites may contact the appropriate network registry, as outlined in table 17-1, which follows.

Table 17-1: Network Address Registry Contact Information

Location	Postal Address	Telephone	Electronic Address	Notes
Asia/ Pacific Rim	Asia Pacific Network Information Center, Tokyo Central Post Office, Box 351, Tokyo, 100-91, Japan	+81-3-5500- 0481	ip-request@ rs.apnic.net	Obtain the file /apnic/ docs/Contents via anonymous ftp from ftp.apnic.net. Determine which form is required, obtain and complete the form, and then submit it electronically.
Europe	RIPE NCC, Kruislaan 409, 1098 SJ Amsterdam, The Netherlands	+31 20 592 5065; +31 20 592 5090 (fax)	hostmaster@ ripe.net	Obtain the files /ripe/ forms/netnum-appl.txt and /ripe/forms/ netnum- support.txt via anonymous ftp from ftp.ripe.net.
United States, Americas, and other locations	Network Solutions, InterNIC Registration Services, 505 Huntmar Park Dr, Herndon, VA 22070 USA	(703) 742-4777; (703) 742-4811 (fax)	hostmaster@ internic.net	Obtain the file / templates/internet-number-template.txt via anonymous ftp from ds.internic.net. Complete the form, and submit it electronically.

➥   *NOTE: Canada will soon have a network registry as well. At present, Canadian sites should obtain the form* /templates/canadian-ip-template.txt *via anonymous* ftp *from* rs.internic.net. *Submit the form to the U.S. registry according to the instructions for U.S. sites. (Do not follow the submission instructions on the Canadian form.)*

Internet Protocol

The process of connecting two computer networks is called *internetworking*. The networks may or may not be using the same network technology, such as Ethernet or token ring. For an internetwork connection to function, a transfer device that forwards datagrams from one network to another is required. This transfer device is called a *router* or, in some cases, a *gateway*.

In order for internetworked computers to communicate, they must "speak the same language." The language supported by most computers is the Transmission Control Protocol/Internet Protocol (TCP/IP). The TCP/IP protocols are actually a collection of many protocols. This suite of protocols defines every aspect of network communications, including the "language" spoken by the systems, the way the systems address each other, how data are routed through the network, and how the data will be delivered. The Internet is currently using version 4 of the Internet protocol (IPv4).

Network Addresses

For internetworked computers to communicate, there must be some way to uniquely identify the address of the computer where data are to be delivered. This identification scheme must be much like a postal mail address. It should provide enough information so that the systems can send information long distances through the network, yet have some assurance that it will be delivered to the desired destination. As with postal addresses, there must be some sanctioned authority to assign addresses and administer the network.

The TCP/IP protocol defines one portion of the addressing scheme used in most computer networks. The Internet protocol defines the Internet protocol address (also known as an IP address, or Internet address) scheme. The IP address is a unique number assigned to each host on the network.

The vendors of network hardware also provide a portion of the addressing information used on the network. Each network technology defines a hardware (media) layer addressing scheme unique to that network technology. These hardware-level addresses are referred to as media access controller (MAC) addresses. The following sections provide details about both IP and MAC addresses, and discuss how each is used to facilitate communications between systems.

Internet Protocol Addresses

Hosts connected to the Internet have a unique Internet Protocol (IP) address. IP addresses consist of a 32-bit hexadecimal number, but they are typically represented as a set of four integers separated by periods. An example of an Internet address is 192.168.3.1. Each integer in the address must be in the range from 0 to 255. IP addresses in this format are also known as "dotted quad" addresses.

There are five classes of Internet addresses: Class A, Class B, Class C, Class D, and Class E. Class A, B, and C addresses are used for host addressing. Class D addresses are called multi-cast addresses, and Class E addresses are experimental addresses.

Class A Addresses

If the number in the first field of a host's IP address is in the range 1 to 127, the host is on a Class A network. There are 127 Class A networks. Each Class A network can have up to 16 million hosts. With Class A networks, the number in the first field identifies the network number, and the remaining three fields identify the host address on that network.

✗   *WARNING: 127.0.0.1 is a reserved IP address called the "loopback*
       address." All hosts on the Internet use this address for their own in-
       ternal network testing and interprocess communications. Do not
       make address assignments of the form 127.x.x.x; nor should you re-
       move the loopback address unless instructed otherwise.

Class B Addresses

If the integer in the first field of a host's IP address is in the range
128 to 191, the host is on a Class B network. There are 16,384
Class B networks with up to 65,000 hosts each. With Class B
networks, the integers in the first two fields identify the network
number, and the remaining two fields identify the host address on
that network. An Internet protocol address space is shown in the
following illustration.

Address Class	Bit 0 | 1 | 2 8 16 24 31	Class Range	# Nets	# Hosts per net
Class A	0 Network | Host	0 – 127	127	15,777,214
Class B	1 0 Network | Host	128 – 191	64	65,534
Class C	1 1 0 Network | Host	192 – 223	32	254

Example 172.168.25.98 = 10101100 . 10101000 . 00011001 . 01100010

Network = 172.168   Host = 25.98

Internet protocol address space.

Class C Addresses

If the integer in the first field of a host's IP address is in the range
192 to 223, the host is on a Class C network. There are 2,097,152
Class C networks with up to 254 hosts each. With Class C networks, the integers in the first three fields identify the network
address, and the remaining field identifies the host address on
that network.

✗   *WARNING: The numbers 0 and 255 are reserved for special use in an IP address. The number 0 refers to "this network." Number 255, the "broadcast address," refers to all hosts on a network. For example, the address 192.168.0.0 refers to the class C network 192.168. The address 192.168.255.255 is the broadcast address for the 192.168.0.0 network and refers to all hosts on that network.*

➡   *NOTE: Due to the immense growth of the Internet, the IPv4 protocol suite is quickly running out of unique addresses. The Internet grew from 6,000 hosts in 1986 to more than 600,000 hosts in 1991. But do not despair: a new version of the IP protocol suite will soon provide more addresses.*

Public Versus Private IP Addresses

At some point the IETF determined it would be prudent to reserve a range of IP addresses that would not be guaranteed to be unique on the Internet. These address ranges could be used by sites that do not connect to the Internet, or by sites that employed a firewall to isolate themselves from the Internet.

Table 17-2, which follows, lists the address ranges that may be used as "private" address space. These addresses are not routed on the Internet, and in fact should never be seen on the Internet because they are not guaranteed to be unique. System administrators should take care that their hosts/routers do not allow these addresses to be sent to the Internet; nor should they allow packets from these addresses to enter their network from the Internet

✓   *TIP: It is good practice to configure the router that provides external connectivity to disallow packets coming in from the Internet that claim to be from the internal network. These packets are often "spoofed" packets generated in an attempt to penetrate the local network security.*

Table 17-2: Private IP Address Ranges

Class	Starting Address	Ending Address	CIDR Notation
A	10.0.0.0	10.255.255.255	10.0.0.0/8
B	172.16.0.0	172.31.255.255	172.16.0.0/12
C	192.168.0.0	192.168.255.255	192.168.0.0/16

A company that is not connected to the Internet may select addresses from the private address space and assign these addresses to their hosts without contacting the Network Information Center for permission. If the site decides to connect to the Internet at a later date, they may choose to renumber their hosts within the address range assigned to them, or them may employ "address translation" for their Internet connection.

Address Translation

Address translation is a technique that allows a router to translate private addresses into public addresses. This allows sites to use private addresses internally, yet communicate with other hosts on the Internet using an assigned public IP address.

Some typical applications where address translation is used include connections to cable-modems, DSL, and other high-speed Internet Service Provider (ISP) networks. The ISP assigns a single IP address to a customer that signs up for cable-modem service. The customer has several computer systems that need to communicate over the Internet. The customer installs Address Translation software on their router. When an internal host needs to communicate with an external host, the translation software goes to work.

The translation software "traps" the outbound packet, and records the destination IP address and service requested in a memory-resident table. It then changes the packet's "source" address to the address of the local router, and sends the packet out on the network. When the remote host replies, the packet is sent to the local router.

The address translation software intercepts the reply packet, determines where it came from, and looks this address up in the memory-resident table. When it finds a match, the software determines which internal host was communicating with this external host, and modifies the packet destination address accordingly. The packet is sent along to the internal host using the private address assigned to that host.

Media Access Controller (MAC) Addresses

In addition to the IP address assigned by the NIC, most networks also employ another form of addressing known as the hardware or media access controller (MAC) address. Each network interface board is assigned a unique MAC address by the manufacturer. In the case of Ethernet interface boards, the MAC address is a 48-bit value. The address is typically written as a series of six 2-byte hexadecimal values separated by colons.

Each network interface manufacturer is assigned a range of addresses it may assign to interface cards. For example, *08:00:20:3f:01:ee* might be the address of a Sun Microsystems Ethernet interface. This address would become a permanent part of the hardware for a workstation manufactured by Sun Microsystems.

Ethernet interfaces know nothing of IP addresses. The IP address is used by a "higher" level of the communications software. Instead, when two computers connected to an Ethernet communicate, they do so via MAC addresses. Data transport over the network media is handled by one system's network hardware communicating with another system's network hardware. If the datagram is bound for a foreign network it is sent to the MAC address of the router, which will handle forwarding to the remote network.

Before the datagram is sent to the network interface for delivery, the communications software embeds the IP address within the datagram. The routers along the path to the final destination use the IP address to determine the next "hop" along the path to the final destination. When the datagram arrives at the destination, the communications software extracts the IP address to determine what to do with the data. Examples of Ethernet address resolution protocol (ARP) and ARP reply packets are shown in the following illustration. The first packet is sent to the network asking "What is the MAC address of *banzai.astro.com*." The second packet contains the reply from *banzai* stating "My MAC address is *0:60:2f:88:da:63*." A more detailed explanation of packet routing and addressing appears later in this chapter.

Sample ARP and ARP reply packets show how hosts exchange MAC addresses.

```
                              Console                              
 Window  Edit  Options                                      Help
 # snoop -d hme0 -v arp
 Using device /dev/hme (promiscuous mode)
 ETHER:  ----- Ether Header -----
 ETHER:
 ETHER:    Packet 1 arrived at 10:11:3.35
 ETHER:    Packet size = 60 bytes
 ETHER:    Destination = ff:ff:ff:ff:ff:ff, (broadcast)
 ETHER:    Source      = 8:0:20:74:26:c2, Sun
 ETHER:    Ethertype = 0806 (ARP)
 ETHER:
 ARP:    ----- ARP/RARP Frame -----
 ARP:
 ARP:    Hardware type = 1
 ARP:    Protocol type = 0800 (IP)
 ARP:    Length of hardware address = 6 bytes
 ARP:    Length of protocol address = 4 bytes
 ARP:    Opcode 1 (ARP Request)
 ARP:    Sender's hardware address = 8:0:20:74:26:c2
 ARP:    Sender's protocol address = 192.88.53.112, legvold.astro.com
 ARP:    Target hardware address = ?
 ARP:    Target protocol address = 129.74.223.81, banzai.astro.com
 ARP:

 ETHER:  ----- Ether Header -----
 ETHER:
 ETHER:    Packet 2 arrived at 10:11:3.42
 ETHER:    Packet size = 60 bytes
 ETHER:    Destination = 8:0:20:74:26:c2, Sun
 ETHER:    Source      = 0:60:2f:88:da:63,
 ETHER:    Ethertype = 0806 (ARP)
 ETHER:
 ARP:    ----- ARP/RARP Frame -----
 ARP:
 ARP:    Hardware type = 1
 ARP:    Protocol type = 0800 (IP)
 ARP:    Length of hardware address = 6 bytes
 ARP:    Length of protocol address = 4 bytes
 ARP:    Opcode 2 (ARP Reply)
 ARP:    Sender's hardware address = 0:60:2f:88:da:63
 ARP:    Sender's protocol address = 129.74.223.81, banzai.astro.com
 ARP:    Target hardware address = 8:0:20:74:26:c2
 ARP:    Target protocol address = 192.88.53.112, legvold.astro.com
 ARP:
 # 
```

Internet Protocol Version 6 (IPv6)

As previously mentioned, Internet Protocol version 4 is running
out of address space as a result of the enormous expansion of the
Internet in recent years. In order to ensure that address space is

available in the future, the Internet Engineering Task Force (IETF) is readying Internet Protocol version 6 (IPv6) for deployment. Beginning with Solaris 8, the Solaris operating system supports IPv6, and Dual-stack (IPv4/IPv6) protocols.   Major differences between IPv4 and IPv6 follow.

- IPv6 addresses are 128 bits long (as opposed to the IPv4 32-bit addresses). A typical IPv6 address consists of a series of colon-separated hexadecimal digits. For example, *0xFEDC:BA98:7654:3210:0123:4567:89AB:CDEF* might be a valid IPv6 host address. Provisions have been made to allow IPv4 addresses on an IPv6 network. These addresses will have hexadecimal numbers separated by colons, followed by decimal numbers separated by periods. Such an address might look like the following:

  ```
  0000:0000:0000:0000:0000:FFFF:192.168.33.44
  ```

This address might be written in a shortened notation as *::FFFF:192.168.33.44*.

- IPv6 implements multicasting to replace the IPv4 broadcast-addressing scheme. The IPv6 multicast scheme allows for several types of "broadcast" addresses (organizational, Internet-wide, domain-wide, and so forth).

- IPv6 does not contain address classes. Some IPv6 address ranges will be reserved for specific services, but otherwise the idea of address classes will vanish.

- IPv6 uses classless Internet domain routing (CIDR) algorithms. This new routing algorithm allows for more efficient router operation in large network environments. Addresses will be assigned regionally to minimize routing tables, and to simplify packet routing.

- IPv6 can encrypt data on the transport media. IPv4 has no facility that allows network data to be encrypted.

- • Current versions of Bind 4.9 and Bind 8 are already able to provide IPv6 name services.

- ↪ *NOTE: IPv6 will employ new packet formats not available in IPv4. These new formats will allow services such as reservation of bandwidth, guaranteed signal quality, and better multicasting capabilities.*

Network Services

Computers are generally networked to take advantage of the information stored on the network's individual systems. By networking many systems, each of which manages a small portion of the data, the corporation has access to all of the data. This is one of the underlying foundations of distributed computing.

For the systems on the network to take advantage of distributed computing, the network must provide network services. A few of these services might be IP address to MAC address resolution, host-name–to–IP-address resolution, network file services, and World Wide Web (WWW) services. The following sections cover the basics of address and name resolution. File services are detailed in Chapter 19, "Network File Systems," and web services are detailed in Chapter 22, "World Wide Web Administration."

Name Services

Computers are very adept at dealing with numbers. In general, people are not capable of remembering the myriads of numbers associated with computer networking. People prefer to work with alphabetic names. The Internet Protocol provides a means of associating a host name with an IP address. The "naming" of computers makes it simple for humans to initiate contact with a remote computer simply by referring to the computer by name.

The Name Server

Unfortunately, computers want to refer to remote systems by IP address. Name services provide a mapping between the host name humans prefer to use, and the IP addresses computers prefer to use. These name services require a host (or hosts) connected to the local network to run special name resolution software. These *name server* hosts have a database containing mappings between host names and IP addresses.

Several name services are available under Solaris. Sun's Network Information Service (NIS+) and the Domain Name Service (DNS) are the two name services used most frequently in the Solaris environment. The NIS+ name service is designed to provide local name service within an organization. The DNS name service is designed to provide Internet-wide name services. Chapter 21, "Network Name Services," provides more information about name services.

Name Resolution Example

When a user on one computer wants to contact another computer, a specific sequence of events must take place. First, the user at host *glenn* enters a command to contact the remote computer (*grissom*). The *glenn* machine searches the */etc/hosts* file for the name *grissom*. If the name is found, *glenn* has the IP address of *grissom*, and name resolution is complete.

If *glenn* does not have an address for *grissom* in the */etc/ hosts* file, it issues a request on the network asking "Does anyone know the IP address of *grissom*?" A name server on the network will receive this request. The name server searches a master database of host names and IP addresses to determine if it knows how to contact the *grissom* machine. If it has an entry for *grissom*, it returns this information to *glenn*, and the name resolution is complete.

If the local name server does not contain an address for the *grissom* machine, it must query other name servers to determine if any of them have an address for the *grissom* machine. The mechanics of name resolution, depicted in the following illustration, is covered more comprehensively in Chapter 21, "Network Name Services."

Sample domain name look-up process.

Step	Action
1	Host requires address of remote host *dilbert.engr.gonzo.com*
2	Local name server polls name server in *.com* domain
3	*.com* name server refers local name server to *gonzo.com*
4	Local name server polls *gonzo.com*
5	*gonzo.com* refers local name server to *engr.gonzo.com*
6	Local name server polls *engr.gonzo.com*
7	*engr.gonzo.com* replies with address of *dilbert.engr.gonzo.com*
8	Local name server passes address of *dilbert.engr.gonzo.com* to host

Address Resolution Protocol (ARP)

As mentioned previously, systems connected to an Ethernet prefer to use MAC-level addresses for communications. But how do the hosts determine the MAC level address of other computers on the network once they know the IP address? The address resolution protocol (ARP) provides a method for hosts to exchange MAC addresses, thereby facilitating communications.

ARP Example

Suppose two computers, host A and host C, are connected to a network, as shown in the following illustration. Before these two computers can exchange data, they must know each other's MAC address. The host software provides a method for the two machines to determine the address information required for communication.

A simple address resolution scenario.

Host requesting connection	Destination requested	Source MAC address	First hop MAC address	Next hop MAC address
A	B	8:0:20:e:ff:37	0:6a:df:3e:22:91	0:3e:7f:11:2:c3
C	A	8:0:20:0:11:8c	8:0:20:e:ff:37	
A	D	8:0:20:e:ff:37	0:6a:df:3e:22:91	0:80:0:7f:f7:77

The ARP software runs on every host on the network. This software examines every packet received by the host, and extracts the MAC address and IP address from each packet. This information is stored in a memory-resident cache, commonly referred to as the *ARP cache*.

If host A needs to communicate with host C, it first checks its own ARP cache to determine if it knows the address for the host C machine. If host A has the MAC address in its cache, it places the datagram on the media with the MAC address of host C in the destination address field.

The Problem

But what happens if the MAC address for host C is not in the ARP cache? Somehow host A must determine the address before it can send the pending datagram. The host A machine resolves the host C MAC address by sending a broadcast ARP packet to the network. Every machine on the network receives the broadcast packet and examines its content. The packet contains computerese asking "Does anyone know the MAC address for the host C machine at IP address W.X.Y.Z?"

Every machine on the network (including host C) examines the broadcast ARP packet it just received. If the host C machine is functioning, it should reply with an ARP reply datagram that (in computerese) says "I am host C, my IP address is W.X.Y.Z, and my MAC address is *8:0:20:0:11:8c*." When the host A machine receives this datagram, it adds the MAC address to its ARP cache, and then sends the pending datagram to host C.

Network Design Considerations

Very few system administrators have the opportunity to design a corporate network. This function is typically managed by a network administrator. Unfortunately, many corporations require the system administrator to perform both functions for the company. In cases where the system administrator also has knowledge of network design and management practice, this situation may not present a problem.

All too often, however, the system administrator does not know how to design a network, or how to go about managing the network that gets implemented. The following sections outline some of the topics the network and/or system administrator must be concerned with when designing a corporate network.

Computer networks may be classified by many methods, but geographic, technological, and infrastructural references are three of the most common classifications. The terms *network geography*, *network technologies*, and *network infrastructure* are used in this book to refer to common classifications. Each term is defined in the following.

- *Network geography:* Physical (geographic) span of the network. Network geography may dictate the technologies used when the network is implemented.

- *Network technology:* Type of network hardware used to implement the network. The network hardware will (in many cases) dictate the infrastructure required to implement the network.

- *Network infrastructure:* Nitty-gritty details of the corporate network wiring/interconnection scheme.

Network Geography

Computer networks can be divided into three general categories: local area networks (LANs), metropolitan area networks (MANs), and wide area networks (WANs). Subsequent sections discuss a few of the characteristics of each of these network geographies.

➡ *NOTE: Network throughput measurements are generally listed in terms of bits per second instead of bytes per second, as is customary with disk subsystems. The difference is attributed to the fact that many network technologies are serial communications channels, whereas disk subsystems are parallel communications channels. The abbreviations Kbps, Mbps, and Gbps refer to kilobits (thousands of bits) per second, megabits (millions of bits) per second, and gigabits (billions of bits) per second.*

Local Area Networks (LANs)

Local area networks are, as the name implies, confined to a small area. These networks typically operate at bandwidths ranging from 4 to 622 megabits per second. LAN networks are typically used to connect systems within a single office building or university campus. Because LAN networks are typically confined to a single building, the underlying cable plant is often owned by the corporation instead of by a public communications carrier.

Metropolitan Area Networks (MANs)

MANs are slower-speed connections that operate at 9.6 Kbps to 45 Mbps but cover larger areas than a LAN. As the name implies, MANs typically provide network services for an area the size of a large city. Many corporations use MANs to connect geographically dispersed offices within a city or state. MANs typically involve a single communications carrier, which makes them much simpler to implement than WAN networks.

Wide Area Networks (WANs)

WANs are generally the most complex network geography because they span vast geographic areas. A WAN typically involves the interconnection of circuits through several commercial communications carriers. Because each carrier's equipment and technology is unique, the complexity of providing an end-to-end connection can be extreme. Most WANs operate at data rates of 56 Kbps to 155 Mbps, but cover large areas such as a state or the entire United States. WANs are used to interconnect metropolitan networks and local area networks. Many corporations use wide area networks to connect worldwide offices into a single corporate network.

Network Technologies

The variety of network technologies available today often leaves new (as well as veteran) system administrators in a fog of confusion. Each technology has advantages and disadvantages that must be considered carefully when selecting equipment for the corporate network. Subsequent sections provide a brief background on several of the more popular network technologies currently available.

➥   *NOTE: Although several common LAN technologies used on Solaris systems are introduced in this section, the discussion is not exhaustive. A few recommended references to help broaden the reader's background on the subject of computer networking are listed in the "References" section at the end of this chapter.*

Ethernet

Ethernet is one of the dominant (if not the most dominant) local area network technologies in use today. A "bus" network (see following illustration) developed in the late 1970s by Xerox Corporation, Ethernet is based on a network signaling method called carrier sense multiple access with collision detection (CSMA/CD), which was originally designed to run on coaxial cable. Each host connected to an Ethernet network constantly monitors the network for the presence of the carrier signal (CS). The carrier is a network control signal that lets hosts know that the network is in use. A host may transmit data onto the network only when the network is idle.

Example of an Ethernet bus infrastructure.

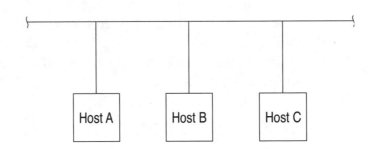

Ethernet has no prearranged order in which hosts transmit data. Any host can transmit onto the network whenever the network is idle. Thus, Ethernet is said to be "multiple access" (MA). If two hosts transmit simultaneously, however, a data collision occurs. Both hosts will detect the collision (CD) and stop transmitting. The hosts will then wait a random period before attempting to transmit again.

It is important to note that Ethernet is a best-effort delivery LAN. In other words, it does not perform error correction or retransmission of lost data when network errors occur due to events such as collisions. Any datagrams that result in a collision or arrive at the destination with errors must be retransmitted. On a highly loaded network, retransmission could result in a bottleneck.

Ethernet networks typically transmit data at 10 Mbps (megabits per second). However, in 1993 Ethernet hardware operating at 100 Mbps became available. Ethernet is best suited for LANs whose traffic patterns are data bursts. NFS and NIS are examples of network applications that tend to generate "bursty" network traffic.

Integrated Services Digital Network

Through the 1950s, the public telephone network was wholly analog; the switching and transmission of voice signals were handled by analog circuitry. In the early 1960s, the telephone industry began implementing digital transmission systems. In the 1970s, the industry began deploying digital switching systems. The Integrated Services Digital Network (ISDN) is a multiplexed digital networking scheme for use over existing telephone facilities. ISDN is a CCITT (Consultative Committee on International Telephony and Telegraphy) standard.

The major advantage of ISDN is that it can be operated over most existing telephone lines. An ISDN connection is typically capable of transmission rates of 64 Kbps (kilobits per second) per channel. The 64-Kbps circuits are referred to as basic rate interfaces or BRIs. A BRI consists of two 64-Kbps bearer (B) channels and a single delta (D) channel. The B channels are used for voice and data, whereas the D channel is used for signaling and X.25 packet networking.

➝　*NOTE: In telephone terminology, 64 Kbps refers to 64,000 bits per second. In computer terminology, 64 Kbps typically refers to 65,536 bits per second.*

ISDN transmission rates can be increased to 128 Kbps by bonding two ISDN B channels, which is still considered "slow" for digital data. It is, however, significantly faster than the typical 9600-baud (9.6-Kbps) or the newer 56-Kbps modems used to connect computer systems via analog telephone lines. The 128-Kbps ISDN circuits are referred to as "2B" circuits. At the customer end of an ISDN link, the connection is a single pair of wires. The maximum length of this link is 5,500 meters (18,000 feet). Due to the length constraints, ISDN is typically used in LANs and MANs.

Digital Subscriber Loop (DSL)

Digital Subscriber Loop (DSL) technology is a relatively new form of long-haul, high-speed networking. The DSL technologies are part of the Broadband ISDN specification that defines ISDN and ATM networking. DSL is available in many "flavors." Each version of DSL requires a special DSL modem. These modems typically have an Ethernet connection on the local side, and a telephone line interface on the remote side.

One of the most common forms of DSL is the Asymmetric DSL, or ADSL. ADSL provides a channel that operates at high speed in one direction, whereas traffic going the other direction is relatively low speed. For example, a typical ADSL link might provide 512 kilobit/second from the service provider to the customer, whereas the customer to service provider link might operate at 128 kilobits per second.

DSL circuits are subject to availability from the local telephone service provider. Some areas offer DSL circuits that operate from 512 Kb/second up to 45 Mb/second. Like ISDN circuits, there are length limits on DSL circuits; the subscriber must be within a small distance of a telephone company central office in order to make use of DSL technology.

Token Ring

Token ring networks are another widely used local area network technology. A token ring network utilizes a special data structure called a token, which circulates around the ring of connected hosts. Unlike the multiple-access scheme of Ethernet, a host on a token ring can transmit data only when it possesses the token. Token ring networks operate in receive and transmit modes.

In transmit mode, the interface breaks the ring open. The host sends its data on the output side of the ring, and then waits until it receives the information back on its input. Once the system receives the information it just transmitted, the token is placed back on the ring to allow other systems permission to transmit information, and the ring is again closed.

In receive mode, a system copies the data from the input side of the ring to the output side of the ring. If the host is down and unable to forward the information to the next host on the ring, the network is down. For this reason, many token ring interfaces employ a dropout mechanism that, when enabled, connects the ring input to the ring output. This dropout mechanism is disabled by the network driver software when the system is up and running. However, if the system is not running, the dropout engages and data can get through the interface to the next system on the ring.

Fiber Distributed Data Interconnect

Fiber distributed data interconnect (FDDI) is a token ring LAN based on fiber optics. FDDI networks, an example of which is shown in the following illustration, typically operate at 100 Mbps. Unlike Ethernet, FDDI is well suited for LANs whose traffic patterns include sustained high loads, such as relational database transfers and network tape backups. The FDDI standards define two types of topologies: single attachment stations (SAS) and dual attachment stations (DAS).

Single-attach FDDI network   **Dual-attach FDDI network**

Example FDDI infrastructures.

An SAS FDDI ring allows a star infrastructure network. A concentrator is placed in the center of the star. A single pair of fibers is connected from the concentrator to each host. If a host needs to send information to another host, the concentrator must forward the information. Single attachment FDDI rings require less fiber than dual attachment FDDI rings, but they contain a serious disadvantage: if a connection between the host and the concentrator is broken, the host is not available to the network. Another disadvantage of a single attachment FDDI ring is the cost of concentrators.

The DAS FDDI ring employs two counter-rotating fiber rings. In a counter-rotating token ring network, one ring sends information in a clockwise direction, and the other ring sends information in a counterclockwise direction. In normal operation, one FDDI ring is used for all communications, while the second ring sits idle. If the primary data ring is broken, the interface reconfigures itself to use the second ring as a loopback to form a single ring between the remaining hosts. This feature provides some fault tolerance to the network at the cost of extra fiber.

Asynchronous Transfer Mode

Asynchronous transfer mode (ATM) networks are rapidly becoming a popular networking technology. Because ATM networks are based on public telephone carrier standards, the technology may be used for local area, metropolitan area, and wide area networks. ATM networks can operate over fiber optic cables at speeds of up to 622 Mbps. Recent additions to the specifications will allow ATM networks to operate at rates over 1 Gbps (gigabits/second), and 2.4-Gbps standards are currently under discussion. Connections over twisted-pair copper media are currently possible at speeds up to 155 Mbps.

ATM, like today's telephone network, is a hierarchical standard that employs a connection-oriented protocol. For two hosts to communicate, one must "place a call" to the other. When the conversation is over, the connection is broken. Most ATM networks are implemented over fiber optic media, although recent standards also define connections over twisted-pair copper cable plants.

ATM is a "cell" networking technology. Whereas Ethernet allows 1,500-byte data packets, and FDDI allows 4,500-byte data packets, ATM operates on 53-byte data cells. A 53-byte cell allows a 5-byte header (which contains control information) and 48 bytes of data. The control information in the header tells how this data cell is to be used in relation to data cells received. The use of small cells reduces wasted bandwidth on the transmission media, and reduces transmission delays on the media.

Network Infrastructures

One of the most difficult (and expensive) problems in network design is defining the network infrastructure. The infrastructure of a network includes the cable plant required to tie the hosts together, the physical topology of the wire within the buildings, the type of electronics employed in the network, and the bandwidth requirements of the hosts connected to the network.

These topics could (and do) fill several books. A few such books are listed in the "References" section at the end of this chapter. The points listed in this section are meant to be general guidelines that should be considered before a network design is implemented.

Bandwidth Requirements

A good rule of thumb for network design is to provide a hierarchical network structure. At the top (root) of the hierarchy is the corporate backbone. Systems connected to this network typically provide corporation-wide services such as routing, name service, and firewall protection. These services are typically in high demand, and therefore require the most network bandwidth. Connections at this level of the network should be the fastest available, typically 100 Mbits/second or faster.

One layer down from the root of the network is the portion of the network that provides connectivity to the other hosts on the network, and other network services. A few examples of the services at this level include network file system servers, e-mail servers, application servers, and license servers. These services, although still in high demand, are typically less bandwidth intensive than the top-level network services. Connections at this level are typically moderate-speed connections (100 Mbits/second).

At the bottom layer of the network are individual workstations. These machines typically require the least bandwidth, and therefore get the slowest links to the network, typically 10 Mbits/second. A typical network bandwidth requirement is shown in the following illustration.

◆◇ *NOTE: These guidelines are meant to be generic suggestions. It is possible, and indeed probable, that some desktop workstations may require more bandwidth than the corporate services backbone. The network designer must determine the typical use for each host connected to the network, and "size" the connection appropriately.*

Typical bandwidth requirements of a corporate network.

Cable Plant

Once the bandwidth requirements have been determined, the next task is to determine the physical implementation of the network cable plant. In many cases, the layout of the cable plant is determined by the architecture of the buildings that house the corporation.

Due to the high bandwidth requirements of the top-level services, the cable plant for this portion of the network should be as compact as possible. If the top-level services can all be located in a single room within the building, the high-speed portion of the network can be built using very short cable segments. These segments could be implemented with fiber optic cabling at a reasonable cost.

The middle layer of the network is responsible for connecting the remote reaches of the network into the top level of the network. Due to the need for this layer to span large distances (100 meters to several kilometers), fiber optic cabling is the media of choice. This layer of the network is typically implemented with a "star" cabling topology. In brief, the cables from the high-speed portion of the network to the low-speed portion of the network are point-to-point links emanating from a central location.

The bottom layer of the network is responsible for connecting individual hosts to the middle layer of the network. These connections typically cover fairly short distances (100 meters). Again, these connections are typically implemented with a star cable topology. Cables emanate from a central point on the floor of a building to individual offices on that floor. A simple star cable topology is shown in the following illustration.

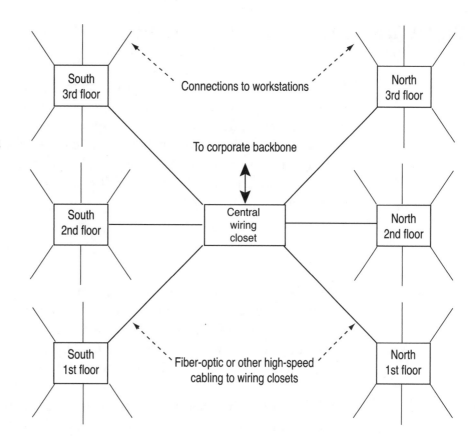

A simple star cable topology.

Network Electronics

With the cable plant complete, the next step in the network design is to determine the type of electronics required to implement the network. For example, should the low-speed connections be implemented via network hubs, network switches, or routers? Should the high-speed connections be implemented via FDDI, ATM, or Fast Ethernet? Should hosts be used as routers, or should dedicated routers be used to handle the routing of network traffic?

Network Implementation Issues

Once the network is physically implemented, the task is not yet complete. Several other issues require the attention of the network designer. For example, how does the manager determine if the network media is overloaded? What can be done to minimize media overload conditions? How should the packets be routed from one host to another? The following sections touch upon these issues.

Netmasks

Netmasks provide a method for segmenting large networks into smaller networks. For example, a Class A network could be segmented into several Class B and Class C networks. The "boundaries" of each network can be defined by providing a "mask" value, which is logically ANDed with the IP address for a host.

Each bit in the netmask with a value of 1 represents a bit of the network address. Netmask bits with zero (0) values represent bits used for the host address. For example, a netmask of 255.255.255.224 provides for 27 bits of network address and 5 bits of host address. Such a netmask would provide a network of 30 hosts.

A Class B network number uses 16 bits for the network address, and 16 bits for the host address on that network. What would happen if a Class B network was treated as a group of Class C networks instead? This would allow the site to have 254 individual networks instead of a single network. The implications of this segmentation are summarized in the following.

- External routers still route packets to the host listed as the gateway for the Class B network. This means that the gateway machine will be required to forward all incoming traffic to other hosts on the internal network.

- Internal routers will be required to keep track of how to forward packets to internal network segments. This implies the existence of a name server, and the use of an internal routing protocol within the internal network. This also suggests that the assignment of IP addresses should be handled geographically such that the routing tables would be minimized within the network.

Subnets

All IP address classes allow more than 250 hosts to be connected to a network. In the real world, this is a problem. Ethernet is the dominant network technology in use in private sector and campus networks. Due to the shared bandwidth design of Ethernet, more hosts on a network means less bandwidth is available to each host. How can this bandwidth problem be overcome?

One method of partitioning the network traffic is through a process called "subnetting" the network. Some organizations place each department on a separate subnet. Others use the geographic location of a host (such as the floor of an office building) as the subnet boundary. By segmenting the network media into logical entities, the network traffic is also segmented over many networks, an example of which is shown in the following illustration.

Typical network segmented into subnets.

For example, the machines on the third floor no longer have to contend with the machines on the second floor for network access. Because the "floor" subnets are tied together by a router (or gateway), the machines may still communicate when required, but otherwise the third floor machines do not "see" traffic bound for the second floor machines. But should subnets be incorporated into a network, and how are subnets implemented?

When to Subnet

There are no absolute formulas for determining when or how to subnet a network. The network topology, the LAN technology being implemented, network bandwidth, and host applications all affect a network's performance. However, subnetting should be considered if one or more of the following conditions exist.

•   There are more than 20 hosts on the network

•   Network applications slow down as users begin accessing the network

- A high percentage of packets on the network are involved in collisions

Obtaining accurate network load statistics requires sophisticated (and expensive) network analysis equipment. However, network load can be estimated by calculating the network interface collision rate on each host. This can be done by using the *netstat –i -n* command, as shown in the following illustration. Note that IPv4 and IPv6 information is listed separately.

Using netstat –I -n *to estimate network load.*

Using the information from the *netstat* command's output, divide the total collisions (*Collis*) by the output packets (*Opkts*), and then multiply the result by 100. For example, on *grissom*, the total collisions are 2,553 and output packets total 242,220. The collision rate on *grissom* is 1.05% (2553/242220 * 100 = 1.05%). To obtain a rough idea of the network collision rate, collect *netstat -i* statistics for all hosts on the network and average them.

Collision rates under 2% are generally considered acceptable. Rates of 2% to 8% indicate a loaded network. Rates over 8% indicate a heavily loaded network that should be considered for subnetting.

✓   *TIP: From a troubleshooting perspective, subnetting allows administra-*
tors to isolate (disconnect) pieces (subnets) of a network when trying to
resolve network problems. By isolating a subnet, it is often easier to de-
termine the source of a network problem. Most routers, gateways, hubs,
and multi-port transceivers have a switch allowing them to operate in
standalone mode. If the switch is set to local mode, the subnet is discon-
nected from the rest of the network. With the switch set to remote mode,
the subnet is connected to the rest of the network.

Subnet Example

Suppose an organization's offices occupy a six-story building. There is a large server room on the first floor that provides services to clients on all six floors. The IP address for the organization is *192.168.44.0*. The corporation would like to subdivide the address space into seven networks (one per floor, plus one for the server room), with an equal number of hosts on each network.

A Class C network may serve 255 hosts. One address on each network is reserved as the broadcast address, and another address on each network is reserved as the "zero" host. Packets sent to the broadcast address are monitored by all hosts on the network. The "zero" address is used by routing protocols to determine how to get information to a particular network.

Dividing 224 by 7 (the number of subnets) yields 36. This means seven subnets of 36 hosts each. To simplify matters (and to provide some future expansion capabilities), the administrator decides that each subnet will be limited to 32 addresses. Five bits are required in order to address 32 entities. From this the administrator determines that the corporation will have eight networks, to be configured as shown in table 17-3, which follows.

Table 17-3: Network Subnet Example

Network	Zero Host	Broadcast	Netmask	Number of Hosts	Serves
One	192.168.44.0	192.168.44.31	255.255.255.224	30	1st floor
Two	192.168.44.32	192.168.44.63	255.255.255.224	30	2nd floor
Three	192.168.44.64	192.168.44.95	255.255.255.224	30	3rd floor
Four	192.168.44.96	192.168.44.127	255.255.255.224	30	4th floor
Five	192.168.44.128	192.168.44.159	255.255.255.224	30	5th floor
Six	192.168.44.160	192.168.44.191	255.255.255.224	30	6th floor
Seven	192.168.44.192	192.168.44.223	255.255.255.224	30	Spare
Eight	192.168.44.224	192.168.44.255	255.255.255.0	29	Servers

With this information in hand, the administrator can install and configure the corporate hosts, resulting in a network topology as shown in the following illustration.

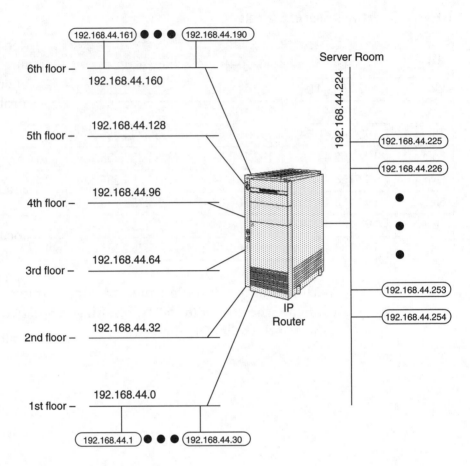

An example of a subnetworked corporate network.

Routing Concerns

All networks require some form of packet routing to ensure that packets get from one host to another. On a simple (single wire) network, the routing is also simple: a source host sends a packet out onto the media, and the destination host picks it up. On a multi-tiered, or subnetted network, routing becomes much more difficult. On the Internet, routing can become a nightmare.

Routing Overview

Network routing is an iterative process. Consider the problem a visitor to a large city would have trying to locate an office in a building in a city that she is unfamiliar with. Generally this problem is broken down into a sequence of smaller problems, as summarized in the following.

- Determine how to get to the destination city.

- Determine which quadrant of the city contains the desired street.

- Determine where the desired building is located on the street.

- Determine where the office is located in the building.

The first two steps in this process are roughly equivalent to external routing. The last two steps are roughly equivalent to internal routing.

In general, an organization should advertise a single routing address. Generally, this address is the network portion of the IP address registered to the organization. Routers on the Internet search the network portion of the IP address, and then forward packets toward the destination based on this information. This process is known as *external routing*.

Once the packet gets to the corporation, the internal routers search the host portion of the address (and any applicable network mask) to determine how to deliver the packet to its final destination. This portion of the routing process is known as *internal routing*.

The following discussion of routing is organized into three areas. First, *internal routing* refers to the routing algorithm used within a corporation's network. *External routing* refers to the routing algorithm used to allow the corporate network hosts to communicate with the outside world (e.g., other hosts on the Internet). Finally, *supernets*, or classless inter-domain routing (CIDR) blocks, are discussed.

External Routing

External routing allows packets from the outside world to be delivered to the local network, and vice versa. Several external routing protocols are available for use on the Internet. Instead of providing details on each of these protocols, this section presents some of the basics of how these protocols operate. For more information of external routing protocols, consult the publications listed in the "References" section of this chapter.

External routing typically involves determining the next hop along the path to the final destination. A distant router that wants to deliver a datagram to a host on the local network does not need to know the exact set of gateways required to get the packet to the final destination. Instead, the external router just needs to know the address of another router one step closer to the final destination.

External routers periodically exchange routing information with their "peers." Each router knows the specifics of packet delivery within its local network. By exchanging information with other routers, they build a routing table that contains information about how to get datagrams to other networks. When the routers exchange routing information they are "advertising" routes to destinations. In other words, the information exchanged between routers includes an implied promise that the router knows how to get datagrams delivered to specific destinations.

External Routing Example

This example presents a simplified view of external routing based on the following illustration. Assume that host A wants to communicate with host B. Host A sends the datagram to the network. Router 1 determines that the destination address is not a local network. Router 1 examines its routing table in an attempt to find a route to the destination. Router 1 determines that router 2 has advertised a route to the destination, so the datagram is forwarded to router 2.

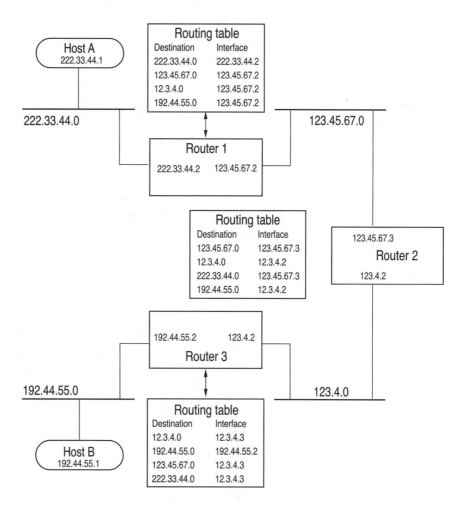

Simplified external routing.

Router 2 receives the datagram and goes through the same process that router 1 followed. Router 2 determines that router 3 has a route to the destination, so the datagram is forwarded to router 3. Router 3 is directly connected to the destination network. Router 3 examines the host portion of the IP address, and forwards the datagram on to host B.

Configuring External Routing

The process of configuring an external routing protocol can be very involved. First, the administrator needs to determine which device on the network will provide external routing services. The decision might be to use a host on the local network, or a dedicated router connected to the local network. The router/host that handles external routing may have immense routing tables stored in RAM memory and on the system disks. In general, it is preferable to use a dedicated router to provide this service, thereby freeing the local hosts to provide other services for the corporation.

➡️   *NOTE: Another reason to use dedicated routers for external routing is the ability to implement traffic filters on a router. Traffic filters allow the administrator to limit access to the local network to outside systems. Although router filters are not a substitute for a firewall, they do allow for better security than most host-router solutions.*

Once the external routing hardware has been chosen, the administrator must determine which external routing protocol to use. This is generally dictated by the routing protocol(s) used by the Internet service provider that provides connectivity to the administrator's site. The administrator must set up configuration files specific to the chosen protocol. If multiple protocols will be used, the administrator may need to set up extra configuration information such that the protocols can share routing information with each other.

In order to enable external routing, the administrator must complete several additional tasks. First, the site must have a host configured to provide name service for the local network. Next, the site's name server must be registered with the Network Information Center (NIC). The NIC adds this host to a master routing database. Any host on the Internet that wants to communicate with a host at this site must contact the site name server to the IP address of the destination machine.

Internal Routing

One host or router on the local network must be configured to run one of the external routing protocols. This router/host will receive all (externally sourced) datagrams bound for hosts on the local network. The router/host will have to make internal routing decisions in order to facilitate the delivery of the data to the local destination host. Likewise, internal network traffic must be routed to destinations on the local network. Therefore, the router that provides external routing must also know how to perform internal routing.

Internal routing may be handled in two ways. First, the network manager may tell each host to send all packets to a central router (or host) for delivery (static routing). Alternatively, the individual hosts may be allowed to determine how to get a packet from point A to point B (dynamic routing). Each routing algorithm has strengths and weaknesses.

Static Routing

Static routing requires that every host on the network send all network traffic to a single central routing authority for data delivery. This routing algorithm is somewhat similar to the U.S. Postal Service delivery method. A host "drops" a packet on the network media destined for the router. Once the host has sent the packet to the router, the host forgets about it, and allows the router to determine how to deliver the data to the final destination. Hopefully, the packet will eventually arrive at the destination.

Static routing is typically used on internal networks where there is little change in the network structure. The administrator creates a */etc/defaultrouter* file on each host on the network. This file contains the IP address of the router that will provide the datagram delivery services. The administrator builds static routing tables in the router's memory to tell the router how to deliver data to any location on the internal network.

Static routing requires minimal host processing support, but it does require intelligent routers and considerable human intervention to build the routing tables. Every time the network configuration changes, the administrator must update the routing tables in each router on the network. Otherwise, if the routing information is allowed to become outdated, a certain subset(s) of hosts may not be able to communicate.

Static routing is typically very efficient. Unfortunately, it is often plagued by a "single-point-of-failure" problem. If the network's static router fails, the entire network is down because all hosts send their data to the router for delivery.

Dynamic Routing

Dynamic routing is typically used in situations where the network configuration is changing rapidly. Dynamic routing is much more processor intensive that static routing, but requires little human intervention once it is configured and working. Dynamic routing requires each router to maintain a local routing table. This table tells the host how to get data from one point to another on the network.

In order to create and maintain these routing tables, each router runs a program that builds the routing table. When one host wants to communicate with another host on the network, the router must determine how to get the data to the final destination. If a dynamic routing algorithm does not know how to get data to the requested destination, it will ask its neighbors if they know a route it can use.

Dynamic routing is typically more robust than static routing. If, for example, a network contains multiple paths between points A and B, the dynamic routing algorithms can adjust the routing tables to survive network failures.

Supernets

The opposite of subnetting is referred to as "supernetting." Because the number of available network addresses is dwindling under IPv4, some entities end up with multiple Class C network numbers when they really need a Class B network number. The process of supernetting allows the organization to maintain a single gateway to the outside world by combining the Class C network addresses into a single addressable supernetwork. These supernetworks are also known as classless inter-domain routing (CIDR) blocks.

Assume that an organization with 400 hosts wanted to connect to the Internet, and applied for an IP address. The Internet service provider may not have a free block of contiguous addresses to assign to the organization. The ISP could assign two Class C network numbers to the organization, and offer to provide the external routing for the organization by advertising a single supernet address to the rest of the network.

For example, suppose the organization was assigned the network numbers 192.168.4.0 and 192.168.5.0. Expressing these two addresses as binary numbers shows that they share a common 22-bit prefix. The ISP could advertise a route to the network 192.168.4/22. This tells the rest of the world that the first 22 bits of the address are used as the network address, with the remaining 10 bits used as the host address.

The external routers know to forward all datagrams bound for hosts on the corporation's two networks to the ISP's router based on this information. The ISP's router forwards the datagrams to the corporate router for final delivery. An example of supernet routing is shown in the following illustration.

Example of supernet routing.

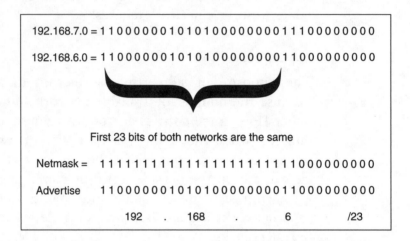

Host Configuration

The following sections outline the steps required to configure a system for use on a network. Although these sections cover a lot of material, they are not an exhaustive guide to configuring a Solaris system for use on a network. Consult the Solaris manual pages and the System Administration *AnswerBook* for more information about configuring systems for use on the network. Above all, consult the local network manager before plugging a new system into the network.

General Setup Procedure

In order to connect a host to the network, the administrator needs the following information.

- Host name for the system
- Domain name for the system
- IP address for the system
- Netmask for the network (if applicable)
- Name service used on the network
- Name or address of the machine providing name service

Once you have this information, you can configure the machine and "plug it in" to the network. The next section explains how this information is used to complete this task.

Configuration Files

Network configuration information is stored in several files on the system. Some of these files specify information about the host's address and host name, or unique setup parameters. Other files specify which network services the host will allow, and which other hosts on the network provide services the host may require.

/etc/inet/hosts *File*

One of the most frequently used network administration files is */etc/inet/hosts*. The file is a registry of IP addresses and associated host names known to a system. At a minimum, it must contain the loopback address (*127.0.0.1*) and the IP address for the host, as indicated in the following illustration. Many networking commands consult the host file in order to resolve the IP address when communications are requested. The format of host file entries follows.

```
IP address<Tab>Fully.Qualified.Name<space>host_alias
```

Minimum required entries in the hosts *file.*

```
# cat /etc/inet/hosts
#
# Internet host table
#
127.0.0.1         localhost
192.168.0.1       grissom.astro.com grissom dhcp-serve loghost
#
# DHCP assigned addresses listed below
#
192.168.0.2       astro-dhcp2
192.168.0.3       astro-dhcp3
192.168.0.4       astro-dhcp4
192.168.0.5       astro-dhcp5
192.168.0.6       astro-dhcp6
192.168.0.7       astro-dhcp7
192.168.0.8       astro-dhcp8
192.168.0.9       astro-dhcp9
192.168.0.10      astro-dhcp10
#
```

All hosts connected to the corporate network need not be listed in the *hosts* file. Typically, only those hosts routinely contacted are listed. For security reasons, *root* is the only user who can edit */etc/inet/hosts*. If a host routinely communicates with many hosts (e.g., more than a dozen) on the network, maintaining a large *hosts* file can be cumbersome. If this is the case, using an on-line distributed name service database system, such as the Domain Name System (DNS) or Network Information Service (NIS+), should be considered. These name services are discussed in Chapter 21, "Network Name Services."

✗   *WARNING: Trailing blank characters on a line in the* hosts *file will cause that entry to fail.*

/etc/hostname.hme0 *File*

Solaris uses */etc/hostname.device/instance* to simplify system configuration at boot time. For a host connected to a single network, the */etc/hostname.hme0* file would contain the host name of the machine. Machines connected to multiple networks would have multiple */etc/hostname.xxx* files.

Each network interface will probably have a unique host name assigned to it. For example, a dual-homed machine with an *hme0* and *le0* interface on separate networks might be called *thishost* on one network interface, and *thishost-that_network* on the other. The */etc/inet/hosts* file should list both addresses for the host in order to facilitate mapping the unique interface names to an IP address.

/etc/nodename *File*

Solaris employs another file to maintain an "alias" for the host on the network. The */etc/nodename* file contains the "alias" name from */etc/hosts*. For a multi-homed host, this allows the host to respond to service requests from all connected networks by a single host name. This allows users to remember only one name for the host, no matter which network interface they use to contact the host.

/etc/services *File*

The */etc/services* file contains a list of network ports and services that correspond to those ports. For example, port 25 is defined as the *sendmail* port, whereas port 80 is reserved as the hypertext transport protocol daemon (*httpd*) port. To add a new service to a host, the administrator must add a port number and service name pair to the */etc/services* file.

In general, ports numbered 0 through 512 are reserved Internet protocol ports. These port numbers are assigned to particular services by the NIC. Port numbers between 512 and 1024 are generally recognized UNIX network service ports. These ports are not reserved, but are generally treated as though they were reserved port numbers. Port numbers above 1024 are unassigned, and may be used for local network services as required by the site.

➡ *NOTE: The use of ports not assigned by the NIC may vary from site to site. Some sites may reserve a group of ports above 1024 for internal use. Checking with the local network administrator before assigning local ports and/or services is recommended.*

/etc/inetd.conf *File*

The inet daemon (*inetd*) monitors incoming network packets for service requests. When a request is received, *inetd* examines the network packet to determine which service is being requested. It then consults the */etc/inetd.conf* file to determine which application to invoke to service the request. You may want to limit the service requests a machine answers. This is done by removing the service from the */etc/inetd.conf* file and restarting *inetd*. Conversely, to add a new service to a host, add information to the */etc/inetd.conf* file and then restart *inetd*.

✓ *TIP: It is sometimes possible to send a signal to the* inetd *process to cause it to re-read the* /etc/inetd.conf *file. This is achieved by sending a* kill -HUP *signal to the process ID (PID) of the running* inetd *process. For example, if the* inetd *process is running with a PID of 73, the command* kill -HUP 73 *should cause the* inetd *process to re-read the configuration file.*

Configuration Commands

Solaris provides several ways for the administrator to enter system network configuration information. During the installation of the operating system, the SunInstall procedure prompts you for network information and builds the files as required for the installation. Once the system is up and running, there are other methods available for updating network information.

/usr/sbin/sys-unconfig

The */usr/sbin/sys-unconfig* command allows the administrator to change the network information of the host. This command removes all network-specific information files, and halts the machine. When the machine is rebooted, a portion of the SunInstall program is invoked to prompt the administrator for the new network connection information.

ifconfig

The */usr/sbin/ifconfig* command is used to bind the IP address, system name, netmask, broadcast address, and other network configuration parameters to a particular network interface. The *ifconfig* command is run automatically at boot time by the *init* process. You can use the *ifconfig* command to examine and/or modify interface parameters while the system is up and running. When issued with the *–a* flag, as shown in the following illustration, *ifconfig* prints the configuration information for all interfaces. Note, that IPv4 and IPv6 information is listed separately on dual-stack hosts. Consult the *ifconfig* manual page for more information on this command.

Using ifconfig *to list interface configuration information.*

Multi-homed Hosts

Hosts with more than one network interface are referred to as *multi-homed hosts*. Such hosts may be configured to act as routers on the internal network. Because these hosts are connected to multiple networks, it may be desirable to have them forward packets from one network to another to facilitate communications. This process is sometimes referred to as *packet forwarding*.

➽   *NOTE: Packet forwarding will require considerable processor overhead, and is usually best left to routers designed to perform this function.*

/etc/ndd

If you want to use a host as a packet forwarding system, the kernel must be configured to provide this service. The appropriate kernel configuration is accomplished via the */etc/ndd* command. In particular, you should use *ndd* to turn on packet forwarding at the IP protocol stack level. To accomplish this task, use the following command.

```
# /etc/ndd -set /dev/ip ip_forwarding 1
```

Conversely, the following command will disable packet forwarding.

```
# /etc/ndd -set /dev/ip ip_forwarding 0
```

Network Metrics

Hosts with multiple network connections have another unique setup situation. These hosts have multiple paths to the rest of the network. The routing algorithms may decide that one path is better than another without paying attention to underlying network technology. For example, a network that offers a 10-Mbit connection from host A to host B might be considered "better" than a 100-Mbit connection between the same hosts.

In such circumstances it may be desirable for the administrator to be able to force the hosts to use the high-speed network link in lieu of the low-speed link the routing algorithms discovered. This may be accomplished by setting the network "metric" value on each network link.

Many routing algorithms determine the "best" route by counting the number of hops (or gateways) the packet has to pass through between points A and B. The route with the fewest hops is considered the "best" route. The network metric value is a way to "add" hops to the route such that the routing algorithm will choose another route to deliver the data. You can set the metric for each network interface with the */usr/sbin/ifconfig* command.

Using DHCP to Assign IP Addresses

A tool that simplifies the IP address assignment configuration is the Dynamic Host Configuration Protocol (DHCP). DHCP automatically assigns an IP address to a host at boot time. Beginning with Solaris 2.6, a DHCP server is shipped with the operating system.

To configure the DHCP service, the administrator should use the *dhcpconfig* command. The *dhcpconfig* command is a front end to the *dhtadm* and *pntadm* utilities that build and maintain the

DHCP configuration tables. The *dhcpconfig* command allows the administrator to configure DHCP, Boot Relay services (required for remote DHCP), or unconfigure the DHCP service, as shown in the following illustration.

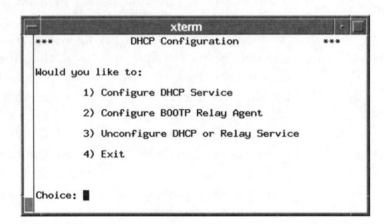

The *dhcpconfig* command will guide the user through several questions, collecting the information necessary to set up DHCP on the server. Once the information is collected by *dhcpconfig*, it will invoke the *dhtadm* and *pntadm* commands to create the DHCP database files, and start the DHCP service.

If this DHCP server were required to provide DHCP service for remote networks, the administrator would have to set up/configure the BOOTP Relay service. The DHCP registration requests are sent as broadcast packets. By default, routers will not forward broadcast packets on to other networks. For the DHCP registration request packets to get to the server, a relay service has to forward them through the router.

As an example of a simple DHCP configuration, the *astro.com* administrators created a DHCP server named (aptly) *dhcpserve*. This machine provides DHCP service to the clients on the 192.168.0.0 network. The network has 10 client machines that obtain IP addresses via DHCP. A completed *dhcpconfig* session for this server is shown in the following illustration.

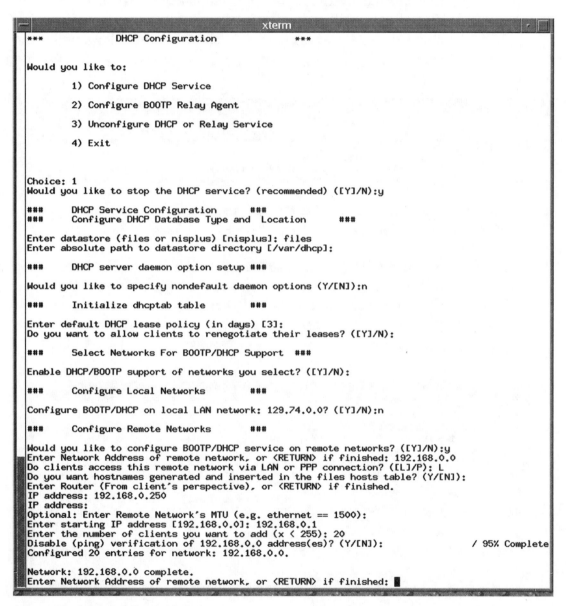

```
***                DHCP Configuration              ***

Would you like to:

        1) Configure DHCP Service

        2) Configure BOOTP Relay Agent

        3) Unconfigure DHCP or Relay Service

        4) Exit


Choice: 1
Would you like to stop the DHCP service? (recommended) ([Y]/N):y

###      DHCP Service Configuration     ###
###      Configure DHCP Database Type and  Location      ###

Enter datastore (files or nisplus) [nisplus]: files
Enter absolute path to datastore directory [/var/dhcp]:

###      DHCP server daemon option setup ###

Would you like to specify nondefault daemon options (Y/[N]):n

###      Initialize dhcptab table       ###

Enter default DHCP lease policy (in days) [3]:
Do you want to allow clients to renegotiate their leases? ([Y]/N):

###      Select Networks For BOOTP/DHCP Support   ###

Enable DHCP/BOOTP support of networks you select? ([Y]/N):

###      Configure Local Networks       ###

Configure BOOTP/DHCP on local LAN network: 129.74.0.0? ([Y]/N):n

###      Configure Remote Networks      ###

Would you like to configure BOOTP/DHCP service on remote networks? ([Y]/N):y
Enter Network Address of remote network, or <RETURN> if finished: 192.168.0.0
Do clients access this remote network via LAN or PPP connection? ([L]/P): L
Do you want hostnames generated and inserted in the files hosts table? (Y/[N]):
Enter Router (From client's perspective), or <RETURN> if finished.
IP address: 192.168.0.250
IP address:
Optional: Enter Remote Network's MTU (e.g. ethernet == 1500):
Enter starting IP address [192.168.0.0]: 192.168.0.1
Enter the number of clients you want to add (x < 255): 20
Disable (ping) verification of 192.168.0.0 address(es)? (Y/[N]):            / 95% Complete
Configured 20 entries for network: 192.168.0.0.

Network: 192.168.0.0 complete.
Enter Network Address of remote network, or <RETURN> if finished: █
```

A completed dhcpconfig *session.*

When the administrator has completed the configuration process, the *dchpconfig* command will create two database files. The files created by the *dhcpconfig* command are stored in the */var/dhcp* directory. The */var/dhcp/dhcptab* file contains information about the server environment, as shown in the following illustration.

```
# cat /var/dhcp/dhcptab
Locale      m       :UTCoffst=-18000:
dhcp-serve  m       :Include=Locale:Timeserv=192.168.25.109:LeaseTim=259200:LeaseNeg:
192.168.0.0 m       :Broadcst=192.168.0.255:Subnet=255.255.255.0:MTU=1500:Broadcst=192.168.0.255:
#
```

Content of the /var/adm/dhcp/dhcptab *file.*

The */var/dhcp/[IP_with_underscores]* file contains the information regarding addresses leased out, and available addresses. If the network number for the DHCP network is 192.168.0.0, this file name would be */var/dhcp/192_168_0_0*, as shown in the following illustration.

```
# cat /var/dhcp/192_168_0_0
0100E09806571C  00      192.168.0.2     192.168.0.1     955756672   dhcp-serve
00              00      192.168.0.3     192.168.0.1     0           dhcp-serve
00              00      192.168.0.4     192.168.0.1     0           dhcp-serve
00              00      192.168.0.5     192.168.0.1     0           dhcp-serve
00              00      192.168.0.6     192.168.0.1     0           dhcp-serve
00              00      192.168.0.7     192.168.0.1     0           dhcp-serve
00              00      192.168.0.8     192.168.0.1     0           dhcp-serve
00              00      192.168.0.9     192.168.0.1     0           dhcp-serve
00              00      192.168.0.10    192.168.0.1     0           dhcp-serve
#
```

Content of the /var/dhcp/192_168_0_0 *file.*

Network Monitoring and Troubleshooting

As with most computer operations, networks are bound to have problems. Troubleshooting networks can be a tedious process. One method of troubleshooting is to use monitoring tools that determine how the network is being used. In some cases, it may not be possible to monitor a network because physical connections may be damaged or gateways may be down. Another method of monitoring the network is to watch the console error messages generated by the machines connected to the network.

Console Error Messages

The error messages Solaris sends to the system console may provide you with a lot of information about the health of the system and the network. Unfortunately, many administrators do not pay attention to the console messages (or worse yet, leave the console window iconized).

Fortunately, Solaris provides the administrator with a simple facility for capturing all console messages in a single location. The Solaris kernel uses the *syslog* facility to log messages to the system console. However, *syslog* also allows these messages to be sent to remote hosts.

/etc/syslog.conf *File*

The */etc/syslog.conf* file controls the action of the *syslog* facility. The Solaris kernel defines several levels of message severity. Entries in the *syslog.conf* file configure the *syslog* facility to handle these different message categories. For example, a simple informational message may be ignored, whereas a system hardware error may be reported to the console with much fanfare. A typical */etc/syslog.conf* file is shown in the following illustration.

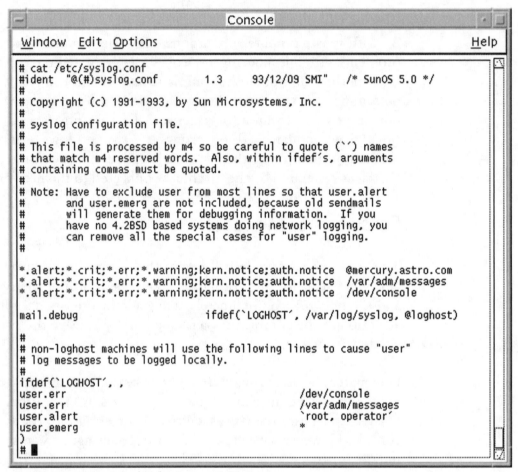

```
                              Console

Window  Edit  Options                                    Help

# cat /etc/syslog.conf
#ident  "@(#)syslog.conf     1.3     93/12/09 SMI"    /* SunOS 5.0 */
#
# Copyright (c) 1991-1993, by Sun Microsystems, Inc.
#
# syslog configuration file.
#
# This file is processed by m4 so be careful to quote (`') names
# that match m4 reserved words.  Also, within ifdef's, arguments
# containing commas must be quoted.
#
# Note: Have to exclude user from most lines so that user.alert
#       and user.emerg are not included, because old sendmails
#       will generate them for debugging information.  If you
#       have no 4.2BSD based systems doing network logging, you
#       can remove all the special cases for "user" logging.
#

*.alert;*.crit;*.err;*.warning;kern.notice;auth.notice  @mercury.astro.com
*.alert;*.crit;*.err;*.warning;kern.notice;auth.notice  /var/adm/messages
*.alert;*.crit;*.err;*.warning;kern.notice;auth.notice  /dev/console

mail.debug                        ifdef(`LOGHOST', /var/log/syslog, @loghost)

#
# non-loghost machines will use the following lines to cause "user"
# log messages to be logged locally.
#
ifdef(`LOGHOST', ,
user.err                                      /dev/console
user.err                                      /var/adm/messages
user.alert                                    `root, operator'
user.emerg                                    *
)
# █
```

Typical content of the /etc/syslog.conf file.

Logging Messages to a Remote Host

In addition to controlling the local disposition of messages, the */etc/syslog.conf* file also allows the administrator to send error messages to a central log host on the network. This procedure enables you to monitor one log file that contains messages from all machines on the network. In order to enable remote *syslog* capabilities, you should add a line similar to the following to the */etc/syslog.conf* file.

```
*.alert;*.crit;*.err;*.warning;kern.notice;auth.notice @mercury.astro.com
```

Once the previous directive has been added to the */etc/syslog.conf* file, you can restart the *syslogd* process on the client machine by sending a *kill -HUP* signal to the process ID of the syslogd process. The *kill -HUP* signal will cause the *syslog* process to re-read the configuration file, and restart with the new parameters. Once you have set up all hosts on the network to log errors to a single location, monitoring the error log is as simple as using the *tail -f /var/adm/messages* command on the log host, as shown in the following illustration.

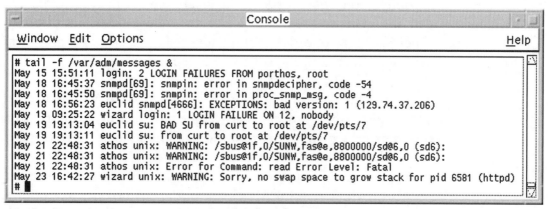

Monitoring syslog *messages using* tail -f /var/adm/messages.

Network Monitoring

Although kernel *syslog* messages may provide some information about the health of the network, it is sometimes necessary to more closely examine the network. The following sections cover selected utilities used for inspection of network status, and tips for using these utilities to troubleshoot network connectivity problems.

Simple Network Management Protocol (SNMP)

A more sophisticated method of monitoring networks is to use a network management tool based on the simple network management protocol (SNMP). The SNMP package allows a network administration host to constantly monitor the network. Information available to the SNMP software includes network utilization, host error counts, host packet counts, and routing information. SNMP allows you to determine normal usage of the network and implement alarms to warn of impending problems.

SNMP operates in a client/server mode. One machine on the network is designated as the SNMP network monitor station. It is configured to poll hosts on the local network in order to collect data into a central repository. The data may be made available to other packages in order to implement alarms, generate graphs of network utilization, and other off-line processing.

The other hosts on the network are configured as SNMP clients. These hosts run a process that watches for polling requests from the network management station. When such a request is received, the SNMP agent code collects the data from the workstation's kernel and forwards it to the management station. Solaris provides the SNMP server and agent utilities as optional software packages in the SUNWconn software cluster.

Remote MONitor (RMON)

Extensions to the SNMP software packages have recently permitted more advanced monitoring of network usage. The Remote MONitor (RMON) package allows the administrator to monitor which applications are utilizing the network and which users are running these applications, giving you the ability to perform bandwidth utilization scheduling for the corporate network.

RMON also provides you with the capability of monitoring disk space usage, processor utilization, memory usage, and other system monitoring functions. RMON is not bundled with the Solaris kernel, but may be obtained from several commercial third-party sources.

Solstice Domain Manager

Tools such as Sun Microsystem's Solstice Domain Manager package use the data collected by the SNMP software to perform higher-level network monitoring. These packages typically include utilities that have the capacity to "discover" systems connected to a network. The discovered systems are then used to construct an interactive schematic of the network. Alarms (thresholds) on numerous network "health" statistics can be set. If a system on the network fails, an alarm for that condition would be tripped, alerting the network manager of the failure by flashing the icon of the failed system on the schematic.

Sun's Solstice Domain Manager, shown in the following illustration, is not bundled with Solaris. It is a separate product marketed by SunConnect, a division of Sun Microsystems. Hewlett-Packard's OpenView product provides capabilities similar to Domain Manager, and is available for a wide range of platforms, including UNIX, Windows NT, and Windows 95.

Monitoring network traffic using Solstice Domain Manager.

Network Troubleshooting

Tracking down network problems is similar to detective work. In time, administrators develop an instinct for where to seek clues. The following discussion is by no means an exhaustive troubleshooting scenario. It will, however, provide some basic tools and techniques that all seasoned network managers use.

snoop *Utility*

There are several ways to monitor network activity and usage. One basic method is to use the */usr/sbin/snoop* command. The output from *snoop* is generally voluminous and somewhat cryptic. However, it is a good command to use when trying to determine whether two hosts on a network are communicating, or to identify the type of network traffic being generated. Some of the more common uses of *snoop* follow.

- *Monitoring traffic between two hosts*. For example, snoop can be very useful when troubleshooting two systems that cannot communicate. For example, a diskless client might issue *tftpboot* requests that are never answered. You could use *snoop* to capture packets from the diskless client to determine if the client is issuing "proper" requests. Next, you could monitor the boot server network traffic to determine if the server is replying to the boot requests, as follows.

  ```
  snoop -d hme0 between client_name server_name
  ```

- *Monitoring traffic using a particular protocol*. For example, a host on the network may not be able to resolve IP-to-MAC addresses. This typically means that the address resolution protocol (ARP) process is failing. You could use the *snoop* command to watch the network for ARP requests and the ensuing replies from the name server, as follows.

  ```
  snoop -d hme0 arp
  ```

- *Monitoring traffic containing specific data*.

- *Monitoring a network to determine which host is creating the most network traffic*. The *snoop* command even provides a "cheap and dirty" means of determining which hosts are generating the most traffic on the network. You could start *snoop* and visually monitor which hosts are generating traffic (*snoop -d hmeo*). Alternatively, you could generate some shell scripts to invoke *snoop* to capture a specified number of packets, and then analyze which hosts were sending and receiving packets during the sample window.

  For example, *snoop -d hmeo -o /tmp/snapshot -c 10000* will collect 10,000 packets and save them in the */tmp/snapshot* file. You can then *use snoop -i /tmp/snapshot | grep hostname | wc -l* to determine how many of the 10,000 packets *hostname* contributed. The illustration that follows shows *snoop* used to monitor network traffic.

✗   *WARNING: By default,* snoop *is a* root*-restricted command because the network interface device special file is created with* root *read permission. Some applications require that the network interface run in promiscuous mode. This changes the read/write modes of the device special file such that any user on the host can use the* snoop *command to capture packets. You may want to have the* root *crontab invoke a utility such as* ifstatus *periodically to determine the mode of the network interface(s) on a host.*

Monitoring network traffic using snoop.

ping *Utility*

If a host cannot be contacted on a network, the first question to answer is whether the problem is specific to the host or the network. A very useful command for making this determination is *ping*. The command name stands for "packet Internet groper," and is used to test reachability of hosts by sending an ICMP echo request and waiting for a reply (see following illustration). The command name is often used as a verb, such as in "Ping *cooper* to see if it's up."

```
┌─────────────────────────────────────────────────────────────────────┐
│ ─                          Console                            · □    │
├─────────────────────────────────────────────────────────────────────┤
│ Window  Edit  Options                                         Help   │
├─────────────────────────────────────────────────────────────────────┤
│ # ping -s mail.astro.com 8192 10                                     │
│ PING columbia.edu: 8192 data bytes                                   │
│ 8200 bytes from mail.astro.com (192.88.53.143): icmp_seq=3. time=135. ms │
│ 8200 bytes from mail.astro.com (192.88.53.143): icmp_seq=4. time=138. ms │
│ 8200 bytes from mail.astro.com (192.88.53.143): icmp_seq=6. time=143. ms │
│ 8200 bytes from mail.astro.com (192.88.53.143): icmp_seq=8. time=144. ms │
│                                                                      │
│ ----mail.astro.com PING Statistics----                               │
│ 10 packets transmitted, 4 packets received, 60% packet loss          │
│ round-trip (ms)  min/avg/max = 135/140/144                           │
│ #                                                                    │
└─────────────────────────────────────────────────────────────────────┘
```

Pinging a host to determine reachability.

By logging on to several hosts on a network and "pinging" other hosts, you can quickly determine whether the problem is network or host related. For example, if *mercury* is able to "ping" all hosts in its network except *cooper*, you can then assume that the network is functioning and the problem centers around *cooper*. The lack of a reply from *cooper* could be the result of a variety of difficulties. Some of the more common problems follow.

• Faulty physical connection

• Faulty host interface

• Corrupted */etc/inet/hosts* file

• Missing or corrupted */etc/hostname.leo* file

• Interface is turned off

• Host is down

Rebooting the non-communicating hosts will frequently solve the problem. If rebooting does not work, examining the files and physical components previously listed would be the next step. In the absence of network diagnostic equipment, replacing compo-

nents may be the only way to track down the failure. A simple
network is shown in the following illustration.

A simple network.

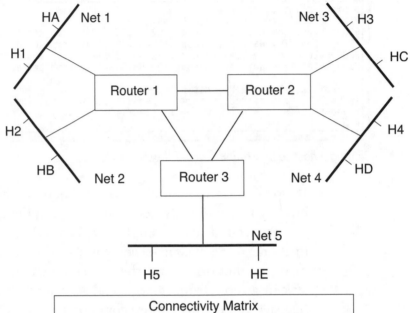

If, on the other hand, no hosts reply to pings, the problem is specific to the network. Some of the more common network faults
follow.

- Missing or damaged terminators
- Broken or damaged cable (may not be visible)
- Faulty connectors
- Damaged transceiver (may be jamming the network)

Once again, in the absence of network diagnostic equipment, replacing components may be the only way to precisely identify the cause of failure. If a host on one subnet is unable to contact a host located on a different subnet, and there is no difficulty "pinging" hosts on the local subnet, the problem could be one of the following items.

- Malfunction of the router connected to the subnet
- Malfunction of the network on the other side of the router
- Malfunction of a distant router not directly connected to the subnet
- Malfunction of the remote host
- Failure of a component or connection between local host and the remote host

To locate the problem, use the techniques discussed earlier in this section for each subnet in question. With subnetted networks, *ping* is often all that is required to locate (or at least narrow down) the source of a network failure.

traceroute *Utility*

The *traceroute* utility does exactly what its name implies: it traces the route from point A to point B. This utility can be very useful when trying to determine why two hosts cannot communicate. For example, to trace the route from a specific local host to a distant host, the administrator could invoke the following command.

```
traceroute fully_qualified_hostname
```

The *traceroute* command prints an informational message about each "hop" along the route from the local host to the remote host. Routing loops and inoperative hosts are very easy to spot in the *traceroute* output, an example of which is shown in the following illustration.

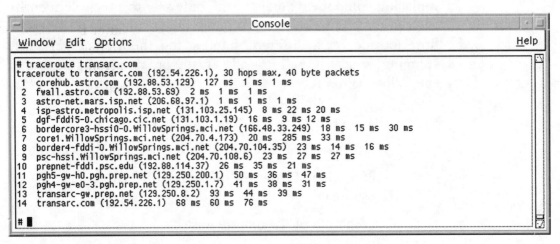

```
# traceroute transarc.com
traceroute to transarc.com (192.54.226.1), 30 hops max, 40 byte packets
 1  corehub.astro.com (192.88.53.129)  127 ms  1 ms  1 ms
 2  fwall.astro.com (192.88.53.69)  2 ms  1 ms  1 ms
 3  astro-net.mars.isp.net (206.68.97.1)  1 ms  1 ms  1 ms
 4  isp-astro.metropolis.isp.net (131.103.25.145)  8 ms 22 ms 20 ms
 5  dgf-fddi5-0.chicago.cic.net (131.103.1.19)  16 ms  9 ms 12 ms
 6  bordercore3-hssi0-0.WillowSprings.mci.net (166.48.33.249)  18 ms  15 ms  30 ms
 7  core1.WillowSprings.mci.net (204.70.4.173)  20 ms  285 ms  33 ms
 8  border4-fddi-0.WillowSprings.mci.net (204.70.104.35)  23 ms  14 ms  16 ms
 9  psc-hssi.WillowSprings.mci.net (204.70.108.6)  23 ms  27 ms  27 ms
10  prepnet-fddi.psc.edu (192.88.114.37)  26 ms  35 ms  21 ms
11  pgh5-gw-h0.pgh.prep.net (129.250.200.1)  50 ms  36 ms  47 ms
12  pgh4-gw-e0-3.pgh.prep.net (129.250.1.7)  41 ms  38 ms  31 ms
13  transarc-gw.prep.net (129.250.8.2)  93 ms  44 ms  39 ms
14  transarc.com (192.54.226.1)  68 ms  60 ms  76 ms
#
```

Sample traceroute *output.*

References

The following references are listed in an attempt to provide the interested reader with more information on the topics covered in this chapter. The list is not intended to be exhaustive, but should provide the reader with a knowledge base on the topic of computer networking.

- Rich Stevens, *TCP/IP Illustrated*, Volumes 1-3, Addison-Wesley Publishing, 1994 (ISBN 0-201-63346-9).

- Craig Partridge, *Gigabit Networking*. Addison-Wesley, 1994 (ISBN 0-201-56333-9).

- Scott Ballew, *Managing IP Networks with Cisco Routers*. O'Reilly & Associates, Inc., 1997 (ISBN 1- 56592-320-0).

- William Stallings, *SNMP, SNMP-v2, and RMO, Second Edition*. Addison-Wesley, 1996 (ISBN 0-201-63479-1).

Summary

This chapter examined selected aspects of computer networking. The coverage contained in the chapter is meant to be a primer on networking for new administrators. Topics discussed include the different classes of Internet IPv4 addresses, as well as coverage of the new IPv6 protocol addresses.

This chapter also discussed how a corporation could obtain a registered Internet address, the basics of network routing, some simple network configuration commands, as well as coverage of several of the network configuration files. Examples of these configuration files were presented to illustrate the information required in order to configure a system for network operation.

Finally, the chapter explored ways to monitor and troubleshoot local area networks. The next few chapters focus on how the network is used for network file system (NFS) file sharing, as well as a few of the popular network name services.

CHAPTER 18
Network Security

Connecting Solaris systems on a network or to the Internet brings with it a new level of security concerns. Primary among these concerns is the trust placed in the management of each of the interconnected systems and the control of access to each system. In some instances, you may want to make certain services widely available to systems that may not be trusted, while preventing access to other services. New security flaws in operating systems and application software are continually being uncovered. Meanwhile, automated tools used to search for and exploit these flaws are being widely distributed and used.

In this chapter, the basics of securing a networked Solaris system are discussed. The types of potential security problems are outlined, along with some typical solutions.

✘   *WARNING: This chapter is intended as a primer on network security. System administrators who intend to connect systems to the Internet securely will need to read further from the list of sources at the end of the chapter.*

Local Versus Network Security

The principal difference between network security and local security (discussed in Chapter 7, "Managing System Security") is the focus on access methods, and the control and monitoring of access. Network services employ a wide variety of access methods. Each type of service has different methods for access control, monitoring of usage, and the potential for unauthorized usage.

One way to view the security of a system is to compare it to the layers of an onion. Security tools and techniques provide layers of protection from unauthorized use such that the inner layers may stop an attack that managed to get past outer layers. This is often called "defense in depth."

From a security standpoint, defense in depth of a networked computer begins with a secure local system (the core), followed by a layer of checks using tools such as *tripwire* and ASET, then a layer of password controls and monitoring, and finally an outer layer of carefully configured services. Making a Solaris system highly secure is a matter of carefully walking outward through the layers.

While securing the system, the administrator needs to remember that a secure computer is less convenient for user interaction than a wide-open system. The administrator needs to make sure that the system security meets with the corporate security requirements, while allowing users to perform their daily duties without undue hardship. Securing the innermost layers is covered in Chapter 7, "Managing System Security." To begin securing the outer layers, first look at the ways a Solaris may be accessed from the network.

Services and Servers

Solaris, like other systems using TCP/IP networking, provides access via server processes that "listen on" connection points known as ports. A *port* is an addressable location a remote machine can contact in order to obtain a service from a local machine. Ports are sometimes called *sockets*, based on the name given to the programming library interface made popular by Berkeley UNIX.

To understand ports, think about a large building. The name of the building is similar to the *hostname* of a machine. The building's street address is similar to the IP address of the machine. The loading dock and the mailroom are similar to ports on a machine. The mailroom and loading usually have a room number associated with them, as do the ports on a networked machine. Each numbered port comes in two flavors: a UDP port for datagram connections and a TCP port for stream connections.

Some services use one flavor or the other; some use both. Ports with agreed-upon names and numbers are referred to as "well-known" ports. These ports are used for services such as e-mail, web service, *telnet*, and *ftp*. These ports are numbered in the range 0 to 255 according to the Assigned Numbers standard, RFC 1700.

Solaris, like most UNIX operating systems, includes the well-known ports as well as a larger range of ports known as "reserved ports." These reserved ports are numbered 0 to 1023. The reserved ports are only usable by service daemons running with *root* privileges. Other ports, up to port 65536, are defined as required, and as agreed upon by cooperating machines.

The following illustration depicts a system as a building. The building's name corresponds to the machine name, and the IP address corresponds to the street address. For each port on which an active service daemon is connected, a door is shown and labeled with the port number and service name. It is easy to see that systems running many services provide many possible entry points. Each of these entry points must be secured for the entire system to be secure. Turning off an unneeded service (the most basic form of network security) removes the corresponding door, leaving fewer entry points to monitor and secure.

Here the system named glenn *at IP address* 192.168.1.34 *is depicted as a building. The active service ports (*telnet, smtp, finger, *and* pop3*) are shown as doors.*

A typical server process accepts a connection on its well-known port and then starts a new process to handle the service request. The new process informs the connecting client to switch to another port to conduct the service. This procedure frees the well-known port for another connection, as shown in the facing illustration.

/etc/inet/services

The collection of well-known ports is listed in the */etc/inet/services* file as discussed in Chapter 17, "Network Configuration and Management." This file cross-references the port number of a given service to a name.

A typical network service request.

1. Client process contacts server process on well-known port.

2. Server process starts a copy of itself to handle request.

3. Client and copy of server process communicate.

For example, the mail transport agent, *sendmail*, uses the simple mail transport protocol (*smtp*) to move electronic mail between systems. The *smtp* port (TCP port number 25) is a well-known port where a *sendmail* daemon listens to accept incoming mail.

Methods of Providing Network Services

There are three basic methods of providing a network service, each with its own security implications. These are described in the following.

- *Daemons:* These are background programs that listen on a specific port for service requests.

- Inetd *services:* The *inetd* daemon is a switchboard daemon that listens on a group of ports and starts the appropriate service daemon as needed. The *inetd* daemon performs its duties based on the rules found in its configuration file */etc/inet/inetd.conf.*

- *RPC services:* The Remote Procedure Call method of providing a service relies on a special look-up daemon, *rpcbind*. RPC requests are processed by first consulting the *rpcbind* daemon to obtain the port number for the required service, followed by connecting to the appropriate port as directed by *rpcbind*.

Daemons

In the first method, server processes such as *sendmail* are started when a Solaris system boots, and run continuously, listening for and processing service requests arriving on a particular port. These processes are part of a group of processes known as *daemons* to denote their continual background activity.

New daemon services are added by installing new server daemon programs and implementing scripts to start/stop the daemon. An example of this might be a web server or a software license server. If the daemon uses a port already defined in the */etc/inet/services* file, the program is simply installed, configured, and started. The startup function is often automated as part of the system boot process, as described in Chapter 5, "Refining System Boot and Shutdown Procedures."

If the new service daemon uses an unregistered port number, the administrator will assign a port by adding a line to */etc/inet/services*. The line contains three fields and an optional comment. The first field gives the service name, the second field describes port number and protocol (UDP or TCP), and the third field lists an "alias" or alternative name the service may be known by. The following is an example for the two ports used by the network time service.

```
time              37/tcp              timeserver
time              37/udp              timeserver
```

inetd

Infrequently required services may be started on demand by a special daemon known as *inetd*. The *inetd* daemon listens to a collection of ports listed in its configuration file, */etc/inet/inetd.conf*. When *inetd* detects a connection on a port, it starts the server process for that port as listed in the configuration file. This process is depicted in the following illustration.

The inetd *daemon answers a service request by starting the appropriate service.*

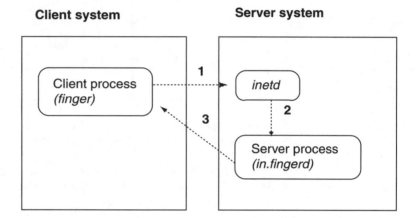

1. A client process, such as the *finger* command, contacts the server.
2. The *inetd* daemon accepts the connection and in turn starts the *in.fingerd* service daemon to process the request.
3. The *in.fingerd* daemon returns the requested information to the client process.

rpcbind *and RPC Services*

The Network File System (NFS) and other services are built on a third remote access technique known as the remote procedure call (RPC). Access to RPC services are made via sockets assigned dynamically by a service named *rpcbind*. To access an RPC service, the client first checks with the server's *rpcbind* daemon to determine the port number being used by the required service. Once it knows the port number, the client connects to the service using the port number returned by the *rpcbind* daemon, as depicted in the following illustration.

Client system   **Server system**

A typical RPC request. The mount command contacts rpcbind, and in turn the in.mountd daemon.

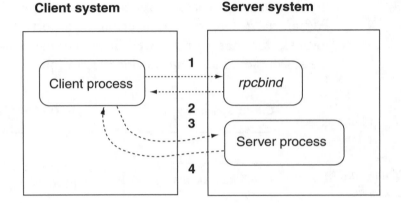

1. An RPC client process contacts the *rpcbind* daemon on the server.
2. The *rpcbind* daemon replies with the port number of the requested RPC service daemon.
3. The client contacts the RPC service daemon on its port.
4. The RPC service daemon processes the request and returns the requested information to the client.

Threats

Each service offered by a system has the potential to be used by an attacker for unauthorized purposes. The three primary types of attack are described in the sections that follow.

Entry Attacks

Entry attacks are situations in which the attacker uses a service to gain access to (and perhaps control of) a system via a flaw in the service's configuration, or the service daemon binary. Flaws in service configuration might include granting trust to untrusted remote systems, or indiscriminately allowing access to services such as file sharing. Flaws in service daemons are most often design or programming errors that allow uncontrolled access.

Often carefully constructed input data from a remote system can exploit a flaw in the daemon, causing it to perform functions not intended by the daemon's author. A typical flaw is a buffer (data storage area) in the daemon program that is undersized. These undersized (or otherwise unprotected) buffers may be over-flowed by lengthy input from a remote system. The result is that the daemon's normal processing is altered, and code may be inserted that causes the daemon to perform unwanted actions.

Denial of Service (DoS) Attacks

"Denial of service" attacks similarly make use of known services and flaws in the daemons that provide the services. The main difference is that "denial of service" attacks usually do not include accessing a system. Instead, the perpetrator just disables the system or service. Some "denial of service" attacks make use of other aspects of TCP/IP network protocols, including malformed packets. By themselves, these packets do not harm a system, but may consume resources to such an extent that other network communications with the system are not possible.

Snooping

The third form of attack is snooping information as it traverses the network. Most network communications contain plain (un-encrypted) *ascii* text data. All of the information is readable if you intercept the data packets as they go by. For example, e-mail, remote terminal session protocols (such as *telnet* and *rlogin*), and web transactions are easily identified, and transmit password information in the clear. Although switched network technologies may help limit the information available to a network sniffer, a better solution to this problem is to use encrypted data transmission.

The most troubling aspect of these attacks on the present-day Internet is that their exploitation is easily automated. Tools that can automatically search for and exploit flaws in service daemons are widely distributed. Services that exist on known ports or that can be determined by consulting a service such as *rpcbind* can be identified very quickly using automated tools. These tools can scan an entire IP address domain in a matter of minutes. At the time of this writing, automated tools are available that combine searching for vulnerable services and, if such a vulnerability is found, automatic exploitation of the service.

✗   *WARNING: No system connected to the Internet is so obscure as to never be scanned. It is just a matter of time before a system will be discovered and probed for possible vulnerabilities.*

Access Methods, Monitors, and Controls

A standalone system, not connected to any network, is secure from any network attack. Unfortunately, such systems are also of limited use. A system connected to a network, but which provides no networked services, is somewhat less secure that a non-networked system. But once again, the system may be of limited value. Most Solaris systems are either a client or provider of some network service.

The problem of securing these systems is one of access control. How can a system provide access to authorized users while denying access to unauthorized users? To open up a system by enabling services means that the system offers more possible avenues of attack. There are several ways to address this problem and manage the threat posed by providing network services.

Firewalls and Filtering Routers

One approach to limiting access is to pass all network traffic through a filtering router and/or firewall. This approach constitutes an attempt to deal with all possible avenues of access by carefully controlling the network traffic as it passes through the network electronics. An advantage of this method is that it provides some measure of security for all services, including those added later or services created by individual users. A disadvantage of this approach is that the inspection of packets impedes the traffic flow across the network. An example of the use of a firewall is shown in the following illustration.

All network traffic filtered by a firewall.

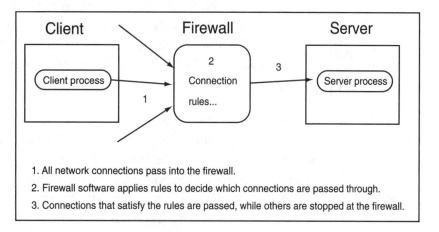

1. All network connections pass into the firewall.
2. Firewall software applies rules to decide which connections are passed through.
3. Connections that satisfy the rules are passed, while others are stopped at the firewall.

A filtering router conditionally passes network traffic between networks. The router filters are rules developed by the network manager. These rules tell the router which packets should be allowed to pass, and which packets should be rejected. Depending on the particular router, it may pass traffic based on the source or destination IP address, the source or destination port number, or other information contained in the packet. Access to a service on a system protected by the filtering router is controlled by allowing (or disallowing) the network traffic to enter (or leave) the network.

There are situations in which it is desirable to exercise finer control over the network traffic than is available using filtering routers. A firewall provides additional control over the network traffic passing through it. Depending on design, firewalls may provide the services of a filtering router, as well as hide the IP addresses of machines behind the firewall. This process is referred to as IP masquerading, or Network Address Translation.

Some firewalls are also capable of monitoring packets for specific content, and preventing objectionable material from passing to the protected network. Setting up an effective firewall requires careful consideration of a large number of factors. If you are working at a site requiring this form of access control, refer to the firewall information listed in the references at the end of the chapter.

Even at their best, firewalls do not provide a complete answer to the network security problem. Firewalls do not provide security to individual hosts. They do not protect hosts on the "secure" network from "inside" attacks. A single hole in the firewall may allow attackers access to all hosts on the "protected" network. Mistakes in the configuration of a firewall may go unnoticed without other security measures in place.

Monitoring Access

No matter how tightly the administrator controls access to a given machine, breakdowns in access control occur. Software bugs and configuration errors in any of the service daemons or a firewall are two common ways in which access control can break down. To cope with such problems and to evaluate the quality of the access control, host-based monitoring is needed.

Log Files

The security-conscious system administrator should routinely monitor system log files. In addition to pointing out security violations, these files can also give you clues to potential problems in other areas (such as failing hardware). Some of these files are under control of the *syslog* service daemon, *syslogd*.

su *Command and Log File*

The *su* command, an example application of which is shown in the following illustration, provides a means of gaining the authority of another user by supplying the user's password when prompted. Invoked without a user name, the *su* command offers access to the *root* account. Because of the authority given to the *root* user, the *su* command maintains its own log file, */var/adm/sulog*.

The su *command in action.*

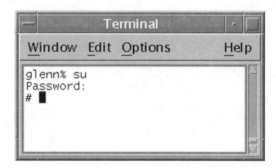

The log file produced by the *su* command is one of the first places a system administrator should check on a routine basis. The command logs both successful and unsuccessful attempts to obtain *root* privileges. is the following illustration shows a sample *su* log from the machine named *glenn*. The successful *su* attempts have a plus symbol (+) listed after the date and time; unsuccessful attempts show a hyphen or minus sign (-) after the date and time. Unsuccessful *su* attempts should be investigated thoroughly. Any attempts made by persons not authorized to perform system administration tasks might be a sign of a security breach.

Last five lines of the /var/log/sulog file.

System log files that should be routinely checked, and a description of their content, are listed in table 18-1, which follows.

Table 18-1: System Log Files

File	Content
/var/adm/sulog	Log of the *su* command. Entries containing a minus sign (-) after the date and time indicate a failed password entry.
/var/adm/log/ asppp.log	PPP software logs errors and connections here. Watch for failed connections as indicators of problems.
/var/log/syslog	Here, by default, the *syslog* service writes error messages produced by *sendmail*. Watch for errors of all types.
/var/adm/ messages	By default, the *syslog* service writes error messages from a number of different daemons to this file. Watch for errors of all types.
/var/log/maillog	Optional *sendmail* log file. Check this file for signs of unauthorized e-mail relay, or attempts to gain access to the system via the SMTP service.
/var/cron/log	By default, the *cron* daemon logs a message when it runs a scheduled *cron* job. Watch here for proper operation of scheduled jobs.

➥   *NOTE: The* syslog *daemon is controlled by a configuration file,* /etc/ syslog.conf, *which can be modified to increase the detail of recorded log file messages, and optionally to display messages on the system con-*

sole, or broadcast to the terminals, of specific users. Syslog is also a network service and logging can be directed to another machine such as a central logging server.

By default, Solaris does not record unsuccessful login attempts. To enable this logging, add the following line to */etc/syslog.conf.*

```
auth.info/var        /log/authlog
```

This statement informs the *syslog* daemon to record the events tagged as "LOG_AUTH" with a severity of "INFO" or higher in the file */var/log/authlog.* The next step is to create this file and the corresponding *loginlog* file. The following commands accomplish this task.

```
# touch /var/adm/authlog /var/adm/loginlog
# chown root /var/adm/authlog /var/adm/loginlog
# chmod 600 /var/adm/authlog /var/adm/loginlog
# chgrp sys /var/adm/loginlog
```

✗   *WARNING: Be aware that an attacker who has gained unauthorized access to a machine you manage will most likely know about these log files and will modify them to remove any indication of his/her actions. Log file monitoring needs be performed in concert with tools such as ASET and* tripwire *to provide a high level of assurance that unauthorized access is discovered.*

Another feature of *syslog* is remote logging. Any of the information logged by *syslog* can be transmitted to another *syslog* process on another system. This allows for log information to be centralized. This has several benefits, including ease of monitoring, and making it more difficult for an attacker to hide his or her tracks. The following is an example of a line from *syslog.conf* that sends the log messages to the mail facility at level *debug* or greater to a remote system. Note that instead of a file name, the notation @ followed by a host name is used.

```
mail.debug        @log.astrocorp.com
```

One final addition to the *syslog* service worth consideration is the use of tools for summarizing data collected by *syslog*. At a moderately large site, the log information collected can be quite lengthy and difficult to visually parse. One such tool is *swatch*. *Swatch* can actively monitor a log file and look for patterns. When *swatch* finds a matching pattern, it can display messages or run commands, such as sending e-mail to alert the system administrator. *Swatch* can be found at the following address.

http://www.stanford.edu/~atkins/swatch/

Inetd Services

Inetd is a switchboard daemon that starts less frequently used services. The most secure method for dealing with services provided by *inetd* is to disable them all by removing */etc/inetd.conf* and */etc/inet/inetd.conf* and commenting the start of *inetd* out of */etc/init.d/inetsvc*. This may be appropriate for workstations that do not provide network services, or systems for which primary access occurs via the system console. This approach may also be reasonable for a server that needs limited remote access, and for which primary access occurs via the console, or via secure methods such as the *ssh* service, described in material that follows.

TCP Wrapper Software (tcpd)

What can be done for situations in which there is a need to provide *inetd*-controlled services but access logging and host-level control is required? First, turn off (disable) unnecessary services listed in */etc/inetd.conf* by commenting them out of the */etc/inet/inetd.conf* file.

For the remaining (required) services, consider using software to control access to the service. The *tcp wrapper* package provides a program called *tcpd* for evaluating every access before the service is invoked, as shown in the following illustration. The *tcp wrapper* is available from the CERT web site listed at the end of the chapter.

TCP-based service protected by the tcpd *wrapper successfully connects and is logged.*

Client system   **Server system**

1. A client process, such as the *finger* command, contacts the server.
2. The *inetd* daemon accepts the connection and in turn starts the *tcpd* daemon.
3. The *tcpd* daemon consults its rules files to decide what to do with the connection.
4. The *tcpd* daemon logs the connection by contacting the *syslog* daemon.
5. The *syslog* daemon writes the log entry to the log file.
6. If the connection is allowed, the *tcpd* daemon starts the server process.
7. The server process responds to the client request.

Once the *tcp* wrapper program is installed, a given service is protected by modifying its entry in */etc/inetd.conf* to invoke the *tcpd* program instead of the service daemon. The *tcpd* program consults two "rules" files, */etc/hosts.allow* and */etc/hosts.deny*, to determine if this contact should be allowed or denied. If the packet "passes" the tests and is allowed by the rules, the *tcpd* program logs the connection via *syslog*, and invokes the service daemon, as shown in the previous illustration. An example of this configuration for the *finger* service is shown in the following illustration.

```
# grep finger /etc/inetd.conf
#finger  stream  tcp      nowait  nobody  /usr/sbin/in.fingerd      in.fingerd
finger   stream  tcp      nowait  nobody  /usr/local/etc/tcpd      in.fingerd
# cat /etc/hosts.allow
ALL: LOCAL grissom.net-kitchen.com apollo.net-kitchen.com
#
```

Protecting the finger *service using* tcpd.

The two lines extracted from */etc/inet/inetd.conf* using *grep* show that the original configuration line for *finger* has been commented out (the line begins with a pound sign) to disable it. The second line is used in place of the original entry. The sixth field in the second line shows *tcpd* as the program to be run, with the seventh field showing the program name as *finger*.

The *tcpd* binary consults */etc/hosts.deny*, and then */etc/hosts.allow*, to determine if this connection is acceptable. By default, the */etc/hosts.deny* file denies access to all services by any host. The */etc/hosts.allow* file then opens up access holes by listing specific hosts/domains that may access the service. If the connecting host does not match a "deny" rule or matches one of the "allow" rules, *tcpd* will start the *finger* process and allow the request to be completed.

If a connection does not meet the criteria specified in the rules files, it can be handled in a number of ways, as specified in the rules files. One way is to simply drop the connection by failing to respond, as shown in the following illustration. The remote system denied a connection receives no answer from the server, the failed attempt to connect is logged via *syslog*, and the service daemon is not started. For more information on configuring *tcpd*, see the documentation that comes with the *tcpd* distribution.

TCP service protected by tcpd *is denied access from a client.*

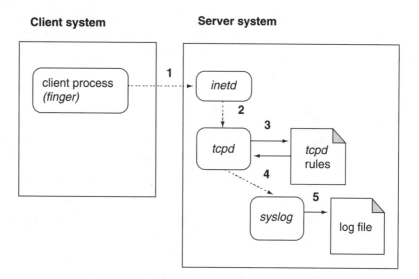

1. A client process, such as the *finger* command, contacts the server.
2. The *inetd* daemon accepts the connection and in turn starts the *tcpd* daemon.
3. The *tcpd* daemon consults its rules files and finds this client is not allowed to connect this service. The response listed in the rule file is to not respond.
4. The *tcpd* daemon logs the connection by contracting the *syslog* daemon.
5. The *syslog* daemon writes the log entry to the log file.

Although the *TCP wrapper* program works well, it is not a comprehensive answer to access security. Wrappers involve an extra process invocation, as well as work on the part of the *syslog* daemon, resulting in a performance loss. Building and installing *tcpd* requires a compiler and other development tools because the program is provided in source code form. The *tcpd wrapper* program is also only usable for TCP-based services; it does not handle UDP- or RPC-based services. Wrapper programs also require the system administrator to update the access rules from time to time.

✗   *WARNING: The* TCP wrapper *program does not protect a system from any flaws in the wrapped service daemon. If a remote system is granted access by the* TCP wrapper, *it is placed in contact with the service daemon and an attack can be made. Sites using any* inetd *service will need to stay up with current patches to correct flaws in service daemons in use at their site.*

The "r" Services

The commonly used Berkeley remote access commands *rlogin rcp* and rsh require some additional consideration if you plan to enable *inetd* entries for these services. The *"r"* commands present additional concerns because they have access controls on a per host and per user basis, as well as the usual log-in password protection.

Users can subvert the password protection by creating a *.rhosts* file in their home directory. The *.rhosts* file contains pairs of system names and user names. If the source machine and user are listed in the *.rhosts* file on the target account, the user is logged in without the need to enter a password! The content of the *root*'s */.rhosts* file needs careful consideration, as indicated in the following illustration.

A secure /.rhosts file has the fewest possible entries.

Placing a machine and user pair in a *.rhosts* file confers a high level of trust on the listed machines and users by *removing the need for passwords when establishing remote connections.* Note that there is also a machine-wide control file for the "*r*" commands, */etc/ hosts.equiv.* This file should be removed and not used.

If the administrator desires to completely disable the use of *.rhosts* files, this can be easily done. The Pluggable Authentication Module architecture (or simply PAM) configuration file, */etc/pam.conf,* controls the authentication methods used by various parts of Solaris. To remove the *.rhosts* authentication scheme, comment out the lines that contain the *pam_rhosts_auth* authentication module.

➥ *NOTE: The PAM service is only available for Solaris 2.6 and later releases.*

Although it is very convenient to allow such easy access between machines, the security implications of such access are considerable, especially if the *root* account is supplied with a *.rhosts* file. Allowing access between machines without requiring passwords makes the collection of machines as secure as the least secure machine in the group. Furthermore, unless the system administrator is certain of the accessibility to each machine, allowing access via *.rhosts* may open a system up to much more access than is intended. In general, the *.rhosts* file should contain the fewest number of entries possible. Better still, it should not be used.

Another problem with the "*r*" commands is the fact that these commands pass all of their information across the network in clear-text. Anyone with a packet sniffer may intercept packets containing e-mail messages, log-in names, passwords, or other sensitive information. This has led many administrators to disable these commands in favor of more secure access methods.

ssh

One solution to the clear-text problems of the "*r*" commands and the threat of snooping is to use a secure communications program that encrypts all data for transit. The *ssh* package is used at many sites as a drop-in replacement for the "*r*" commands and telnet. Because there are also *ssh* clients for Windows and MacOS systems, the *ssh* package is also useful for employees who require access to corporate computing facilities from remote sites (which may not be secure). The *ssh* package provides a secure remote communications channel. The *ssh* package can be obtained from the following address. The package includes installation directions.

http://www.ssh.fi/sshprotocols2/download.html

➡ *NOTE: The licensing agreement for* ssh *is not very clear. Your organization may need to purchase the commercial version, or may want to use* ssh *version 1 or the new OpenSSH. Review the licensing information carefully to ensure that your organization is in compliance with the terms of the* ssh *license agreement.*

The primary benefit of *ssh* is to prevent passwords and other information from being intercepted en route to a remote machine. The *ssh* package employs strong encryption and long keys to prevent this, as well as encrypted access keys on top of the Solaris password mechanism. The *ssh* package also provides key management facilities that greatly reduce the need to use access control files, such as *.rhosts*, for most remote connections. It is also possible to tunnel other sessions through an *ssh* connection, thereby protecting these services. Some sites provide secure tunnels for services such as IMAP mail systems, the X-windows system, and FTP services.

Other Daemon-provided Services

The most secure method of dealing with the services proved by daemons such as *sendmail* is to simply not start them in the first place. As described in Chapter 5, "Refining System Boot and Shutdown Procedures," Solaris uses files in the */etc/rc?.d* directories to specify the order of actions taken as the system enters the various run levels. Removing or renaming these files and directories effectively disables these services.

The following command disables any service from starting in run levels other than single user and multi-user. Solaris should not normally enter these states, and removing these links prevents daemons from being started unexpectedly.

```
rm —f /etc/rc[013].d/*
```

Next, the following sequence of commands disables the Solaris auto configuration tools, NFS and RPC daemons, *sendmail*, and the preservation of *vi* editor files.

```
cd /etc/rc2.d
for file in S30sysid.net S71sysid.sys \
S72autoinstall S60nfs.server \
S73nfs.client S74autofs \
*cache* S71rpc S76nscd \
S88sendmail S80expreserve ; do
```

```
mv $file .NO$file
done
```

Making RPC Services Secure

As with the *inetd* services, a system such as *file server* may require that the RPC service daemons and NFS be run to provide file service. Chapter 20, "Automating NFS with *automount*," provides details on setting up RPC services and NFS for using encryption to lessen the risks involved in running those services.

The *rpcbind* daemon itself can also be replaced with a version of *rpcbind* that works in a fashion similar to the TCP wrapper software described earlier in this chapter. The replacement *rpcbind* logs denied connections using *syslog*. This program is also available on the CERT and COAST sites listed at the end of this chapter.

➡ *NOTE: The replacement* rpcbind *program is fairly dated and has not been recently updated.*

Dealing with Denial of Service

Denial of service via an attack on a specific service (such as a web service daemon) is difficult to prevent. Specific approaches are beyond the scope of this chapter, but are covered in some of the references listed at the end of the chapter.

Solaris can be "hardened" (made less vulnerable) to attacks that employ malformed TCP/IP packets by making some minor adjustment to the TCP/IP parameters in the kernel. The following are some common adjustments that reduce the effect of common attacks. These command strings use the *ndd* command to set kernel variables for the *arp* and *ip* drivers. The commands can be added as a group to the */etc/init.d/inetinit* start-up script, or put in a separate script that is called directly after *inetinit*.

- *ARP cache attacks:* The following commands reduce the time address-resolution information is retained, reducing the effect of an attack that aims to "poison" the target machine's ARP cache.

  ```
  ndd -set /dev/arp arp_cleanup_interval 60000
  ndd -set /dev/ip ip_ire_flush_interval 60000
  ```

- *Packet forwarding:* Systems that allow packet forwarding are targets for attackers because they provide access to other systems and networks. Unless needed, packet forwarding should be shut off using the following command.

  ```
  ndd -set /dev/ip ip_forwarding 0
  ```

- *Multi-homed packet spoofing:* The following commands prevent hosts with multiple network interfaces from forwarding spoofed packets.

  ```
  ndd -set /dev/ip ip_strict_dst_multihoming 0
  ```

- *Directed broadcasts:* By default, Solaris systems acting as routers forward broadcast packets for remote networks. This can create large amounts of unwanted network traffic and form the basis of the so-called "smurf" attack. The following command disables forwarding of directed broadcast packets.

  ```
  ndd -set /dev/ip ip_forward_directed_broadcasts 0
  ```

- *Source-routed packets:* These packets contain a specific path the datagram should follow. Normally, routing should be handled by the routers along the path and not specified by the packet. An attacker may use these packets to bypass specific routers or firewalls. The following command disables the forwarding of source-routed packets.

  ```
  ndd -set /dev/ip ip_forward_src_routed 0
  ```

- *Redirect errors:* Routers use ICMP redirect messages to inform hosts to use a different router. However, attackers may also use them to forge a bogus route. These two commands tell Solaris to ignore these error messages (first line) and prevent a Solaris system from sending these messages (second line).

  ```
  ndd —set /dev/ip ip_ignore_redirect 1
  ndd —set /dev/ip ip_send_redirects 0
  ```

- *SYN floods:* This attack takes advantage of a weakness in the TCP protocol by exhausting the resources used to maintain a list of new connections. This list can be expanded using the following command to reduce the effect of this attack.

  ```
  ndd —set /dev/tcp tcp_conn_req_max_q0 10240
  ```

➥ *NOTE: On versions of Solaris prior to 2.6, use* ndd—set/dev/tcp TCP_ CONN_REQ_MAX 1024.

One attack type not handled by modifying driver parameters is IP spoofing. These attacks make use of a trust relationship between two systems and the ability to predict the sequence numbers used in a TCP connection. To reduce the predictability of the sequence numbers, Solaris 2.6 and later versions implement a better system of creating initial sequence numbers, as recommended in RFC 1498. To enable this better method, change the line in */etc/default/inetinit* that contains the parameter TCP_STRONG_ISS to the following.

```
TCP_STRONG_ISS=2
```

Analyzing Network Security

As described in Chapter 7, "Managing System Security," Solaris provides a local security evaluation tool named ASET. The *aset* tool provides a good first step toward securing individual hosts. For systems connected to the Internet, it is recommended that a tool such as *tripwire* also be employed. These tools can be used to monitor a system for changes that may indicate the system has been accessed in an unauthorized manner. Consider running ASET and *tripwire* and checking the log files they produce on a routine basis.

Systems can also be checked for potential network security problems by employing tools to examine available network services. The *satan*, *nmap*, *cops*, *tiger*, and *nessus* tools are among those most popular for this purpose. These tools will automatically search the network services of a selected system or network for known problems.

These tools can check the security of passwords, look for *world*-writable files and directories, detect and probe network services, and check for secure file sharing as they run. Not surprisingly, hackers developed some of these tools! They have become public-domain mainstays of many security professionals due to their security testing capabilities. References to these programs can be found in the "References" section that follows.

References

The discussion of network security in this chapter has been brief. If you require more detailed information, consider the sources listed in tables 18-2 and 18-3, which follow.

Table 18-2: Books

Source	Comments
Simson Garfinkle and Gene Spafford, *Practical UNIX and Internet Security*, 2d edition. O'Reilly & Associates, 1996 (ISBN 1-56592-148-8).	A detailed and complete guide to security for UNIX systems connected to the Internet.
Simson Garfinkle and Gene Spafford, *Web Security and Commerce*. O'Reilly & Associates, 1997 (ISBN 1-56592-269-7).	Targets web server security and security of transactions executed over the Web.
Brent Chapman and Elizabeth D. Zwicky, *Building Internet Firewalls*. O'Reilly & Associates, 1995 (ISBN 1-56592-124-0).	Covers the setup and maintenance of firewalls.
Hal Pomeranz, editor, Solaris Security: Step by Step. SANS Institute, 1999; (719) 599-4303 or *info@sans.org*.	A very thorough checklist of procedures for tightening the security of a Solaris system.

Table 18-3: Web Sites and Software

Contact Information	Comments
Computer Emergency Response Team (CERT); *http://www.cert.org/*; *cert@cert.org*; 1-412-268-7090.	The national contact for computer security. This web site contains a wealth of information on security matters, and pointers to the security software mentioned in this chapter.
Computer Incident Advisory Capability (CIAC); *http://ciac.llnl.gov/*; *ciac@llnl.gov*; 1-510-422-8193.	Government and education contact for computer security. Provides an excellent mailing list, with updates on the latest security problems and their solutions.

Contact Information	Comments
http://www.cs.purdue.edu/homes/spaf/ hotlists/ csec-top.html	Professor Gene Spafford's well-maintained list of security-related web sites.
ftp://coast.cs.purdue.edu/pub/	A well know FTP archive for security tools, including TCP wrappers, *tripwire, nessus, nmap,* and *satan.*

Summary

Securing networked systems is a challenge, especially for systems connected to the Internet. The current Internet provides an environment rich in potential threats, including automated scans and attacks. Solaris provides numerous network services, each with the potential of offering unauthorized access.

This chapter discussed several methods of securing a networked system. The use of firewalls and/or filtering routers can provide a site with a first line of defense against unauthorized access. Turning off services wherever possible affords the best security. When services are required, careful attention to access methods and controls, and continual monitoring for unauthorized access, can help make a networked system more secure.

CHAPTER 19

Accessing Network Resources with NFS

The network file system (NFS) is an integral component of many Solaris installations. Developed by Sun Microsystems in the mid-1980s, NFS is a published standard, or protocol, that allows file sharing among hosts over a network. It is a standard feature on most UNIX systems and is available for most operating systems, including VMS, MS-DOS/Windows, and Macintosh (MacOS). Solaris provides a secure version of NFS called Secure NFS, which provides enhanced security features to protect NFS operations from network security breaches. A discussion of secure network communications is covered later in the chapter.

NFS usefulness derives from its ability to transparently access files located on remote hosts via a network connection. For example, NFS allows hosts on a network to access data or programs from a central location. This reduces system maintenance requirements by limiting the number of hosts, or locations, where programs and data files must be installed.

Furthermore, in environments where users regularly work on different hosts, NFS can automatically mount their centrally located home directories to any host they log on to. To users, it appears as though respective home directories follow them from host to host. NFS can also be used to share network devices such as CD-ROMs, or files from different operating systems such as the aforementioned VMS, UNIX, Macintosh, or MS-DOS.

NFS Function and Terminology

The most basic NFS configuration consists of a single host operating as a server and another operating as a client. The server *shares* or makes available some or all of its files by running a special set of processes. These processes are run in the background and are often called *daemons*.

The *share* command is used to instruct the daemons which directories to share with other hosts. By placing configuration information in a special file, */etc/dfs/dfstab*, the *share* commands will run automatically when the server is booted. As detailed in Chapter 5, "Refining System Boot and Shutdown Procedures," the system will execute the run control script */etc/rc3.d/S15nfs.server* during the boot sequence. This script is responsible for starting the NFS subsystem on the server.

The client attaches or *mounts* one or more of the file system sections made available by the server. The *mount* command incorporates the identified file system section into the local host's file tree in the same manner that local disk-based file systems are mounted. The client also runs a special set of daemon processes and uses configuration files that list the file systems to mount from the server. The client's NFS daemons are started at boot time when the */etc/rc2.d/S72nfs.client* script is executed.

Servers, Clients, and Sharing

A host that shares file systems via NFS is called a *server*. A host that mounts NFS file systems from other hosts is called a *client*. A host can be both client and server. Networks on which there are servers and clients are called *client-server networks*. Networks where hosts are both client and server to each other are often called *peer-to-peer networks*. Most large networks have a mix of clients, servers, and peers.

A file system, which is offered for mounting by a server via NFS, is said to be *shared*. This term comes from the *share* command used to offer file systems for mounting. Similar to other file system types, a remote file system is mounted on a client using the *mount* command. The shared portion of the server's file systems can be mounted in the same location, or in a different location in the UNIX file tree, on the client.

Remote Procedure Call (RPC)

A key to NFS operation is its use of a protocol called *remote procedure call* (RPC). The RPC protocol was developed by Sun Microsystems in the mid-1980s along with NFS. A *procedure call* is a method by which a program makes use of a subroutine. The RPC mechanism extends the procedure call to span the network, allowing the procedure call to be performed on a remote machine. In NFS, this is used to access files on remote machines by extending basic file operations such as read and write across the network.

How RPC Works

To understand the purpose of RPC, first consider how the computer system interacts with the files stored on its local disk. When a Solaris program is read into memory from disk, or a file is read or written, system calls are made to the read or write procedures in the SunOS kernel. As a result, data is read from or written to the file system.

When a file is read from or written to an NFS file system on a network, the read or write calls are made over the network via RPC. The RPC mechanism performs the read or write operation needed on the server's file system on behalf of the client. If the NFS client or server is a different type of host or operating system, such as VMS or a PC running Windows, the RPC mechanism also handles data conversion.

The result is that the information returned by the read or write call appears to be the same as if the read or write had occurred on a local disk. Although this appears complex, it all occurs at the file system level and requires no changes to normal operations on files or internal programs. It is similar to reading and writing a floppy disk versus a hard disk or a CD-ROM. As far as the programs that need to read and write the files are concerned, NFS is just another file system. The transparent nature of NFS is one of its best features. A typical NFS operation is shown in the following illustration.

*Typical NFS
operation.*

1. Client program reads or writes a remote file.

2. Client makes RPC call to server.

3. Server performs read or write to file on disk.

4. Server returns data or acknowledgment of write to client.

5. Client RPS returns data, acknowledgment, or write to client program

NFS File System Differences

The major difference between an NFS file system and other file system types derives from the network connection between NFS clients and servers. Several clients can use NFS file systems simultaneously. For example, it is possible to share the */usr/share/man* portion of a server's file system among many clients. Sharing a single file system among many clients allows the administrator to reduce the amount of disk space required for every system on a network. It also reduces the effort required to install and maintain software because a single copy of a program is shared among many clients.

Although NFS is quite handy and versatile, it does have certain drawbacks. NFS is not as fast as local disks. Current disk technology allows transfers at speeds up to 160 megabytes/second. Typical Ethernet connections range in speed from 5 to 80 megabits per second! NFS can cause problems on clients when the server of a mounted NFS file system becomes unavailable or network problems occur. NFS is only as good as the underlying network connection. These problems and some preventive measures are discussed later in this chapter and in Chapter 20, "Automating NFS with *automount*."

Setting Up an NFS Server

The setup of an NFS server involves dealing with the administrative issue of UID and GID numbering, editing the *dfstab* file, and verifying that the required daemon processes are started. Once an NFS server has been set up, the necessary *share* commands and daemons will be automatically started when the server is booted.

Match UID and GID Numbers for All Users

Setting up a host as an NFS server is fairly easy. First, the administrator must ensure that the clients and servers use identical UIDs and GIDs for all users and groups. In other words, a user must have the same UID number on all hosts on which he or she will be sharing files via NFS. Mismatches can result in files being inadvertently accessible to users who might not normally have access to them, or inaccessible to users who should have access to them.

If the site is using NIS or NIS+ to handle the */etc/passwd* and */etc/group* files, the UID and GID numbers should already match up across the hosts that share the same NIS or NIS+ server. NFS uses the UID and GID numbers when identifying file access permissions on remote file systems. See Chapter 21, "Network Name Services," for information on setting up name services.

Insert share Commands in /etc/dfs/dfstab

With uniform UID and GID numbering in place, the next step is to add lines to the */etc/dfs/dfstab* file. Each line is a *share* command specifying the portion of the server's file tree to be shared, which hosts are allowed to mount it, and how they are allowed to mount it. It is also possible to issue the *share* command on the command line to allow a file system to be shared. Placing the command in */etc/dfs/dfstab* causes the file system to be shared every time the host is booted. A simple example appears in the following illustration.

```
# cat /etc/dfs/dfstab

#       Place share(1M) commands here for automatic execution
#       on entering init state 3.
#
#       Issue the command '/etc/init.d/nfs.server start' to run the NFS
#       daemon processes and the share commands, after adding the very
#       first entry to this file.
#
#       share [-F fstype] [ -o options] [-d "<text>"] <pathname> [resource]
#       .e.g,
#       share  -F nfs  -o rw=engineering  -d "home dirs"  /export/home2

share   -F nfs  -o ro=grissom.astro.com -d "Manual Pages"       /usr/man
#
```

A simple /etc/dfs/dfstab *file.*

The final line of the *dfstab* file is the actual *share* command. The lines preceding it begin with a pound sign (#) and are treated as comments. The first option to *share*, *-F nfs* specifies that the portion of the file system is to be shared via NFS.

The next flag (reading from left to right) is the *-o* flag, which specifies how the portion of the file tree is to be shared and by whom. Several options are available. The *ro=* and *rw=* options specify a list of hosts allowed to access the file system on a read-only or read-write basis, respectively. Other options, discussed later in the chapter, may be used to control secure access to a file system.

In the previous example, the *ro=grissom.astro.com* option allows the host named *grissom* to mount the shared resource, */usr/man*, as read-only. Additional hosts can be added by separating their names with a colon (e.g., *ro=grissom.astro.com:glenn.astro.com*). This procedure can become unwieldy for large numbers of hosts. Creating a net group using NIS+ allows the administrator to group collections of hosts together and refer to them by a single name. See Chapter 21, "Network Name Services," for information on setting up NIS+ and creating a net group.

✓   *TIP: When specifying host access lists, the fully qualified domain must be used when employing the DNS name service.*

In the example, the *-d* option appears in the *share* command after the *-o* option flag and its arguments. This option allows a description of the shared item to be displayed when the server is queried concerning available shared items. The final item on the line is the path for the portion of the file system to be shared.

Starting Daemons, and the nfs.server Script

The next step is to start the required daemon programs. The daemons will start automatically during the boot sequence only if *share* commands are found in */etc/dfs/dfstab*. When setting up an NFS server for the first time, it will be necessary to start them manually after configuring the */etc/dfs/dfstab* file. A simple way to start these daemons is to run the shell script that would have been run during the boot sequence, as shown in the following illustration.

```
 ┌─                              Terminal                            · ┌─┐
 │ Window  Edit  Options                                         Help  │
 │ # /etc/init.d/nfs.server start                                    ▲ │
 │ # ps -ef | egrep "nfs|mount"                                        │
 │    root  3754     1  1 07:16:44 ?        0:00 /usr/lib/nfs/nfsd -a 16│
 │    root   133     1  0   May 24 ?        0:00 /usr/lib/nfs/statd     │
 │    root   135     1  0   May 24 ?        0:01 /usr/lib/nfs/lockd     │
 │    root   150     1  0   May 24 ?        0:17 /usr/lib/autofs/automountd│
 │    root  3756     1  2 07:16:45 ?        0:01 /usr/lib/nfs/mountd    │
 │    root  3758 29018  1 07:16:58 pts/3    0:00 egrep nfs|mount       │
 │ # ▮                                                                 │
 │                                                                   ▼ │
 └─                                                                   ─┘
```

Manual start of nfs server daemons.

In the previous example, the *nfs.server* script was run, the *ps* command was used, and then its output was filtered with the *egrep* command to produce a list of NFS daemons. The *statd* and *lockd* daemons provide for server crash recovery and file locking across the network. If the host will be running only as an NFS client, these are the only daemons started.

When a server crashes and comes back into service, the *statd* daemon handles the resolution of any locks pending at the time of the crash. The *statd* on the client informs the system that a server has crashed and rebooted pending locks; information about file locking and status is not maintained across server crashes. The *lockd* daemon enforces file locking on the shared file system by communicating with the *lockd* daemon on clients that mount and use a shared NFS file system.

Next, the *nfs.server* script starts the *mountd* and *nfsd* daemons. The *mountd* daemon handles requests from clients to mount shared items. It uses the information provided by the *share* command to help it determine permission for mounting, as well as access allowed a particular shared item. The *nfsd* daemon sets up the

NFS mechanism inside the SunOS kernel and provides its interface to the network.

➡️   *NOTE: The* nfsd *process does not directly handle RPC requests, but instead passes them on to the SunOS kernel. As a result, the* nfsd *process may not accumulate CPU time while NFS operations are occurring. Avoid using CPU time accumulated by* nfsd *as a measure of NFS operations.*

The number 16 as provided to the *nfsd* is used to allocate threads in the SunOS kernel. Each thread can handle one RPC operation at a time. A thread is a separate line of execution or control, allowing each request to be handled asynchronously without the overhead of separate processes handling each request. The number of threads set depends on the amount of NFS traffic the server is expected to handle. The amount of traffic is a function of the number of clients and how frequently RPC requests will be made.

More threads will improve service under heavy loads, up to the point at which the server's resources are fully utilized. The proper setting for a specific environment can be determined only by experimentation, and depends on factors such as the server's CPU type, CPU speed, disk system performance, network interconnections, and so forth. The default of 16 works well for most small to moderately sized sites.

Checking the NFS Server

In addition to using the *ps* command, it is possible to check on an NFS server by using the *showmount* and *share* commands, an example of which is shown in the following illustration. With *showmount*'s *-e* option, it is possible to specify the name of a remote host and view the list of file systems it is currently sharing. Alternatively, the *share* command, when used without arguments, will list the shared file systems on the local system.

Checking the NFS server with the showmount *and* share *commands.*

NFS server setup is a one-time activity. Once a server is configured, the *share* commands and daemon processes will automatically be started as part of the system boot procedure. Adding shared directories is a matter of inserting them in the *dfstab* file and running the *share* command.

Setting Up an NFS Client

Configuring an NFS client is even easier. All the administrator need do is specify remote file systems to mount in */etc/vfstab* and then issue the requisite *mount* commands. An NFS file system entry looks similar to entries for other file system types. Like other file systems listed in */etc/vfstab*, NFS file systems are automatically mounted at system boot. One major difference between NFS and other file systems is the fact that NFS-mounted file systems are not *fsck*ed by the client machines. An example */etc/vfstab* file is shown in the following illustration.

```
╔═══════════════════════════════════════════════════════════════════════╗
║  ─                           Terminal                          ⌐  □    ║
╠═══════════════════════════════════════════════════════════════════════╣
║  Window   Edit  Options                                        Help    ║
╠═══════════════════════════════════════════════════════════════════════╣
║ # cat /etc/vfstab                                                      ║
║ #device            device          mount          FS    fsck   mount    mount    ║
║ #to mount          to fsck         point          type  pass   at boot options  ║
║ #                                                                      ║
║ #/dev/dsk/c1d0s2 /dev/rdsk/c1d0s2 /usr            ufs   1      yes      –        ║
║ fd        –       /dev/fd fd       –       no     –                    ║
║ /proc     –       /proc   proc     –       no     –                    ║
║ /dev/dsk/c0t1d0s1         –        –       swap   –      no      –      ║
║ /dev/dsk/c0t1d0s0        /dev/rdsk/c0t1d0s0       /     ufs    1        no       ║
║ –                                                                      ║
║ /dev/dsk/c0t1d0s6        /dev/rdsk/c0t1d0s6       /usr  ufs    1        no       ║
║ –                                                                      ║
║ swap      –       /tmp    tmpfs    –       yes    –                    ║
║ glenn:/usr/share/man             /mnt/man        nfs    –      yes      ro       ║
║ # /etc/init.d/nfs.client start                                        ║
║ # df                                                                   ║
║ /                 (/dev/dsk/c0t1d0s0 ):   287898 blocks    96825 files ║
║ /usr              (/dev/dsk/c0t1d0s6 ):   605666 blocks   324084 files ║
║ /proc             (/proc            ):         0 blocks      418 files ║
║ /dev/fd           (fd               ):         0 blocks        0 files ║
║ /tmp              (swap             ):   105760 blocks     5194 files  ║
║ /mnt/man          (glenn:/usr/share/man):  605664 blocks  324084 files║
║ # █                                                                    ║
╚═══════════════════════════════════════════════════════════════════════╝
```

Adding an NFS file system to /etc/vfstab.

In the upper part of the window is the */etc/vfstab* file. The last line mounts */usr/share/man* from the host named *glenn*. It is mounted read-only on */mnt/man*. Running the */etc/init.d/nfs.client* shell script executes the *mountall* command to mount the remote file systems using the same procedure used when the system starts up. The output of the *df* command shows the mounted file system. Note that for NFS file systems the remote host name and shared path take the place of the device name shown for local file systems, and the number of files is shown as *-1*.

NFS File System Mount Options

Like other special file system types, NFS file systems have a variety of specialized *mount* options. Most of these options pertain to the networked nature of an NFS file system. Some options are paired as alternative choices separated by slashes.

mount Command Options

The following are the *mount* command options, with descriptions of their purpose and how they operate.

- *rw/ro*: Read/write and read-only. NFS file systems mounted read/write can block activity on the client when the server providing the file system becomes unavailable. See the descriptions of *intr* and *bg* that follow.

- *hard/soft*: The *hard* option mounts an NFS file system in such a way as to ensure that data is written to the remote file system. If the file server becomes unavailable, a file system mounted with the *hard* option will stop all remote file operations until the file server becomes available again. All file systems mounted with the *rw* option should also use the *hard* option to ensure the integrity of data written to the file system. The *soft* option does not provide assurance of data writes to the remote file system, but does not stop remote file operations in the case of a file server becoming unavailable. This option is useful for file systems that are mounted read-only.

- *suid/nosuid*: The *nosuid* option negates the effect of programs on the remote file system for which respective *setuid* bits are set. *Setuid* programs run from NFS file systems mounted with the *nosuid* option are executed with the normal permissions of the user executing the program, *not* those conferred by the *setuid* bit. This option is used to increase the security of the client by preventing *setuid* programs on remote file systems from being used on the client system.

- *bg/fg*: This option pair controls how to handle a failed mount of an NFS file system. Mounts with the *bg* option are retried in the background, freeing the shell that issued the *mount* command. Use this option when mounting file systems in */etc/vfstab* to prevent a workstation from stopping during the boot sequence because a file server is down.

- *intr/nointr*: The *nointr* option prevents program interrupts when programs cause an NFS operation to occur. This can result in programs being uninterruptible when an NFS file server becomes unavailable. The default is to allow interrupts so that those programs can be aborted in the event of server failures.

- *retry=n*: Number of times to retry a failed mount. The default of 10,000 is usually sufficient.

- *timeo=n*: Time-out value for retrying NFS operations. Increase this value to permit very slow systems, such as near-line file stores, more time to complete basic operations.

- *retrans=n*: Number of retransmissions of a given NFS operation. The setting depends on the network and type of server being used. Some networks where packet loss is high benefit from an increase in the number of retransmissions.

- *rsize=n*: Read buffer size. Some servers and clients (e.g., those with slower or less reliable network connections) perform better when the buffer used for NFS operations is a different size than the default.

- *wsize=n*: Write buffer size. Similar to *rsize* in usage.

- *vers=n*: NFS protocol version number (2 or 3). By default, the *mount* command will attempt to use the highest version number protocol available from the server. This option allows the protocol version to be fixed for use with servers that support only the older version 2 protocol.

- *proto=?*: Controls the network protocol used to transport NFS data. NFS uses IP datagrams by default. By setting *proto=tcp*, NFS will use *tcp*, thereby improving performance when moving data over wide-area networks and the Internet.

- *secure*: Solaris 7 and previous versions. This option enables NIS+ public key cryptography to be used for NFS authentication and encrypted communications.

- *sec=?*: New in Solaris 8. This option replaces the *secure* option of previous versions. It controls the security mode used for NFS. Options are *sys* for UID- and GID-based control, *dh* (Diffie-Hellman) for NIS+ based public key cryptography, *kerb4* for MIT Kerberos authentication, and *none*, which provides no authentication and maps all access to *nobody*, as described in the material that follows.

Using mount Command Options

Choosing the correct mix of *mount* command options for various circumstances is a bit tricky. General guidelines to help select the necessary options for a particular situation are discussed in the sections that follow.

hard *Option*

In general, NFS file systems mounted with the *rw* option should also use the *hard* option. When a problem occurs in writing to an NFS file system that is mounted soft, it is reported when the file is closed. Few programs check the return value from the *close* system call. The *hard* option handles this at the file system level, forcing the client to retry failed writes until they succeed (e.g., the server returns to operation). Data loss is thus prevented. However, using the *hard* option has serious ramifications for overall system functioning, as discussed later in this chapter.

bg *Option*

To prevent a client from hanging before completion of system start-up, use the *bg* option whenever possible. This option allows the boot process to complete by placing any NFS file system mounts that cannot be executed at boot time in the background. The mount will complete automatically once the problem preventing the mount from completing (e.g., network connections or server down) is fixed.

Use rsize *and* wsize *with Care*

Exercise great care when changing the values for *rsize* and *wsize*. NFS puts packets of the largest possible size on a network to improve efficiency. Reducing the size of these by adjusting the *rsize* and *wsize* parameters should be done only when a constraint requires such a change. For example, some older equipment with limited network hardware can cause network problems solved only by changing the NFS buffer sizes.

Use timeo *with Equal Care*

In a similar vein, the *timeo* value should be adjusted with care. Check to be certain that the underlying network is sound before adjusting this parameter. Flooding a problematic network with extra NFS request retransmissions will only make a bad situation worse. However, slow NFS servers may require increasing the value of *timeo* to avoid sending unnecessary retransmission of NFS requests.

In general, take care to examine the network hardware before making adjustments to the various NFS parameters. NFS is very demanding on a network and can bring to light problems not seen under lesser network loads. Network hardware failures can often result in additional network traffic as a result of NFS request retransmissions. See the "Troubleshooting NFS" section that follows for additional information.

Improving NFS Security

When using NFS to share files, the network becomes more like a single large computer system than a group of separate machines. The implications of this for system security are important to note. Access to files shared via NFS depends on the security of each machine involved in the sharing. Because the content of files are transferred over the network, you must consider access to the network itself as well. As the network grows, the ability of other machines on a network to monitor network communications or even alter them becomes more significant as a security concern.

File and Access Security

NFS security is only as good as the security of the individual hosts sharing file systems. Some things can be done to harden NFS against security problems, but the proper place to begin is by evaluating the security of each client and server. See Chapter 7, "Managing System Security," and Chapter 18, "Network Security," for discussions of the basic tools and areas that should be addressed on each host.

Server Sharing Security Options

With the individual hosts secured, move on to evaluate the permissions given by the file server for mounting its file systems. By default, an NFS file server maps access requests made by the *root* user on a client to a special UID. This UID does not have *root* file access privileges, and is normally set to that of the *nobody* account. You can change it by using the *anon=* option to the *share* command. For example, the following is an entry that would be placed in */etc/dfs/dfstab* to share */export/progs* with the *root* user on remote systems mapped to UID 10.

```
share -F nfs -o anon=10,rw=local -d "local programs" /export/progs
```

In some circumstances, you may want to offer *root* access to a server's file system, perhaps to facilitate remote software installation. This can be implemented on a host-by-host basis using the *root=* option. This option accepts a list of hosts in much the same way as the *ro=* and *rw=* options to the *share* command. Hosts listed in the *root=* list will have full *root* access to the file system.

✗   *WARNING: Use the* root= *option with extreme caution. Under normal operations, the NFS server maps requests by* root *from a client to the* nobody *UID. This provides some file security to the server. The* root= *option allows the administrator to bypass this security measure. Shared directories with the* root= *option assigned to them are accessible by the* root *account on any client that may have access to them with full* root *file access.*

In situations where the administrator may not have complete control over the security of client workstations, it is prudent to limit access to file systems offered by the server.

Client Mounting Security Options

On the client side, the *mount* commands should use the *nosuid* option to improve security. This option disables the effect of the *setuid* and *setgid* bits when a program from a remote file system is executed. The programs are run using the normal user permissions, *not* those conferred by the *setuid* or *setgid* bits. This helps to prevent non-secure programs offered by a server from being used by users on clients to affect client security.

Security at the Wire Level

Solaris provides for additional security for NFS operations through a secure version of the base remote procedure call (RPC) mechanism. By default, RPC uses the UNIX UID and GID numbering information as the basis of its authentication method. In Solaris 8 this is called "sys" security. These numbers can be easily seen by network monitoring tools as they go by on the network, and it is impossible for the client and server to determine if the requests they send and receive are genuine.

Secure RPC Uses a Public Key Cryptography System

Secure RPC solves the problem of authenticating the client and server by the use of a public key cryptography system. A cryptographic system uses a mathematical formula to encrypt (or encode) information in such a way as to be unreadable except for those who have the needed numerical keys to decrypt (or decode) the information to a readable form.

A public key cryptography system works as follows: two mathematically related large numbers are generated for each user. One of the numbers (the private key) is kept secret, and the other number (the public key) is published in a public database available to all hosts. Something encrypted using one number as the key can be decrypted using the other number as the key. By using a combination of keys from the client and server, information can be exchanged securely with assurance of the sender's and recipient's identity.

The public and private keys are used in the following ways. First, the client randomly picks a session key to be used to encrypt data being sent to and from a server. The client encrypts the session key using a combination of the current time, the client's private key and the server's public key, and then sends this encrypted information to the server. The server decrypts the message using the current time, the client's public key, and the server's private key. The server now has the session key and is ready to respond. The information flow in this type of system is shown in the following illustration.

Flow of information through crypto system.

Due to the mathematical properties of the keys and the encryption method, the message the server and client exchange can only be decrypted by the specific server whose keys were used, and could only have come from the specific client whose keys were used. This process works to ensure that no other host on the network is masquerading as either the client or the server.

The time stamp and session key are used to avoid "playback attacks." A playback attack is when a host simply records RPC requests and then reuses the encrypted information later in forged RPC requests. A secure RPC server will reject requests containing invalid time information.

➥   *NOTE: Time synchronization between client and server hosts is required*
       for secure RPC to function properly.

Secure RPC Setup

With Solaris 8, the administrator has two options for securing NFS, Kerberos and NIS+, also known as Diffie-Hellman for the name of the inventors of the key exchange method it uses. The administrator should check to see which method is in use at his or her particular site. Choosing a method if none is already in place will involve considering such factors as the availability of suitable NFS clients and servers for all systems in use at the site, as well as the infrastructure needed to support the different authentication methods (such as NIS+ or Kerberos servers).

To configure the more common secure RPC using the NIS+ option, the administrator may first need to assign a domain name to the collection of hosts that will share public keys and establish an NIS+ server. (This procedure is covered in Chapter 21, "Network Name Services.") Once a domain has been established, the following sequence of tasks can be performed.

1.   Generate new public and private keys using *newkey*.

2.   Start the key server by typing */usr/sbin/keyserv*.

3.   Edit */etc/dfs/dfstab* on the servers to add the *sec=dh* option to the *share* commands and NFS file system entries.

➥   *NOTE: In Solaris 7 and previous versions of Solaris, use the secure*
       option to specify NIS+ authentication. To configure NFS to use Ker-
       beros authentication, the administrator may first need to establish a
       Kerberos server and find out what the site's principal KDC and realm
       name is. Next, keys will need to be generated, as described in Sun's
       SEAM Administration guide. Finally, Kerberos authentication is
       enabled on the NFS server by adding the sec=kerb4 *option to the*
       share *command line entries in* /etc/dfs/dfstab.

Troubleshooting NFS

NFS depends on the underlying network to function. If systems are encountering problems mounting a remote file system, check the network first. Refer to Chapter 17, "Network Configuration and Management," for specific items to check. For NFS in particular, check to be sure that large data packets can be reliably sent between client and server. On an Ethernet, check using the *ping* command. Use the following command on the client.

```
/usr/sbin/ping -s server 1500 100
```

This command will send 100 maximum-size packets (1500 bytes) from the client to the server and report statistics for each packet with totals at the end. On a quiet network, packet loss should be minimal. Even on busy networks, losses of more than one or two packets should be examined. Although NFS will continue to function even in the face of severe network problems, performance will suffer greatly. Attempting to adjust NFS mount parameters such as *rsize*, *wsize*, and *retrans* without first checking the network can mask network problems.

Server Problems

Once the network connection is proven to be sound, the next items to check are the various NFS daemon processes. These background processes must be present for NFS to function. Use the *ps* command to look on the NFS server for the following daemons.

- *mountd*: If this daemon is missing, clients will be unable to mount NFS file systems from the server. Restart the daemon by becoming *root* and typing */usr/lib/nfs/mountd*. Check for the daemon again, using *ps*. If it fails to start, check the */etc/dfs/dfstab* file for errors.

- *nfsd*: Absence of *nfsd* means no NFS service. To restart the daemon, become *root* and type */usr/lib/nfs/nfsd -a X*, where *X* is the number of threads used by *nfsd*. The default is 16. If another number is used for a particular server, use it here when restarting *nfsd*.

- *rpcbind*: NFS is built on RPC. The *rpcbind* daemon provides an RPC name look-up service. Without this daemon, RPC calls will not work and NFS and other RPC services will stop. Because several services are dependent on *rpcbind*, the best course of action is to reboot if this daemon is missing.

If the clients still do not seem to be able to get NFS services from the server, try killing the *nfsd* process and restarting it. This can be accomplished by using the *nfs.server* script found in */etc/init.d*, which is reproduced in the following.

```
# /etc/init.d/nfs.server stop
# /etc/init.d/nfs.server start
```

If using the script fails to correct the problem, try rebooting the server.

Client Problems

The problems that can occur on clients are usually less a matter of daemons not running and more a matter of errors in configuration. If a particular client cannot mount NFS file systems that other clients can, carefully check the configuration files. Check for errors in the server's */etc/dfs/dfstab* file and the client's */etc/vfstab* file.

Check to be sure that the server is sharing the directory the client is trying to mount, and that the server has the correct name and IP address for the client. If the site is using NIS or NIS+, verify that the client and server are receiving current information from those services. Refer to Chapter 21, "Network Name Services," for ways to test and verify that the NIS+ service is functioning properly.

NFS Design Pitfalls

As mentioned earlier, the simple NFS configuration described here has a few drawbacks. To mount a file system read/write and maintain file integrity, file systems must be mounted using the *hard* option. However, if the file server of such a file system becomes unavailable, processes on the client that refer to the remote file system will stop working, and wait for the server to return to service. This can happen even in the case of file systems not directly involved in the processing the client is executing at the time.

As networks grow, maintaining the */etc/vfstab* files on all clients becomes increasingly laborious. The */etc/vfstab* on each client will usually require modification as each new file server is added or as file servers are reorganized, renamed, or moved. The likelihood of mistakes in the file escalates, and the fact that all clients need not mount all file servers at any given time becomes increasingly evident. If files are shared from a common file server, the */etc/vfstab* method makes no allowance for adding file servers to help balance the load or to provide redundancy in the face of failures.

To deal with these problems, Solaris uses a special daemon called an *automounter*, and a special file system type known as *autofs*. These tools are covered in Chapter 20, "Automating NFS with *automount*."

Summary

Sun's network file system (NFS) is a widely available mechanism for sharing files between Solaris systems and a large number of other UNIX, VMS, PC, and Macintosh systems. Through the use of specific options to the *share* and *mount* commands, special circumstances arising from the networked nature of NFS can be handled.

This chapter examined the configuration and operation of the NFS subsystem, as well as tips for securing and troubleshooting NFS installations. Unfortunately, a simple NFS setup has drawbacks due to problems that occur when file servers are unavailable. These problems are addressed using the *automount* service, discussed in the next chapter.

Automating NFS with automount

As mentioned in Chapter 19, "Accessing Network Resources with NFS," growing networks present the system manager with many challenges. As new machines are added, correctly maintaining */etc/vfstab* files becomes an increasingly daunting chore. In addition, when a file server becomes unavailable, processes on client workstations may stop even if they do not depend on that file server. The *automount* service was developed to minimize these problems.

General Concept of automount

In most cases, an NFS client does not require continuous access to all NFS file systems that may be available on a network. If you can avoid mounting NFS file systems until they are needed, problems caused by unavailable file servers are greatly reduced. In brief, mechanisms for identifying and using alternative file servers when possible are required for this purpose. The combination of the *autofs* file system type and the *automount* daemon provides an automated NFS file system mounting service as a solution to the unavailable file server problem, while providing an efficient means of managing the mounting of large numbers of shared file systems.

The *automount* service works by intercepting file access requests, mounting the required NFS file system, and then allowing the request to proceed. Assume that the */usr/man* directory, an unmounted NFS file system under control of the *automount* service, is accessed. The access request would be intercepted by the *automount* daemon, which in turn would mount the file system and then allow the request to continue. This mounting process is transparent; the file access occurs as if the file system had been mounted all along. The *automount* service also cleans up (unmounts) NFS file systems no longer being accessed.

Advantages of automount for the Small Network

Although the demands of large networks drove the development of the *automount* service, it provides benefits for even very small networks. In all cases where an unavailable file server can cause problems, the *automount* service is useful. As a small network grows, the *automount* service makes adding and managing additional shared file systems easy.

As additional resources become available, critical programs and data files can be offered on duplicate file servers to further improve reliability and performance. Through the use of the *auto_home* feature, user accounts can appear to have a single home directory, whereas the system manager is free to move the location of the user's files as needed to adapt to changing configurations and disk usage demands.

autofs and automountd

The *autofs* is a special file system type that resides in the SunOS kernel. When the *automount* daemon *automountd* is started, it reads a series of configuration files known as *map files*, and creates *autofs* file systems associated with each of the mount points listed in the maps.

Unlike other mount points, which are attachment points for specific data storage devices in the UNIX file tree, the *autofs* mount points attach the *automountd* process to the UNIX file tree (see the following illustration). Imagine that the mount point for a local disk is a shelf or cubbyhole. An *autofs* mount would be similar to a person who would wheel into place a shelf or cubbyhole as needed based on the request and the rules found in a map file.

A few autofs *mount points in a UNIX file tree.*

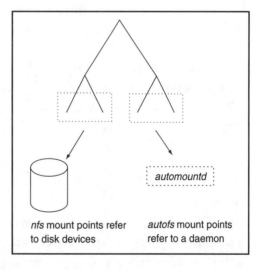

nfs mount points refer to disk devices

autofs mount points refer to a daemon

automountd

When a file or directory access request traverses an *autofs* mount point, the request is intercepted and a message is sent to the *automount* daemon. The *automount* daemon consults the map files for entries pertaining to the mount point and mounts the referenced NFS file system. The access request is then allowed to proceed by being directed to the newly mounted NFS file system. This process is shown in the following illustration.

Steps taken when automount initially mounts an NFS file system.

After the initial access request, all remaining requests are passed on to the remote NFS file system.

automountd

Remote NFS file system.

mount NFS file system...

Ordinary file system access goes directly to a device.

autofs file system access causes the *automount* daemon to mount the remote NFS file system, and then pass along the access request to the remote file system.

All subsequent requests for file access that traverse this particular *autofs* mount point are directed to the mounted NFS file system. The *automount* daemon monitors the *autofs* mount point. If the mount point is not traversed during a predefined period, the referenced NFS file system is automatically unmounted.

The automatic mounting imposes a brief delay when an unmounted NFS file system is first accessed. Little additional overhead occurs after the initial mount, and the automatic unmount occurs in the background. Consequently, there is no interference with ongoing processing on the client. Great flexibility and ease of maintenance are the payoff in return for a tiny increment to overhead.

Starting and Stopping the automount Service

The *automount* service is built on top of the NFS service. To use *automount*, first be sure NFS is properly configured and that the NFS daemons are started, as described in Chapter 19, "Accessing Network Resources with NFS." With NFS configured and running, *automount* is most easily started manually by running the run control or *rc* shell script used by Solaris to start the *automount* service during the system boot process. From the *root* user prompt, type the following.

```
# /etc/init.d/autofs start
```

Conversely, to stop the *automount* service, key in the following.

```
# /etc/init.d/autofs stop
```

These commands are automatically performed as part of the system boot and shutdown procedures, respectively.

automount Maps

The heart of the *automount* service is its map files. These files specify the *autofs* mount points, control the mount options used when mounting specific NFS file systems, and allow for automated selection of NFS file systems based on various parameters. The automatic selection of redundant file servers and the general flexibility of the map files allow for a single collection of map files to be used across an entire network. This process greatly eases the maintenance chore when shared file systems are added, removed, or moved.

Map files are divided into direct and indirect types. Direct maps list the mount points for remote file systems as their keys. Indirect maps list mount points within a specified directory and use symbolic links, which point from the key entry to the remote file system mounted elsewhere. Every map file is listed in the master map file, as shown in the following illustration.

Master automount file and respective map files.

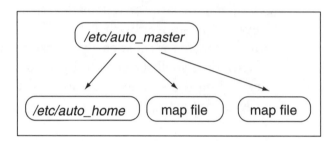

Master Map

The master map (*/etc/auto_master* file) forms the basis of the map system. The format of lines in the file is simple, consisting of a mount point, the name of a map file that specifies NFS file systems to be mounted at that mount point, and any mount options to be applied to the *autofs* file system at the mount point. An example of an auto_master file is shown in the following illustration.

Sample auto_master file.

Each line in the sample *auto_master* file uses a different feature of the master map. The first two lines begin with a pound sign (#)

and are treated as comments. Adding comment lines helps document changes made to the file or to describe what each of the listed map files does. The third line shows a direct map file. The dash (-) listed for the mount point is a placeholder.

Direct maps list the path name of each mount point as the key entry in the map file itself. NFS file systems mounted by a direct map are mounted at the mount point listed as the key in the direct map file. This line also specifies the mount point as *-ro* (read-only). A read-only mount point might be used when mounting a file system containing programs or data that do not change (e.g., manual pages).

Direct Maps

Upon reading each line in the master map file, the *automount* daemon reads the listed map file. Map files are assumed to be in the */etc* directory. An example of a direct map file is shown in the following illustration.

Direct map file.

```
# cat /etc/auto_direct
# mount /usr/man from grissom
/usr/man          -ro      grissom:/usr/man
# mount framemaker binaries from grissom *or* chaffee
/usr/frame        -ro      grissom:/usr/frame \
                           chaffee:/usr/frame2.0
# 
```

Each line in the direct map file consists of a key, a list of mount options, and the location of the NFS file system to be mounted. In a direct map, the key is the path to the mount point on the client where the NFS file system is to be mounted. The second line of this file shows a simple example where */usr/man* is mounted using the *-ro* (read-only) option. The */usr/man* directory is shared by the file server *grissom*.

The fourth and fifth lines show how multiple file servers can be used to redundantly serve applications or data. In this case, the */usr/frame* directory contains a popular text processing program. It is available on two different file servers in two different directories. The map file specifies that either of these file servers can be used to obtain */usr/frame*.

Server Connection

To decide which file server to use, the *automount* daemon consults the client workstation's network address information and tries to use the *closest* file server to reduce network traffic. The closeness measure used is to first try servers on the same network as the client, and then those on other networks. This is done to minimize NFS traffic across routers.

If more than one file server is close, the *automount* daemon will contact all close file servers and use the file server that responds first. If the chosen file server does not respond, the *automount* daemon will automatically try the other file servers in the list. The result is that the nearest available file server will be used.

Note also that the line has different remote directory names being used to locate */usr/frame* (i.e., */usr/frame* and */usr/frame2.0*). When the remote file system is mounted by the *automount* daemon, the name of the key is used regardless of the name of the remote file system. Thus, in this case, if *chaffee* were chosen as the file server, *chaffee:/usr/frame2.0* would be mounted as */usr/frame* on the client.

This allows the map file to mask directory-naming schemes that might be confusing to users on client workstations. System managers should be aware that this feature can be used to provide easily remembered names for users, whereas a different naming scheme can be used on the file server for purposes of labeling the software or data to aid in its management.

✓ *TIP: When setting up a heterogeneous network of NFS servers and clients from different vendors, consider using the* auto_home *map. Vendors have different conventions for the location of user home directories. With the use of* auto_home, *these different locations can be consolidated into a single, easy-to-use scheme.*

�'t *NOTE: The multiple location feature of the* automount *service is the key to building highly reliable NFS file sharing. When designing a network where shared collections of programs and data will be used, consider replicating the shared items and using* automount *to avoid single points of failure.*

Variables Used in Map File Entries

As a further aid to creating maps, the *automount* daemon understands and can substitute special symbols into map entries. These symbols can be used to create map entries that mount different file systems from the same key, depending on the machine type or operating system version in use. Descriptions and examples of map file variables are presented in table 20-1, which follows.

Table 20-1: Map File Variables

Variable	Meaning	Example
ARCH	CPU architecture type	sun4
CPU	CPU type	SPARC
HOST	Host name	glenn
OSNAME	Operating system name	SunOS
OSREL	Operating system release number	5.6
OSVERS	Operating system version	FCS1.0

For example, consider the following direct map file entry.

```
/usr/man     chaffee:/export/manuals/$OSREL/man
```

Assuming that the file server *chaffee* had several versions of the manual pages available, this entry would mount the correct version of the manual pages for the client on which it is being used. The ARCH and CPU variables can be used in a similar manner to create maps used on Solaris systems based on both SPARC and Intel CPUs without modification. For example, to mount the correct version (SPARC or Intel) of a local collection of programs with a direct map, you could use the following statement.

```
/usr/local/bin          grissom:/export/$CPU/local/bin
```

Another map feature is the use of the plus symbol (+). Map files can be included inside each other by using the +*mapfile* notation. For example, to include a map file named */etc/more_direct_maps* in the direct map file, the system administrator could add the following entry to the */etc/direct_maps* file.

```
+more_direct_maps
```

➡️   *NOTE: Where the automount daemon will look for plus sign (+) entries depends on the automount entry in the* /etc/nsswitch *file. By default the* automount *daemon will look in* /etc *for files first, and then consult the NIS service. See Chapter 21, "Network Name Services," for more information.*

The objective of using these map features is to create general-purpose maps that can be used anywhere on a given network. By carefully crafting a general-purpose map, the problem of configuring a large number of NFS file systems on many different clients can be reduced to maintaining a single configuration file.

Indirect Maps

The fourth line of the sample *auto_master* file in the following illustration lists an indirect map. In an indirect map, the NFS file systems are mounted at a different mount point than the one listed in the table, and a symbolic link is made from that location to the directory listed in the master map.

Sample auto_master *file.*

This allows for some more-exotic *automount* tricks. A diagram of an indirect mount with */home* as its entry in *auto_master* and a key of *frank* in the map file appears in the following illustration.

Mount points and symbolic links created by an indirect map.

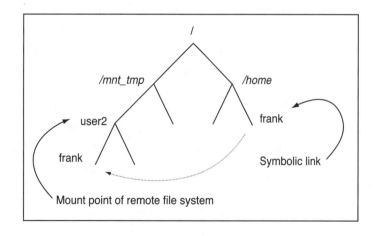

auto_home *Map*

One of the most common indirect maps is *auto_home*. The *auto_home* map has keys in the */home* directory. It is used to collect users' home directories into a single location, and greatly eases the task of maintaining home directory location information.

The following illustration shows a typical *auto_home* map. Indirect maps have the same items in each entry line as found in a direct map: a key, an optional list of mount options, and the location of a remote file system.

An auto_home *indirect map.*

```
# cat /etc/auto_home
# Home directory map for automounter
#
dwight   glenn:/export/home/&
frank    chaffee:/user2/&
tom      grissom:/export/users/&
steve    glenn:/export/home/&
#
```

Note that the line starting with the key *frank* does not contain a full path. The key in an indirect map specifies the name of the symbolic link to be made in the directory specified in the entry for the map in the *auto_master* file. In this case, all keys in the *auto_home* map will create symbolic links in the */home* directory.

The ampersand character (&) stands in place of the key. It is a shorthand notation that can be used in any type of map file. For example, it shortens the entry that begins with the key *tom* to *grissom:/export/users/&* instead of *grissom:/export/users/tom*.

Indirection by use of symbolic links is useful for several reasons. First, the actual mount point can be placed so as to avoid unnecessary mounting of remote file systems. Consider what could happen if the command *ls -F* is invoked in a directory that contained numerous *autofs* mount points controlled by direct maps. All remote file systems would be referenced and mounted.

The symbolic link used in an indirect map avoids unintended mount point traversal and mounting. Mounting the remote file system and creating the connection occur when the mount point is traversed for access to files or directories beyond the link itself. When a mount occurs, the referenced file system is mounted, and then a connection is made from the key to the file system.

Why Use auto_home?

The major administrative reason for using the *auto_home* map is to allow for easier maintenance of users' home directories. Under this system, a user can always refer to his or her home directory as being in */home*. In the previous example, Tom's home directory could be referred to as */home/tom*, Frank's as */home/frank*, and so on, even though they are located on different file servers or in different directory trees. Because this path is an automatically created symbolic link, the actual location of the directory (e.g., */export/users/tom* on *grissom*) can easily be changed if necessary.

Moving user home directories is a common occurrence for the system administrator due to system changes such as new disks, existing file systems filling, and so on. With the use of *auto_home*, the only change required when making a move is to adjust the *auto_home* map entry. The user's password file entry, and programs or shell scripts written by users that refer to their home directory, can remain untouched. This is easier for the system manager and more convenient for users.

➡️   *NOTE: Do not use the /home directory for local file systems. The* auto-mount *service by default will use this directory for the* auto_home *map. By convention, home directories are placed in the /export file system.*

Map Maintenance

How maps are maintained depends on whether or not the site uses NIS+. When using NIS+, the table administration command *nistbladm* is used to make changes to map files. The command updates the NIS+ databases as necessary to make any changes in the *automount* maps available via NIS+.

If changes to the *auto_master* map have been made, the *automount* command must be run on each client to ensure that new mount points are created. Chapter 21, "Network Name Services," provides a more detailed description of *nistbladm* and other commands used to manage NIS+.

Maintaining the auto_home Map File

The *auto_home* map is a special maintenance case. The user account administration tool (*admintool*) described in Chapter 6, "Creating, Deleting, and Managing User Accounts," allows the administrator to easily update the *auto_home* map. If you do not use the *admintool*, you will need to manually edit the map file when moving user home directories around. Be sure to distribute the updated file to all client workstations as soon as the home directory has been moved, to avoid mismatches between the map entry and the actual location of the user's home directory. Mismatches would prevent the user from accessing her/his home directory.

When you are not using NIS+ to manage the map files, the procedure is slightly different. First, the administrator will need to edit the appropriate map files. Then, if changes are made to the *auto_master* map, run the *automount* command. A convenient method for distributing the map files to a large number of clients is the *rdist* tool, as described in Chapter 16, "Automating Administrative Tasks."

Sample rdist File

The following is a sample *rdist* file that will distribute the *auto_master*, *auto_direct* and *auto_home* maps used as examples in this chapter. The *rdist* file will automatically check each file and update files that are out of date on the listed clients. Whenever the *auto_master* file is updated, *rdist* will also run the *automount* command.

```
# Define the clients who will receive the files.
SLAVES = ( grissom shepard )
# Define the files to be distributed
FILES = (
    /etc/auto_master
    /etc/auto_direct
    /etc/auto_home
    )
# Distribute the files to the clients.
# Specify special handling for the /etc/auto_home file.
${FILES}->> ${SLAVES}
    install;
    special /etc/auto_master /usr/sbin/automount;
```

If the foregoing *rdist* instructions were placed in a file named *auto.dist*, the *rdist* command could read and use the file with the following command.

```
rdist -f auto.dist
```

Given this command, *rdist* would compare the date and size of the files on the distribution host with the files on the listed clients. Files on the distribution host would be installed on the clients in instances where differences were encountered. If the */etc/*

auto_master file were updated on a client, *rdist* would also run the */usr/sbin/automount* command on the client.

Troubleshooting automount

When trying to locate and fix problems with the *automount* service, bear in mind that *automount* is built on top of NFS, which in turn is built on top of the basic Solaris TCP/IP networking functions. The first step should be to check that NFS and the network features on which it depends are working. Refer to Chapter 19, "Accessing Network Resources with NFS," and Chapter 17, "Network Configuration and Management," for additional help. Then, if NFS is working, proceed to check the *automount* map files.

Disappearing File Systems

A common problem when beginning to use *automount* is the placement of *autofs* mount points over the top of existing mount points. This is most commonly seen when you start the *automount* service and a file system seems to disappear beneath the *autofs* mount point.

To use features such as the *auto_home* map, the user file systems must be mounted in a location other than */home*. The most common convention is to place local file systems in */export*, and the *autofs* mount points in the final location. For example, it might be desirable to mount a user file system such as */export/home* and then use the *auto_home* map to make the user's home directories appear in */home*. This method allows both the file server and clients to use the same *automount* maps, as indicated in the following illustration.

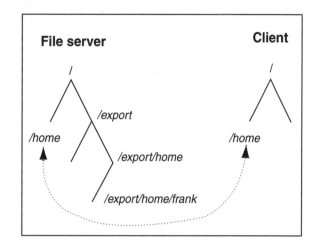

Home directories in /export can be automounted by both clients and servers.

Syntax Errors

Another common problem is syntax mistakes in the map files. When error messages appear or incorrect syntax is suspected, try using the *-v* option flag to the *automount* command to obtain additional messages. The next step you could attempt is using the pound sign comment character (#) to "comment out" lines to help track down the problem.

```
# bob mercury:/export/home/bob
```

Lines that start with a pound sign are ignored by the *automount* daemon. Inserting a pound sign on a suspect map file line disables it for testing whether a problem exists on a particular line.

Summary

Using the *automount* service can add flexibility and greater reliability to the already useful NFS file service. A single set of map files can be created for use across the entire network. This process greatly reduces the effort required to maintain NFS file access in large networks. By using the *auto_home* map, user home directory maintenance can be simplified. This chapter covered the configuration of *automount* map files, the starting and stopping of the *automount* service, and troubleshooting *automount* problems.

CHAPTER 21
Network Name Services

Previous chapters focused on how to manage system operations information (e.g., host and password files) on individual hosts. Maintaining system information stored on individual hosts is cumbersome on networks as small as five or six hosts, and becomes unmanageable on larger networks. For example, the */etc/inet/hosts* file is accessed by a variety of network applications in order to establish communications between hosts. If a new host is added to a network of 100 existing hosts, the */etc/inet/hosts* file on each of the existing hosts would have to be updated with the new host's name and IP address.

Chapter 16, "Automating Administrative Tasks," illustrated how the */bin/rdist* utility could be used to distribute important files to systems on the network. Unfortunately, some files change too frequently to be kept current by an overnight *rdist* run. The problem of keeping system information current is especially troublesome with very dynamic information such as passwords.

Solaris 8 provides three different name look-up services to help the administrator cope with this problem. The NIS+ (Network Information Service Plus) environment addresses this problem by maintaining system information on a central host known as an NIS+ server. The NIS+ server provides its clients with system information (e.g., host and password) upon request through the NIS+ lookup service. The Domain Name Service (DNS) is also a name lookup service. Because NIS+ does not scale to networks the size of the Internet, DNS is often used in conjunction with NIS+ to provide the name services required for efficient network communications.

The newest name lookup service to be included in Solaris is LDAP, the lightweight directory access protocol. It is supported natively in Solaris 8 and can be used in conjunction with or in place of NIS+ or DNS. Sites considering using LDAP should be aware of the complexity of properly securing LDAP transactions. Unlike NIS+, which uses strong encryption for authentication and data transmission, LDAP uses a combination of access control lists (ACLs) and varying security modes.

Although the use of the anonymous authentication and the user authentication methods are common for many LDAP functions, they provide no security from snooping of data as it passes over the network. Sites that use LDAP for sensitive data such as user account passwords should consider using SSL and certificates to provide over-the-wire security and cryptographically secure authentication.

✓   *TIP: See the references at the end of this chapter for books that go into detail on LDAP configuration and security.*

This chapter focuses on configuring the NIS+ environment, the setup of DNS services, and how to use NIS+ in conjunction with

DNS and LDAP to provide efficient name services for the corporate network.

What Is NIS+?

NIS+ is a network lookup service that provides information about the users, workstations, and network resources to hosts that reside within an NIS+ domain. An NIS+ domain is a collection of NIS+ related hosts. NIS+ is a component of ONC+ (Open Network Computing Plus) in Solaris 2, and replaces its predecessor, NIS, which is used in Solaris 1. Like NIS, NIS+ is not proprietary, but may be licensed to any vendor who wants to include the service as part of its operating environment.

➙   *NOTE: NIS is also known as YP (Yellow Pages).*

NIS+ Servers and Clients

NIS+ solves the problem of multiple machines sharing the same information by storing the data in databases on NIS+ server hosts. The server hosts make the data available to NIS+ clients on demand. This arrangement is known as the NIS+ client-server model, shown in the following illustration. A NIS+ client process uses remote procedure calls (RPC) when making information lookup requests. An NIS+ server is a process that services an NIS+ client RPC request, searches for requested information in a database, and returns the information to the client process.

*NIS+ client-
server model.*

Every NIS+ domain is served by one master server, and may also utilize one or more replica servers. An NIS+ master server contains the master set of system information in the form of NIS+ tables. If changes are made to the NIS+ tables on the master NIS+ server, they are automatically pushed to the replica servers. An NIS+ replica server maintains copies of the tables in order to distribute the load in answering client requests and to provide backup sources of information in the event the master server fails.

NIS+ Commands

Solaris provides a suite of commands that allow alteration of NIS+ data. Before proceeding to explore the inner workings of NIS+, a basic understanding of these commands is essential. The names of many of the NIS+ commands are very similar to the names of standard UNIX commands. The NIS+ commands only operate on NIS+ data, but are otherwise very similar in operation to standard UNIX commands.

NIS+ User Commands

The following NIS+ user-level commands may be invoked by any user on the system, but only return data if that user is authorized access to the NIS+ data.

- *nisaddent*: Add */etc* files and NIS maps to NIS+ tables.

- *niscat*: Display NIS+ tables and objects.

- *nischgrp*: Change group ownership of an NIS+ object.

- *nischmod*: Change access rights on an NIS+ object.

- *nischttl*: Change lifetime of NIS+ object.

- *nisdefaults*: Show NIS+ default values.

- *niserror*: Show NIS+ error messages.

- *nisgrep*: Search NIS+ information.

- *nisgrpadm*: NIS+ group administration command.

- *nisln*: NIS+ symbolic link utility.

- *nisls*: List NIS+ directory.

- *nismatch*: NIS+ search utility.

- *nismkdir/nisrmdir*: Create/remove NIS+ directory.

- *nispasswd*: Change NIS+ password.

- *nisrm*: Remove NIS+ object.

- *nisshowcache*: Print content of shared NIS+ cache file.

- *nistbladmin*: NIS+ table administration utility.

- *nistest*: Test the state of NIS+ name space.

NIS+ Administrator Commands

In addition to the user-level NIS+ commands, Solaris provides a suite of administrator commands for NIS+. The NIS+ administrator commands are limited to use by the system administrator or other authorized NIS+ administrators. The following commands are used to initialize and manage NIS+ databases.

- *aliasadm*: Manipulate NIS+ aliases.

- *nis_cachemgr*: NIS+ cache manager utility.

- *nisaddcred*: Create NIS+ credentials.

- *nisclient*: Initialize NIS+ credentials for NIS+ clients.

- *nisinit*: NIS+ client/server initialization utility.

- *nislog*: Display the NIS+ transaction log.

- *nisping*: Send *ping* to NIS+ servers.

- *nispopulate*: Populate the tables of the NIS+ domain.

- *nisserver*: Set up an NIS+ server.

- *nissetup*: Initialize an NIS+ domain.

- *nisstat*: Report NIS+ server statistics.

- *nisupdkeys*: Update public keys in an NIS+ directory.

- *rpc.nisd/rpc.nisd_resolv*: NIS+ service daemon.

- *sysidnis*: System configuration tool.

NIS+ Domains

A collection of related hosts and the information managed and utilized among the hosts is known as an *NIS+ domain*. Every NIS+ host must be associated with a specific NIS+ domain. For example, *astro.com.* might be the domain name issued to a company named Astro Inc. Because Astro is a commercial entity, it would be assigned to the sub-domain named *com* (commercial). Similar to IP addresses, domain names are issued by the NIC. The domain name issued by the NIC is often referred to as the *root* domain or top-level domain. A *root* domain must contain two components ending with a dot, such as *astro.com.*

➥   *NOTE: NIS+ domain names are not case sensitive. Thus,* Astro.com. *and* astro.com. *are equivalent.*

The name of the NIS+ domain a host is associated with is declared in the */etc/defaultdomain* file. Domains can be divided into sub-domains to accurately reflect the hierarchical structure of an organization. For example, a company named Astro Inc. with hardware, software, marketing, and sales divisions might set up the domain hierarchy shown in the following illustration.

NIS+ hierarchical structure for astro.com.

NIS+ Objects

An NIS+ name space is a hierarchical structure in which NIS+ information is stored. The structure includes the *root* domain and all domains below it. Every name space has a *root* master server that serves the *root* domain at the top of the name space. An NIS+ name space is similar in structure to a UNIX file system hierarchy. Directory, table, and group objects are the most common types of objects in the NIS+ name space. The following objects are stored in the */etc/nis* directory on the NIS+ server.

- *Directory objects*: Similar to UNIX directories in that they can contain other objects such as table and group objects.

- *Table objects*: Store information in the NIS+ name space and are analogous to a UNIX plain file. The Solaris 2 environment provides 16 default tables, each of which stores information within a domain about hosts, users, networks, services, and so forth. A set of NIS+ tables stores information for that particular domain only.

- *Group objects*: Used for NIS+ security. Like the group ownership within a UNIX file system, an NIS+ group is a collection of users and hosts identified by a single name and is used to facilitate NIS+ security. The default NIS+ group created in every domain is called *admin*. Although additional NIS+ groups are optional, they are a security convenience allowing an NIS+ administrator to assign access rights to a group of users and hosts.

Directory Objects

Directory objects make up the framework of a name space and are created with the *nismkdir* command. Each sub-domain in an NIS+ hierarchy is analogous to a subdirectory in a UNIX file system. Two default NIS+ directories reside within every domain: the *org_dir* directory stores NIS+ table objects, and the *groups_dir* directory stores NIS+ security group objects (see the following illustration). The directory object at the top of a name space (*root* domain) is called the "root directory." A name space is said to be "flat" if it consists of a *root* domain only. A flat domain can become hierarchical by creating sub-domains.

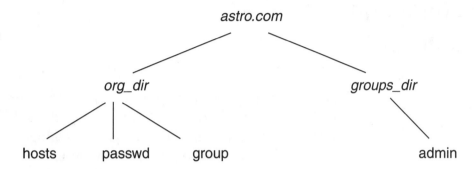

Default objects in every NIS+ domain.

Using NIS+ Object Names

NIS+ object names are formed by appending the *root* directory name (including the period) to the names, known as a "fully qualified name." For example, *sales.astro.com.* and *software.astro.com.* are both fully qualified names that represent the *sales.astro.com.* and *software.astro.com.* sub-domains within the *astro.com.* name space. Like UNIX files and directories, NIS+ objects can be referred to by partially or fully qualified names. Characteristics of these names are summarized in the following.

- Similar to a relative path name in a UNIX file system hierarchy, a partially qualified NIS+ name is simply the name of the component. For example, the *hosts* table's partially qualified name would be *hosts*.

- A fully qualified name resembles an absolute UNIX name in that it is the complete name of the component. Such a name is formed by starting with the name of the component and appending the names of all involved components up to the *root* domain, and is delimited by periods (.). For example, the fully qualified name of the sales division's *hosts* table at Astro Inc. is *hosts.org_ dir.sales.astro.com*. Likewise, the fully qualified name of the software division's *hosts* table would be *hosts.org_dir.software.astro.com.*

Configuring an NIS+ Domain

This section provides a step-by-step procedure for setting up a flat domain named *domain.astro.com.* using the NIS+ installation scripts. The first step in the process is to configure the *root* master server. The second step consists of populating the *root* master server tables with information. The files in the */etc* directory are frequently used to achieve this step. Finally, clients are added to the domain.

Creating a Flat NIS+ Domain Using NIS+ Scripts

Solaris 2 includes the following NIS+ administration shell scripts to facilitate the creation of NIS+ servers, tables, and clients.

- */usr/lib/nis/nisserver*: Used to set up a *root* master or its replica NIS+ servers.

- */usr/lib/nis/nispopulate*: Used to populate NIS+ tables from corresponding files in a specified domain.

- */usr/lib/nis/nisclient*: Used to initialize NIS+ hosts and users.

Configuring a Root Master Server

1.   Log in as *root* on the host that is to become the NIS+ *root* master for the name space *astro.com.*

2.   Run the *nisserver* script with the *-r* and *-d* options (shown in the following). The *-r* option is used to indicate that the server being initialized is a *root* server, and the *-d* option specifies the domain name the server is administering.

```
mercury# /usr/lib/nis/nisserver -r -d astro.com.
```

3. The *nisserver* script will ask for setup verification (as shown in the script that follows). The following script sets up the machine *mercury* as an NIS+ *root* master server for domain *astro.com*.

    ```
    Domainname : astro.com.
    NIS+ Group : admin.astro.com.
    YP compatibility : OFF
    Security level : 2=DES
    Is this information correct? (Y or N) y
    ```

4. The *nisserver* script will then run the *nisinit* and *nissetup* commands and start an NIS server process. You will be prompted to enter an NIS+ network password required by NIS+ for security purposes. Enter either the master server's *root* password or another password. To keep things simple, most sites use the *root* password as the NIS+ password, as shown in the following.

    ```
    adding credential for mercury.astro.com.
    Enter login password: rootpasswd
    ```

✓   *TIP: Whichever password is chosen for NIS+ security, be sure to record it in a safe place. It is easily forgotten and may be required to repair problems in the NIS+ database should they occur.*

Populating a Root Master Server's Tables

To populate the *root* master server's tables, perform the following steps.

1. On the master server for the domain, run the *nispopulate* script to populate the empty tables created by the *nisserver* script. Typically, the system file located in the */etc* directory is used as the source for NIS+ table population.

    ```
    mercury# cd /etc
    mercury# /usr/lib/nis/nispopulate -F
    ```

2.   The *nispopulate* script will ask for setup verification, as shown in the following.

```
NIS+ Domainname: astro.com.
Directory Path: (current directory)
Is this information correct? (Y or N) y
```

3.   The source files are listed and you are prompted once again to verify the creation of the destination NIS+ tables.

```
This script will populate the following NIS+ tables for domain your.do-
    main.name. from the files in current directory:
auto_master auto_home ethers group hosts networks passwd protocols services
    rpc netmasks bootparams netgroup aliases shadow
Do you want to continue? (Y or N) y
```

Setting Up the NIS+ Client

Use the *nisclient* script on the *root* master to create the credentials for the client. In this case, *mercury* is the master server for the domain, and *grissom* is the client being initialized.

```
mercury# /usr/lib/nis/nisclient -c grissom
```

1.   The user is asked to verify the client.

```
You will be adding DES credentials in domain astro.com. for mercury
**nisclient will not overwrite any existing entries ** in the credential
    table
Do you want to continue? (Y or N) y
checking astro.com. domain...
checking cred.org_dir.astro.com. permission...
adding DES credential for grissom...
Adding key pair for unix.grissom@astro.com.
Enter grissom.astro.com. root login
password:rootpasswd
Retype password:rootpasswd
```

2.   For all new NIS+ clients added, you should run the following statement on the client's machine.

```
nisclient -i -h root_master -a root_master_IP_address -d your.domain name.
```

3. Log on to the designated client as *root* and execute the *nis-client* script to initialize the client.

```
grissom# nisclient -i -h mercury -a 154.50.2.2 -d astro.com.
Initializing client grissom for domain astro.com....
Once initialization is done, you will need to reboot your machine.
Do you want to continue? (Y or N) y
setting up domain information astro.com....
setting up the name service switch information...
Please enter the network password that your administrator gave you.
Please enter the Secure-RPC password for root: rootpasswd
Please enter the login password for root: rootpasswd
Client initialization completed!!
Please reboot your machine for changes to take effect.
```

Setting Up an NIS+ Replica Server

An NIS+ client can be configured as a replica server. In this case, the NIS+ client *grissom* created in the previous sections will be converted to a replica server.

1. Log in as *root* and start the NIS+ service daemon, *rpc.nisd*.

    ```
    grissom# rpc.nisd
    ```

2. Run the *nisserver* script on the master server to initialize the replica.

```
mercury# /usr/lib./nis/nisserver -R -d astro.com. -h grissom
This script sets up an NIS+ replica server for domain your.domain.name.
Domainname : astro.com.
NIS+ Server : grissom
```

3. You will be prompted to verify the setup.

    ```
    Is this information correct? (Y or N) y
    ```

4.   This script will set up machine *grissom* as an NIS+ replica
       server for domain *astro.com*. In order for *grissom* to serve
       this domain, you will need to start up the NIS+ server dae-
       mon (*rpc.nisd*), with proper options on *grissom* if they are
       not already running.

5.   As a precaution, you are asked once more if you want to
       carry out this procedure.

```
Do you want to continue? (Y or N) y
Added grissom.astro.com. to group admin.astro.com.
```

Troubleshooting NIS+

Like NFS, the NIS+ service depends on a set of service daemons
to function. When a problem occurs, the first check to perform
is to see if the NIS daemons are running on the server. For an
NIS+ only setup, check for *rpc.nisd*. Also check that the domain
name is correctly set on the server and on the client using the
domainname command to display the domain name.

As with other network services, NIS+ is only as good as the network connection it runs over. Troubleshooting an NIS+ problem
should continue with a check of network connectivity using tools
such as *ping* to check that NIS clients and servers can reliably
communicate over the network. Also check that the *rpcbind* service is running on the NIS+ server. *Rpcbind* provides the critical
service lookup function that enables clients to locate the NIS+
server. Check for proper function of *rpcbind* using the *rpcinfo*
command to list the *rpc* services provided by a server. The NIS+
service should be in the list.

Authentication Problems

Most NIS+ problems are authentication problems. A typical user
authentication error is "Password does not decrypt secret key for
username." This is seen when the user logs in. This is due to the
user's login password and network password not being the same.

The corrective action is two steps. First, use *keylogin* to obtain the NIS+ credentials for this user, and then use *chkey* to synchronize the user's login and network passwords.

A more serious problem occurs when the NIS+ server suffers minor database corruption and fails to properly authenticate to clients. A common situation on a client is the error message "Unable to Authenticate NIS+ Server." Correcting this problem takes several steps, detailed in the following. The basic flow of the procedure is to first reduce the security level of the NIS+ server, and then correct the corruption by reentering the *root* user's keys and propagating them into the database files. Finally, the changes are propagated out to the replica servers and the NIS+ server is restarted in full security mode.

First, log in as *root*, stop the NIS+ server, and then restart it at security level o. The first command shown in the following uses *grep* to filter the output of *ps* to reveal the *process id* (PID) of the NIS+ server. The second command uses that number to stop the NIS+ server via the *kill* command. The final command restarts the NIS+ server in security level o.

```
ps -ef | grep rpc.nisd
kill PID
/usr/sbin/rpc.nisd -r -S 0
```

Next, remove and re-create the *root* user's key using the *keylogout* and *nisaddcred* commands.

```
keylogout —f
nisaddcred des
```

Next, stop the *keyserv* daemon, push the *root* user's credentials into the database files, and then restart the *keyserv* daemon. This is done as previously, using *grep* and *ps* to determine the *process id*

(PID) of the *keyserv* daemon, and using that number with the *kill* command to stop the daemon. The trio of *nisupkeys* commands pushes the *root* user's key into the data files. The last command simply restarts the *keyserv* daemon, completing this step.

```
ps -ef | grep keyserv
kill PID
nisupdkeys `nisdefaults -d`
nisupdkeys org_dir.`nisdefaults -d`
nisupdkeys groups_dir.`nisdefaults -d`
/usr/sbin/keyserv
```

If the NIS+ configuration includes replica servers, these changes need to be propagated to those servers. The *nisping* commands shown in the following cause the replica servers to note the change and update their copies of the NIS+ databases.

```
/usr/lib/nis/nisping `nisdefaults -d`
/usr/lib/nis/nisping org_dir.`nisdefaults -d`
/usr/lib/nis/nisping groups_dir.`nisdefaults -d`
```

Finally, the NIS+ server is stopped and restarted in secure mode. Here again, *ps* and *grep* are used to identify the *process id* of the NIS+ server, and the *kill* command is used to stop it. The last command restarts the NIS+ server.

```
ps -ef | grep rpc.nisd
kill <pid>
/usr/sbin/rpc.nisd -S 2
```

✓   TIP: *The primary method of avoiding NIS+ data corruption problems is to make sure the NIS+ server is adequately configured. The NIS+ server performs best on systems for which there is sufficient memory to eliminate swapping. How much space is needed is a function of the size of the NIS+ data files.*

Domain Name Service

The NIS+ name service allows the local administrator to maintain data about the machines in the local domain. The information in the NIS+ tables is only available to hosts within the NIS+ domain. In addition to IP address to hostname mappings, NIS+ stores information about several site-specific parameters, as described in the following.

- */etc/ethers*: Contains MAC addresses for use by the *tftp* protocol (e.g., diskless client booting).

- */etc/inet/hosts*: Local host name to IP address lookup table.

- */etc/inet/networks*: List of local networks and aliases.

- */etc/inet/netmasks*: List of local *netmask* values.

- */etc/inet/protocols*: List of protocols in use at the local site.

- */etc/inet/services*: List of services available at the local site.

- */etc/mail/aliases*: List of local e-mail aliases.

- */etc/netgroup*: List of hosts allowed access to file systems and services at the local site.

- */etc/resolv.conf*: List the *nameserver*, domain, search list, and other name resolution options.

Much of the information stored by NIS+ is useless to systems outside the local network; or worse yet, could be used by hackers to compromise the security of hosts on the local network. But what happens when a remote host (a host outside the NIS+ domain) wants to communicate with a host within the NIS+ domain? The remote host needs a way of resolving the host name to an IP address, but the NIS+ data are not available to it. Because NIS+ was never intended to service the entire Internet, other name services have been devised to provide this service.

➥   *NOTE: The following sections explore the workings of the Domain
      Name Service (DNS) and provide a limited overview of DNS.
      Because entire publications have been devoted to DNS, the reader is
      encouraged to obtain one of the reference books listed in the "References"
      section of this chapter for more information on DNS.*

DNS Terminology

The Domain Name Service (DNS) is the name service used on
the Internet. The DNS duplicates some of the information
stored in the NIS+ tables, but DNS information is available to all
hosts on the network. The DNS software uses many of the same
terms used in NIS+. In particular, the concepts of domains, clients, servers, and databases are found in both name services.
However, the operation of DNS is very different from NIS+.
Before delving into how DNS operates, understanding how DNS
defines these terms is recommended.

Domains

Under NIS+, a domain is defined as an NIS+ server and its collection of NIS+ clients. NIS+ allows for hierarchical domains to a
maximum depth of 128 levels. These hierarchical domains allow
the corporation to partition the network services in whatever
fashion the corporate data management group desires.

Like NIS+, DNS allows hierarchical domains. Unlike NIS, DNS
uses the domain name to describe a generic "ownership" of the
hosts.

Root Domain

Under DNS, the domain name space (see following illustration) is
very similar to the UNIX file system. One domain at the *root* of
the name space contains all other domains. Known as the *root*

domain, this domain is usually written as a single period (.) at the end of the domain name.

DNS domain name space.

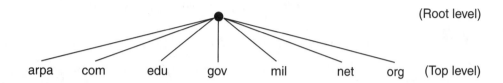

DNS domain name space.

Top-level Domains (U.S.)

DNS defines several "top-level" domains. These domains are global in nature. For example, in the United States, the most common top-level domains are *edu, com, net, org, gov,* and *mil.* The names of these top-level domains make an attempt to describe the type of organizations they contain. For example, the *edu* domain contains educational institutions; *com,* commercial (private sector business) organizations; and *mil,* military institutions.

Under each top-level DNS domain there may be any number of sub-domains. For example, under the *edu* domain are sub-domains for organizations such as The University of Notre Dame (*nd.edu*), Purdue University (*purdue.edu*), Stanford (*stanford.edu*), Cornell (*cornell.edu*), and most other universities. These second-level domain names are registered by the Network Information Center (NIC) such that they are unique throughout the Internet. See Chapter 17, "Network Configuration and Management," for information about registering a domain name.

Under each of these second-level domains there may be other sub-domains. For example, the Computer Science and Engineering Department at The University of Notre Dame is the *cse.nd.edu* domain, whereas the Engineering Computer Network at Purdue University is the *ecn.purdue.edu* domain. The assignment of third-level domain names is delegated (by the NIC) to the "owner" of the registered second-level domain. Consequently, the local administrator has some flexibility in assigning sub-domain names, and can simplify the operation of the DNS name resolution protocol, as shown in the following illustration.

Top-level Domains (Outside the U.S.)

Outside the United States, the top-level domains usually look a little different than they do within the United States. This difference is a historical artifact that requires a bit of explanation. When the Internet was first put into operation, it only served hosts and organizations within the United States. The top-level domains described in the previous section made sense in the context of the network's planned scope.

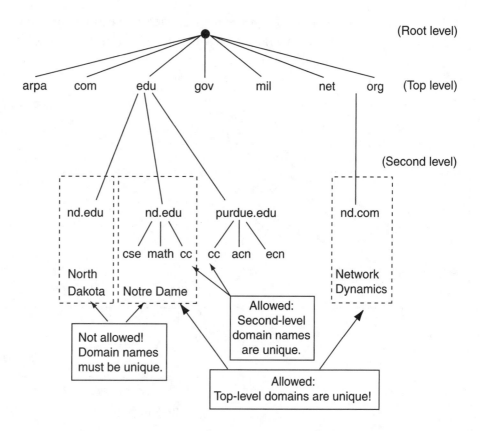

Extended DNS domain space.

Over time the Internet expanded to locations outside the United States. The International Standards Organization (ISO) developed a series of standard two-letter country codes (ISO standard 3166). These two-letter domains were inserted under the *root* domain, with special provisions ensuring that existing U.S. top-level domains could remain unchanged.

Most locations outside the United States subdivide their top-level domain in a manner similar to the U.S. top-level domains. For instance, *edu* and *ac* are educational or academic organizations, *com* and *co* are for commercial organizations, and so forth. In the United States, ISO standard 3166 defines second-level domains for each U.S. state and territory, permitting these entities the flexibility to name further sub-domains as they see fit (see the following illustration).

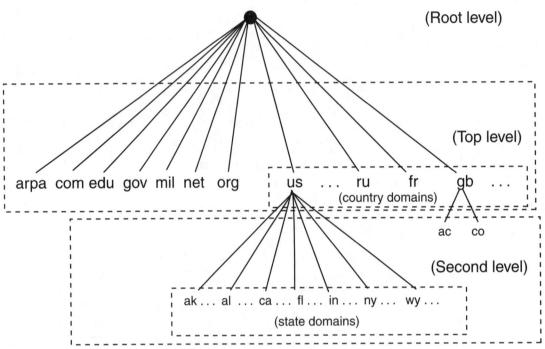

A politically correct rendition of the DNS name space.

Traversing the Domain Tree

Domain names read from left to right, where the leftmost component of the name is the furthest point from the *root* of the name space. For example, the name *hostname.store21.kmart.com* might describe a host located in store 21 in the *kmart* domain under the *com* top-level domain. Similarly, the names *grumpy. cse.nd.edu* and *mischief.ecn.purdue.edu* describe hosts within the *edu* top-level domain. The *grumpy* machine is located under the *cse* domain, which is under the Notre Dame (*nd*) domain, whereas the *mischief* machine is located under the *ecn* domain, which falls under the Purdue domain (see the following illustration).

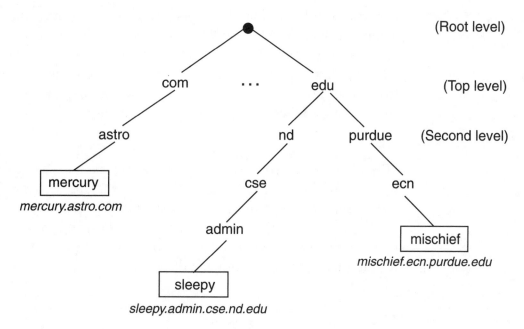

Reading DNS domain names.

To be precise, there should be a dot at the end of any domain/ host name (e.g., *grumpy.cse.nd.edu.*). The dot signifies that the domain is fully qualified (specified) all the way to the *root* domain. When the trailing dot is left off the name, the name is said to be partially qualified.

DNS Servers

Under DNS, name servers are programs that store information about a domain name space. A name server maintains database files that contain the master information for a domain. The client machines within the domain are configured in such a way that they will query the local name server when name-to-address mapping is required.

When a request is made for a name not in the local name server's database, a slightly different approach is taken. The host will query its local name server to get the name to address mapping. Then the local name server queries another name server in an attempt to resolve the query. This (often recursive) process will continue until the local name server gets a valid mapping (query is resolved), or until all appropriate name servers have been queried and have failed to return the requested information (host not found).

When the local name server makes a request of another name server and obtains a valid answer, it saves the answer for a designated period. The time period is specified in the answer itself. This is known as caching, which reduces the frequency of name service requests between name servers. An example of name resolution is shown in the following illustration.

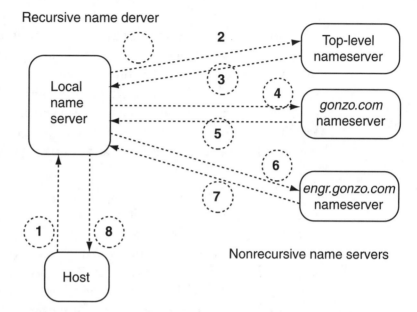

Recursive name derver

Name resolution example.

Nonrecursive name servers

Step	Action
1	Host requires address of remote host *dilbert.engr.gonzo.com*
2	Local nameserver polls nameserver in *.com* domain
3	*.com* nameserver refers local nameserver to gonzo.com
4	Local nameserver polls *gonzo.com*
5	*gonzo.com* refers local nameserver to *engr.gonzo.com*
6	Local nameserver polls *engr.gonzo.com*
7	*engr.gonzo.com* replies with address of *dilbert.engr.gonzo.com*
8	Local nameserver passes address of *dilbert.engr.gonzo.com* to host

In order to ensure that name lookups will succeed, each domain should have more than one name server configured to answer queries about hosts within that domain. In many cases, organizations may cooperate such that each corporation provides the primary name service for its own domain, but also agrees to provide backup name service for the other organization's domain.

Primary Name Servers

The primary name server obtains information about the domain from disk files contained on the name server host. The domain may actually have several primary name servers in an attempt to spread the workload around and provide better service. The DNS specification refers to primary name servers as "primary masters." Primary master name servers are defined as name servers that contain the master information about hosts within their domains.

Secondary Name Servers

Each domain can also define a number of secondary name servers. Secondary name servers obtain their databases from primary name servers through a process called "zone transfer." The DNS specification refers to these secondary name servers as "secondary masters." Secondary master name servers are name servers that provide backup name service for a domain. These secondary name servers are queried in the event the primary name server(s) do not respond to a query.

Caching Name Servers

Caching name servers have no direct access to any authoritative information about the domain. These name servers query primary and secondary name servers for all DNS requests, and store the results temporarily, as is done for requests outside the local domain. Caching name servers usually contain valid data, but because they do not load information directly from primary name servers, the data can become stale. Caching name servers are not considered authoritative name servers.

Root Name Servers

In order to provide a master list of name servers available on the Internet, the Network Information Center (NIC) maintains a group of *root* name servers. These name servers provide authoritative information about specific top-level domains. When a local name server cannot resolve an address, it queries the *root* name server for the appropriate domain for information that may allow the local name server to resolve the address.

DNS Clients and the /etc/resolv.conf File

DNS clients invoke a series of library calls that issue RPC requests to a DNS server. These library routines are referred to as "resolver code." Whenever a system requires a name to IP address mapping, the system software executes a *gethostbyname()* library call. The resolver code contacts the appropriate name service daemon to resolve the query.

In order to make DNS operate, the administrator must configure the */etc/resolv.conf* file on each host. This file tells the name service resolution routines where to find DNS information. The format of the entries in the */etc/resolv.conf* file is a keyword value. The following are valid keywords, with descriptions.

- *nameserver*: Valid values for name servers are IP addresses in standard dot notation. The administrator may specify a maximum of three name servers to query on separate lines.

- *domain*: The domain value will be appended to any host name that does not end in a dot. For example, if a user types *telnet mercury*, the resolver code will automatically append *.astro.com* to the request such that the telnet session will actually be issued as *telnet mercury.astro.com*.

- *options inet6*: Tells the resolver to look for IPv6 records.

- *search searchlist*: Tells the resolver to look for the *hostname* in the list of domains listed in *searchlist*.

A typical */etc/resolv.conf* file for a client in the *astro.com* domain might look like the following.

```
domainastro.com
nameserver192.168.1.1
```

On the name server itself, the *nameserver* keyword is followed by the reserved address *127.0.0.1*, also known as *localhost*. This causes the resolver code to direct queries to the local machine.

DNS Database Files

The DNS service consults a series of database files in order to resolve name service queries. Such database files are often referred to as *db files*. Unfortunately, there are two versions of the BIND software, and each version uses its own *db* files. In order to provide scalability and modularity, the database information is split into several files, described in the following sections.

Bind 4 db Files

The following are Bind 4 *db* files.

- */etc/named.boot*: Start-up file for the name server. This file contains information specific to the named program. Information in this file determines whether the name server is a primary name server or a secondary name server, where the *db* files are located, and other local setup information.

- *db.domainname*: This database provides the host name to IP address mapping for the hosts within a domain name. There may be multiple *db.domainname* files on a server (one file for each domain for which this server provides name service).

- *db.address*: This database provides the IP address to host name mapping (reverse mapping) for the hosts on a network address. There may be multiple *db.address* files on the name server (one file for each network within the domain).

Bind 8 db Files

The following are bind 8 files.

- */etc/named.conf*: Start-up file for the name server. This file contains information specific to the named program. Information in this file determines whether the name server is a primary name server or a secondary name server, where the *db* files are located, and other local setup information.

- *db.domainname*: This database provides the host name to IP address mapping for the hosts within a domain name. There may be multiple *db.domainname* files on a server (one file for each domain for which this server provides name service).

- *db.address*: This database provides the IP address to host name mapping (reverse mapping) for the hosts on a network address. There may be multiple *db.address* files on the name server (one file for each network within the domain).

Name Service Resource Records

The information stored in the database files is organized into resource records, and the format of these records is determined by the DNS design specifications. Subsequent sections explain the types of records, the format for each record, and the information provided by the record.

Comment Records

As with any programming language, DNS provides a method for embedding comments in *db* files. You should place comments in these files in an attempt to explain the local site setup. A comment record begins with a semicolon (;). Everything that follows a semicolon (on a line of the *db* file) is considered a comment. The following is an example.

```
; The following lines define local host addresses with A records.
```

SOA Records

Start of authority (SOA) records provide information about the management of data within the domain. The SOA record must start in column 1. The format of an SOA record is shown in the following, followed by an explanation of each field.

```
domain class SOA primary_server rp (
        serial_number          ; comment about serial number
        refresh_value          ; comment about refresh value
        retry_value            ; comment about retry value
        expiration_value       ; comment about expiration value
        TTL )                  ; comment about time to live value
```

- The *domain* field is the domain name (e.g., *astro.com.*).

- The *class* field allows the administrator to define the class of data. Currently, only one class is used. The *IN* class defines Internet data.

- The *SOA* field tells the resolver that this is a start of authority record for *domain*.

- The *primary server* field is the fully qualified host name of the primary name server for this domain. For example, *mercury.astro.com.* would be the name server for the *astro.com.* domain.

- The *rp* field gives the fully qualified e-mail address of the person responsible for this domain. Note that the usual "at" symbol (@) is replaced by a dot (.) in the *rp* record. For example, *curt.mercury.astro.com.* might be the responsible person for the *astro.com.* domain.

- The *serial number* field is used by the secondary name servers. This field is a "counter," which gives the version number for the file. This number should be incremented every time the file is altered. When the information stored on the secondary name servers expires, the servers contact the primary name server to obtain new information. If the serial number of the information obtained from the primary server is larger than the serial number of the current information, a full zone transfer is performed. If the serial number from the primary name server is smaller than the current information, the secondary name server discards the information from the primary name server.

- The *refresh_value* field is used by the secondary name servers. The value in this field tells the secondary servers how long (in seconds) to keep the data before they obtain a new copy from the primary name server. If the value is too small, the name server may overload the network with zone transfers. If the value is too large, information may become stagnant and changes may not propagate efficiently from the primary to the secondary name servers.

- The *retry_value* field is used by the secondary name servers. The value in this field tells the secondary servers how long (in seconds) to wait before attempting to contact a nonresponsive primary name server.

- The *expiration_value* field is used by the secondary name servers. The value in this field tells the secondary servers to expire their information after *value* seconds. Once the secondary server expires its data, it stops responding to name service requests.

- The *TTL* field tells name servers how long they can "cache" the response from the name server before they purge the information obtained in response to a query.

The following is an example of an SOA record.

```
astro.com IN SOA mercury.astro.com curt.mercury.astro.com. (
    2000033001  ; Serial - format YYYYMMDD##
    10800       ; Refresh every 3 hours
    3600        ; Retry after an hour
```

```
60480           ; Expire data after 1 week
86400 )         ; TTL for cached data is 1 day
```

NS Records

NS records define the name servers within a domain. The NS record starts in column 1 of the file. The format of an NS record follows: *domain class NS fully_qualified_hostname*. The following is an example of an NS record.

```
astro.com       IN      NS      mercury.astro.com.
```

A Records

A (Address) records provide the information used to map host names to addresses. The A record must begin in column 1 of the file. The format of an A record is *fully_qualified_hostname class A address*. The following are examples of A records.

```
localhost.astro.com. IN      A       127.0.0.1
glenn.astro.com.     IN      A       192.168.5.1
```

AAAA Records

AAAA (IPv6 Address) records provide the information used to map host names to IP version 6 addresses. The AAAA record must begin in column 1 of the file. The format of an AAAA record is *fully_qualified_hostname class AAAA address*. The following is an example of an AAAA record.

```
glenn IN AAAA     3ffe:0c00:8002:0001:4800:5aff:fefc:2b36
```

PTR Records

PTR (pointer) records provide the information used for reverse mapping (looking up a host name from an IP address). The PTR record must begin in column 1 of the file. The format of the PTR record is *address class PTR fully_qualified_hostname*.

The PTR record does have one quirk: the address portion appears to be written backward, and contains information the user typically does not see. The information is presented in this format in order to simplify the lookup procedure, and to maintain the premise that the information at the left of the record is furthest from the *root* of the domain. An example of a PTR record follows.

```
1.5.168.192.in-addr.arpa. IN PTR glenn.astro.com.
```

The *Ipv6* protocol utilizes a new domain, *IP6.INT*, for reverse mappings, as shown in the following example.

```
6.3.2.b.c.f.e.f.f.f.a.5.0.0.8.4.1.0.0.0.2.0.0.8.0.0.c.0.e.f.f.3.IP6.INT
    IN PTR glenn.astro.com.
```

MX Records

MX (Mail eXchanger) records provide a way for remote hosts to determine where e-mail should be delivered for a domain. The MX records must begin in column 1 of the *db* files. The format of MX records is *fully_qualified_hostname class MX preference fully_qualified_mail_hostname*.

There should be an MX record for each host within a domain. The *preference* field tells the remote host the preferred place to deliver the mail for the domain. This allows e-mail to be delivered in the event the primary mail server is down. The remote host will try to deliver e-mail to the host with the lowest preference value first. If that fails, the remote host will attempt to deliver the e-mail to the host with the next lowest preference value. The following are examples of MX records.

```
glenn.astro.com. IN MX 1 mail.astro.com.
glenn.astro.com. IN MX 10 www.astro.com.
```

CNAME Records

CNAME (Canonical NAME) records provide a way to alias host names. For example, the host name *saturn-apollo-13.astro.com* might be a little difficult for some people to type. Users may have an easier time typing *apollo13.astro.com*. To provide this alias mapping, the administrator would have to enter a CNAME record in the appropriate DNS *db* files. The following is an example of a CNAME record.

```
apollo13.astro.com. IN CNAME saturn-apollo-13.astro.com.
```

TXT Records

TXT records allow the administrator to add text information to the *db* files, giving more information about the domain. The TXT records must start in column 1 of the *db* file, and the format of TXT records is *hostname class TXT "Random text about hostname in this domain"*. The following is an example of a TXT record.

```
mercury IN TXT "Mercury is in room G-101 of Bldg 3"
```

RP Records

RP (responsible person) records define the person responsible for the domain. The RP records must start in column 1 of the *db* files. The format of an RP record *is hostname class RP fully_qualified_email_address fully_qualified_hostname*. The following is an example of an RP record.

```
truax IN RP  curt.mercury.astro.com  vonbraun.engr.astro.com.
vonbraun IN TXT "Engineering Division of Astro.com (900) 555-1212"
```

Setting Up a DNS Server

Solaris 2 includes the DNS software as a standard portion of the operating system. This limits the process of setting up a primary name server to building the database files, then starting the */usr/sbin/in.named* daemon. Prior to building the database files, the administrator should understand the various pieces of software that perform the DNS function.

✗   *WARNING: Old versions of the DNS software have known security vulnerabilities that can be exploited to gain* root *access. Be sure to obtain and apply the current set of patches to the DNS service* before *it is put into service on the Internet. See Chapter 7, "Managing System Security," for details on obtaining and applying patches.*

BIND

The BIND software is the Berkeley Internet Name Daemon. BIND is the current implementation of the DNS software, and is available for most UNIX ports currently on the market. BIND consists of the following components.

- *in.named*: Process that answers name-to-address resolution queries.

- *nslookup*, *dig*, and *host*: These utilities allow users to query the DNS database(s) to determine what information is available.

- DNS resolver libraries: Routines that perform the name-to-address query resolution for *in.named*.

Recent versions of Solaris ship with a variant of Bind 8, whereas earlier versions shipped with Bind 4 variants. The examples that follow show Bind 4 implementations, with Bind 8 information presented as applicable. These examples refer to *IPv4* addresses in private address space. Because these files use private address space, they would not be suitable for name service on the Internet, but could be used on an internal name server. Consult Chapter 17, "Network Configuration and Management," for more information on private addressing.

Example DNS Configurations

A simple example of using DNS is to set up a caching-only name server. This configuration might be appropriate for a small department or group that wants to have a local name server on a network segment. The administrator will need to create the following configuration files.

- *   */etc/named.boot (/etc/named.conf* for Bind 8)

- *   *db.cache (root.hint* for Bind 8)

- *   /etc/resolv.conf

Bind 4 /etc/named.boot File

For this configuration, the */etc/named.boot* file simply contains the name of the directory used to store the named configuration files and the name of the data file containing information about the top-level DNS servers, as shown in the following.

```
directory/usr/local/named
cache.db.cache
```

Bind 8 /etc/named.conf File

The Bind 8 *named.conf* file is considerably more verbose than its Bind 4 counterpart. The Bind 8 file contains global information used by the bind binary, as well as information about the zone files required for this server. The following example is a minimal *named.conf filr* for use with Bind 8.

```
options {
    directory ".";
    statistics-file "named.stats";
    named-xfer "/usr/sbin/named-xfer";// _PATH_XFER
    dump-file "named_dump.db";        // _PATH_DUMPFILE
    pid-file "named.pid";             // _PATH_PIDFILE
    statistics-file "named.stats";    // _PATH_STATS
    notify yes;
    check-names master warn;          // default. //
    check-names slave warn;
    auth-nxdomain yes;
    interface-interval 60;            // scan for new or deleted interfaces
                                      // every 'interface-interval' minutes

    statistics-interval 60;           // log statistics every
                                      // 'statistics-interval' minutes
};
zone "astro.com" {
    type master;
    file "astro.com";
    check-names warn;
    allow-update { none; };
    allow-transfer { any; };
    allow-query { any; };
    max-transfer-time-in 120;
};
 zone "168.192.in-addr.arpa" IN {
    type master;
    file "d.168.192.in-addr.arpa";
    check-names fail;
    allow-update { none; };
    allow-transfer { any; };
};
zone "." {
    type hint;
    file "root.hint";
};
```

```
zone "localhost" IN {
    type master;
    file "localhost.zone";
    check-names fail;
    allow-update { none; };
    allow-transfer { any; };
};
zone "0.0.127.in-addr.arpa" IN {
    type master;
    file "127.0.0.zone";
    check-names fail;
    allow-update { none; };
    allow-transfer { any; };
};
```

db.cache *File* (root.hint *file for Bind 8)*

The *db.cache* file (Bind 4) contains hints the name server can use to contact top-level name servers. This file should be the same for every name server on the Internet. A master copy of the *db.cache* file is available via anonymous *ftp*, as *named.root*, from *ftp.rs.internic.net* (198.41.0.5). Bind 8 refers to this file as *"root.hint"*. The *root.hint* file contains the same information as the Bind 4 *db.cache* file. The *db.cache* file for the example would contain the following directives.

```
;
; Data file for initial cache data for root domain servers.
;
.6D IN NSG.ROOT-SERVERS.NET.
.6D IN NSJ.ROOT-SERVERS.NET.
.6D IN NSK.ROOT-SERVERS.NET.
.6D IN NSL.ROOT-SERVERS.NET.
.6D IN NSM.ROOT-SERVERS.NET.
.6D IN NSA.ROOT-SERVERS.NET.
.6D IN NSH.ROOT-SERVERS.NET.
.6D IN NSB.ROOT-SERVERS.NET.
```

```
.6D IN NSC.ROOT-SERVERS.NET.
.6D IN NSD.ROOT-SERVERS.NET.
.6D IN NSE.ROOT-SERVERS.NET.
.6D IN NSI.ROOT-SERVERS.NET.
.6D IN NSF.ROOT-SERVERS.NET.
;
G.ROOT-SERVERS.NET.5w6d16h IN A192.112.36.4
J.ROOT-SERVERS.NET.5w6d16h IN A198.41.0.10
K.ROOT-SERVERS.NET.5w6d16h IN A193.0.14.129
L.ROOT-SERVERS.NET.5w6d16h IN A198.32.64.12
M.ROOT-SERVERS.NET.5w6d16h IN A202.12.27.33
A.ROOT-SERVERS.NET.5w6d16h IN A198.41.0.4
H.ROOT-SERVERS.NET.5w6d16h IN A128.63.2.53
B.ROOT-SERVERS.NET.5w6d16h IN A128.9.0.107
C.ROOT-SERVERS.NET.5w6d16h IN A192.33.4.12
D.ROOT-SERVERS.NET.5w6d16h IN A128.8.10.90
E.ROOT-SERVERS.NET.5w6d16h IN A192.203.230.10
I.ROOT-SERVERS.NET.5w6d16h IN A192.36.148.17
F.ROOT-SERVERS.NET.5w6d16h IN A192.5.5.241
```

/etc/resolv.conf *File*

The */etc/resolv.conf* file tells the DNS resolver code where to find the DNS name server, and the name of the current domain. The following sample */etc/resolv.conf* file contains two name servers and sets the domain to *astro.com*. A caching name server might be one of the listed name servers.

```
domain astro.com
nameserver 192.168.4.1
nameserver 192.168.4.4
```

With these files in place, the name server is started, as shown at the end of the example in the following section.

◆ *NOTE: On the system running* in.named, *the name server line in* /etc/ resolv.conf *can simply point to the special "localhost" address,* 127.0.0.1.

A More Complex Example

In order to set up a primary name server for a domain, the administrator must create the *db* files for that domain. The following illustration shows a sample domain, which will be examined in subsequent sections. The sample domain will not employ sub-domains initially, but as the company grows, logical sub-domain boundaries might be *engr.astro.com* (engineering), *mkt.astro.com* (marketing), *corp.astro.com* (corporate services), and *adm.astro.com* (administration).

The astro.com *DNS domain configuration.*

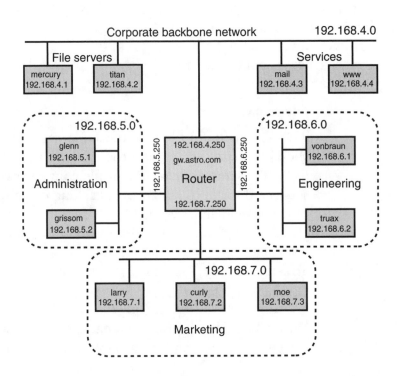

For the example domain, the administrator must create the following files.

- */etc/named.boot*

- *db.astro (db.domain)*

- *db.192.168.4*

- *db.192.168.5*

- *db.192.168.6*

- *db.192.168.7*

- *db.127.0.0*

- *db.cache*

- */etc/nsswitch.conf*

- */etc/resolv.conf*

- ➡ *NOTE: The DNS administrator may decide to build one db.192.168 file that contains all of the subnetwork information, instead of separate files for each subnet.*

/etc/named.boot *File*

As mentioned previously, the *named.boot* file customizes the DNS information for a particular site. To simplify matters, the *directory* instruction is used to inform the named software to look in */usr/local/named* for the *db* files for this server. The *primary* instruction tells the named software where to find individual *db* files required to resolve DNS queries. The *cache* instruction tells the named software where to find the file that contains information about the *root* name servers. For the example network, the *named.boot* file would contain the following directives.

```
directory /usr/local/named
primary astro.com db.astro
primary 4.168.192.in-addr.arpa db.192.168.4
primary 5.168.192.in-addr.arpa db.192.168.5
primary 6.168.192.in-addr.arpa db.192.168.6
primary 7.168.192.in-addr.arpa db.192.168.7
primary 0.0.127.in-addr.arpa db.127.0.0
cache . db.cache
```

db.domain *File*

The *db.astro* file contains information mapping host names within the *astro.com* domain to IP addresses for the hosts in the *astro.com* domain. It also contains an SOA record, as well as NS, MX, CNAME, PTR, and other records about these hosts. The *db.astro* file for the example would contain the following directives.

```
astro.com.   IN   SOA  mercury.astro.com.  curt.mercury.astro.com. (
                 1           ; Serial Number
                 10800       ; Refresh every 3 hours
                 3600        ; Retry after 1 hour
                 604800      ; Expire after 1 week
                 86400 )     ; TTL is 1 day
;
; Astro.com name servers
;
astro.com.        IN       NS        mercury.astro.com.
astro.com.        IN       NS        www.astro.com.
;
; astro.com A records
;
localhost.astro.com.      IN       A        127.0.0.1
mercury.astro.com.        IN       A        192.168.4.1
titan.astro.com.          IN       A        192.168.4.2
mail.astro.com.           IN       A        192.168.4.3
www.astro.com.            IN       A        192.168.4.4
gw.astro.com.             IN       A        192.168.4.250
glenn.astro.com.          IN       A        192.168.5.1
```

```
grissom.astro.com.          IN      A       192.168.5.2
gw.astro.com.               IN      A       192.168.5.250
vonbraun.astro.com.         IN      A       192.168.6.1
truax.astro.com.            IN      A       192.168.6.2
gw.astro.com.               IN      A       192.168.6.250
larry.astro.com.            IN      A       192.168.7.1
curly.astro.com.            IN      A       192.168.7.2
moe.astro.com.        IN      A     192.168.7.3
gw.astro.com.         IN      A     192.168.7.250
;
; Aliases for astro.com hosts
;
gw-4.mercury.com.   IN   CNAME       gw.astro.com.
gw-5.mercury.com.   IN   CNAME       gw.astro.com.
gw-6.mercury.com.   IN   CNAME       gw.astro.com.
gw-7.mercury.com.   IN   CNAME       gw.astro.com.
;
; MX records for astro.com
;
mercury.astro.com.  IN   MX    1       mail.astro.com.
titan.astro.com.    IN   MX    1       mail.astro.com.
mail.astro.com.     IN   MX    1       mail.astro.com.
www.astro.com.      IN   MX    1       mail.astro.com.
gw.astro.com.       IN   MX    1       mail.astro.com.
glenn.astro.com.    IN   MX    1       mail.astro.com.
grissom.astro.com.  IN   MX    1       mail.astro.com.
vonbraun.astro.com. IN   MX    1       mail.astro.com.
```

db.192.168.4 *File*

The *db.192.168.4* file contains information mapping IP addresses on the 192.168.4.0 network to host names. It also contains an SOA record, as well as NS, PTR, and other records about these hosts. The *db.192.168.4* file for the example would contain the directives that follow. The *db.192.168.{5,6,7}* files are not shown here, but will be very similar to the *db.192.168.4* file.

```
4.168.192.in-addr.arpa.  IN  SOA mercury.astro.com. curt.mercury.astro.com.
   (
                    1           ; Serial Number
                    10800       ; Refresh every 3 hours
                    3600        ; Retry after 1 hour
                    604800      ; Expire after 1 week
                    86400 )     ; TTL is 1 day
;
; Astro.com name servers
;
4.168.192.in-addr.arpa.      IN      NS      mercury.astro.com.
4.168.192.in-addr.arpa.      IN      NS      www.astro.com.
;
; astro.com hosts on 192.168.4 net
;
1.4.168.192.in-addr.arpa.    IN      PTR         mercury.astro.com.
2.4.168.192.in-addr.arpa.    IN      PTR         titan.astro.com.
3.4.168.192.in-addr.arpa.    IN      PTR         mail.astro.com.
4.4.168.192.in-addr.arpa.    IN      PTR         www.astro.com.
250.4.168.192.in-addr.arpa.  IN      PTR         gw.astro.com
.
```

db.127.0.0 *File*

The *db.127.0.0* file contains information mapping IP addresses on the 127.0.0 network to host names. It also contains an SOA record, and NS, PTR, and other records about these hosts. The *db.127.0.0* file for the example would contain the following directives.

```
0.0.127.in-addr.arpa. IN SOA  mercury.astro.com. curt.mercury.astro.com.
   (
                    1           ; Serial Number
                    10800       ; Refresh every 3 hours
                    3600        ; Retry after 1 hour
                    604800      ; Expire after 1 week
                    86400 )     ; TTL is 1 day
0.0.127.in-addr.arpa.        IN      NS          mercury.astro.com.
```

```
0.0.127.in-addr.arpa.      IN    NS      www.astro.com.
1.0.0.127.in-addr.arpa.    IN    PTR     localhost.
```

db.cache *File*

The *db.cache* file contains hints the name server can use to contact top-level name servers. This file is the same as in the first example, and should be the same for every name server on the Internet.

/etc/resolv.conf *File*

The */etc/resolv.conf* file tells the DNS resolver code where to find the DNS name server, and the name of the current domain. The following sample */etc/resolv.conf* file contains two name servers and sets the domain to *astro.com*.

```
domain astro.com
nameserver 192.168.4.1
nameserver 192.168.4.4
```

Starting the DNS Daemon

Once all of the configuration files are set up and checked, the administrator should start the DNS daemon on the name server machine, as follows.

```
# /usr/etc/in.named
```

Troubleshooting the DNS Daemon

If the configuration files have been set up correctly, the name service will start without a hitch. The operation of the name service can be tested using the *nslookup* command to query the DNS name server for information on a local host. For example, the command # *nslookup glenn* should return the following information about the *glenn* machine in the *astro.com* domain.

```
Server: mercury.astro.com
Address: 192.168.4.1
Name: glenn.astro.com
Address: 192.168.5.1
```

If *nslookup* does not return the proper information for a host in the database, the name service daemon may not have started correctly. Consult the */var/adm/messages* file to determine if the named program wrote any error messages to the log file.

Once the *nslookup* command returns correct information for local hosts, the administrator should create start-up files in the appropriate */etc/rc.d* directories such that the */usr/etc/in.named* process starts when the name server is booted. Consult Chapter 5, "Refining System Boot and Shutdown Procedures," for more information on creating start-up scripts.

Using NIS+ with DNS

To provide a robust name service environment, it is sometimes necessary to run more than one name service. For example, an Internet-connected corporation will need to run DNS for Internet name service, but the local administrator may want to maintain local network information in NIS+ maps.

Still other corporations may have a need to use other name services such as the Federated Name Service (FNS). The following section covers the basics of configuring hosts to use multiple name services. The discussion is limited to NIS+ and DNS name services, but the concepts presented may be extended to other name services.

Configuring Multiple Name Services

Many sites will opt to use the NIS+ name service for local network information, and DNS for external name service. This creates a minor logistics problem for the name resolution libraries. For instance, which name service should be contacted first? In order to facilitate the use of multiple name services, the administrator should configure the system by editing the */etc/nsswitch.conf* file.

/etc/nsswitch.conf *File*

The */etc/nsswitch.conf* file contains directives that instruct the name service resolution routines on where to look for name resolution information. A few valid options for where the resolver should look are files (search */etc/inet/hosts*, */etc/passwd*, */etc/group*, and so on), DNS (consult a DNS server), LDAP (consult an LDAP server or LDAP -> DAP gateway) and NIS+ (consult an NIS+ server).

The administrator can configure the name service to search a single source, or to search all sources in a specific order. In addition, you can instruct the system to search one source for host name to IP address mappings, another source for user password information, and yet another source for other information (e.g., Ethernet MAC addresses for network boot requests).

To provide the most reliable name service for the corporate network, many sites configure the machines to consult NIS+ first. If a resolution is not found from NIS+ information, contact DNS. If there is still no resolution to the query, consult the local files in */etc*.

An Example /etc/nsswitch.conf File

The following sample */etc/nsswitch.conf* file will implement an NIS+, DNS, and then local files search in order to resolve name service queries. Note that for information other than host name to IP address, or IP address to hostname lookups, only NIS+ and local files are consulted. If an LDAP server is available, the keyword *ldap* could be put in place of the *nisplus* or *dns* keywords, or added to the list of lookup services.

> ✓   TIP: *Although many combinations are possible, it is often desirable to have a files entry for each service. This provides a fallback service for the system to use in the case that the other name lookup services are unavailable.*

The *[NOTFOUND=return]* directive tells the name service not to return an error if the query cannot be resolved by the name service. If the request cannot be resolved once all of the name service information has been searched, and if the information is not available in local files, an error will be returned for this query.

```
# /etc/nsswitch.conf
# uses NIS+ (NIS Version 3) in conjunction with files.
# "hosts:" and "services:" in this file are used only if the
# /etc/netconfig file has a "-" for nametoaddr_libs of "inet" transports.
# the following two lines obviate the "+" entry in /etc/passwd and
# /etc/ group.
passwd: files nisplus
group: files nisplus
# consult /etc "files" only if nisplus is down.
# Use DNS and NIS+. Must also set up /etc/resolv.conf file for DNS
# nameserver lookup.
hosts: nisplus dns [NOTFOUND=return] files
services: nisplus [NOTFOUND=return] files
networks: nisplus [NOTFOUND=return] files
protocols: nisplus [NOTFOUND=return] files
```

```
rpc: nisplus [NOTFOUND=return] files
ethers: nisplus [NOTFOUND=return] files
netmasks: nisplus [NOTFOUND=return] files
bootparams: nisplus [NOTFOUND=return] files
publickey: nisplus
netgroup: nisplus
automount: files nisplus
aliases: files nisplus
sendmailvars: files nisplus
```

References

The following references are listed in an attempt to provide the interested reader with more information on the topics covered in this chapter. These sources are not exhaustive, but should provide the reader with a knowledge base on the topics of computer name services.

- Paul Albitz, Cricket Liu, and Mike Loukides, *DNS and BIND*. O'Reilly & Associates, 1998 (ISBN 1-56592-5122).

- Hal Stern, *Managing NFS and NIS*. O'Reilly & Associates, 1991 (ISBN 0-937175-75-7).

- Auburn University's *NIS+ Frequently Asked Questions* web page, *http://www.eng.auburn.edu/users/rayh/solaris/NIS+_FAQ.html*, maintained by Ray W. Hiltbrand.

- W. Richard Stevens, *Introduction to IPv6*, available at *http://www.kohala.com/start/classes/classes.html*.

- Timothy Howe, Mark Smith, and Gordon Good, *Understanding and Deploying LDAP Directory Services*. Macmillan Technical Publishing, 1999 (ISBN 1-57870-070-1).

Summary

This chapter discussed the basic concepts of network name services. Two name services were examined in detail: NIS+, a local name service, and DNS, an Internet-wide name service. Examples outlined the steps required to configure an NIS+ environment in a local area network. In addition, basic step-by-step procedures for creating and initializing an NIS+ master server, NIS+ client, and NIS+ replica server, as well as the steps required to populate NIS+ tables, were presented. Some common NIS+ problems and solutions were shown.

Similarly, examples illustrated DNS database information setup, the configuration of a Bind 4 name server, and the configuration of a Bind 8 DNS name server in both a simple and complex configuration. Finally, the chapter examined methods of making LDAP, NIS+, and DNS name services cooperate in order to provide a robust, reliable name service for corporate network users.

CHAPTER 22
WWW Administration

Perhaps no other area of computing holds more interest and promise for the future than the World Wide Web (WWW). Solaris provides an excellent platform for providing WWW services. However, the Web is complex and makes unique demands on the server providing services, as well as on system administrators. The complexity of web service also adds new challenges in the area of securing the server system.

What Is the World Wide Web?

The World Wide Web is a client-server based application originally developed by Tim Berners-Lee at CERN to distribute documentation. Researchers at various locations, notably the National Center for Supercomputer Applications at the University of Illinois, extended the original design to include the distribution of a wide variety of media, including graphics, audio, video, and small applications (or applets).

707

WWW clients, known as browsers, are used to make requests from WWW servers and display the results in the form of a *page*. These pages may contain the addresses of other pages, in the form of links that can be selected. The World Wide Web name derives from the interconnected nature of these page links.

Pages and other resources are referenced using a universal resource locator (URL). The format of a URL is a resource type tag, followed by the name of the system holding the resource, followed by the path to the resource, which may include option flags and other data. For example, the URL for the Dilbert comic strip follows. This URL refers to a web page as noted by *http* (hypertext transport protocol). The referenced page is stored on a system named *www.unitedmedia.com* in a directory named */comics/ dilbert/*.

http://www.unitedmedia.com/comics/dilbert/

In typical usage, a browser connects to a server to obtain pages as requested by the user. A single page may contain information obtained from more than one server. The communications protocol or language used to communicate between WWW browsers and servers, as well as the language that describes how a page is rendered, is standardized, allowing browsers functioning on different operating systems to access servers operating on different operating systems. The ability to publish information in a form viewable independent of operating systems used to serve and view the information is one key to WWW's success.

Another key to the success of the WWW are forms and secure transactions. Web pages can contain fill-in-the-blank forms. The responses entered into these forms can be transmitted back to the server for processing. In addition, pages and the responses from a form can be encrypted and securely transmitted.

Web servers have the ability to run external programs through a feature known as the Common Gateway Interface (CGI). This feature allows for pages to be constructed dynamically and for processing of information received via forms. The combination of these features makes a wide variety of web-based applications possible, such as on-line product ordering. An example of a web page is shown in the following illustration.

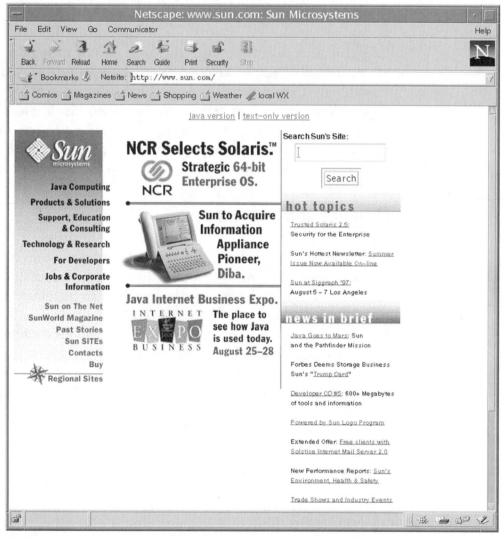

Typical WWW page with graphics, text, and underlined links to other pages.

WWW Clients

Administrating WWW clients is primarily a matter of keeping up to date with browser and page content development. At present, leading browsers such as Netscape Communications and Microsoft products are undergoing rapid development. New versions of a browser every six months are common.

Likewise, new page content in the form of new media data types are continually being developed. Not all media types are directly viewable by a given browser. Additional software may be needed to view certain content types, such as video. Such additions to the browser come in two flavors: (1) extensions to the browser program itself, often called *plug-ins*, and (2) separate applications started under the browser's control, known as *helper applications*.

Plug-ins and Helpers

Solaris version 8 includes Netscape Communicator. Netscape Communicator is a suite containing browser, e-mail, calendar, and discussion tools. Additional media formats can be added to Navigator through the *mime.types* and *mailcap* files. Some are added as part of the installation of a plug-in or helper application, whereas others are added separately. For example, the *xanim* video player is a helper application that requires a line to be added to either the user's *.mailcap* file or the system-wide files (see the following illustration).

Adding the xanim *player definition to the user's* .mailcap *file and starting Netscape.*

Client Security Issues

Web browsers present several security problems. Problems with downloaded applets and scripts are currently receiving the most attention. Java is a language designed to run within a portable virtual machine. Java applets are downloaded mini-programs that are run inside the browser's Java virtual machine. Although the designers of the Java language built in numerous security protections, some implementations of the Java virtual machine (and the local libraries used by running Java applets) are not secure.

Fortunately, most browsers allow the user to optionally turn off the execution of Java applets. Likewise, the less capable, but more frequently found, JavaScript can also be turned off. Turning off these features will disable certain interactive features of web pages that utilize JavaScript or Java applets while protecting the user from problems caused by bugs in the Java virtual machine or libraries. The desirability of turning off JavaScript and Java applets depends on the specific pages being viewed and the applets they contain. These features may be required for some applications.

The controls for deactivating these features can be found in Navigator's Preferences dialog box, obtained by selecting the Preferences item from the Edit menu. The Java language controls are found by clicking on Advanced in the list at the left of the dialog box. An example of Netscape Navigator security preferences is shown in the following illustration.

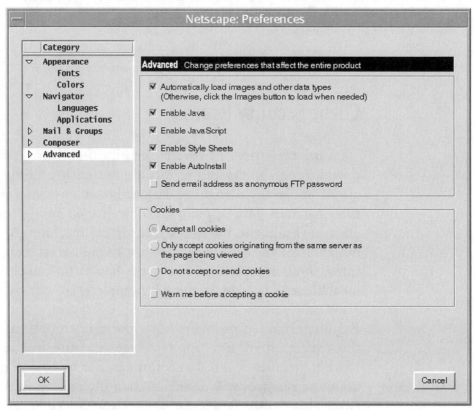

Netscape Navigator's security preferences for languages and "cookies."

Bugs in the browser itself constitute a more common problem. Browsers are complex, often including their own Java virtual machine, as well as internal versions of *ftp* and other network tools. System managers at sites concerned about security should continually monitor the browser vendor web pages for updates that address security problems.

✗   *WARNING: There are numerous security vulnerabilities associated*
       with downloaded applets and scripts on Microsoft Windows platforms
       that affect the security of other systems on a network. These include the
       unintended installation of malicious software that may examine or dis-
       rupt network traffic or adversely affect the operation of servers and oth-
       er networked systems. Security-conscious sites need to consider not only
       the security of their servers but the risks involved in their choice of client
       platforms and software.

Referring page information that some browsers pass along to a
web server as they switch from page to page while following links
is a less discussed problem. This information can be used by a
web server to identify the page a user comes from when accessing
a new page. The problem derives from the additional information
on the referring page that is also passed along. Such additional
data may include information believed to be secure if the browser
moves from a secure page to an unsecured page.

Many web sites avoid this problem by "wiping the browser's feet"
via directing the browser to a blank or unrevealing page after
requesting secure information. By default, the Netscape browser
will alert users to this problem by posting an alert message when
the user moves from a secure page to an unsecured page.

➡   *NOTE: Modern browsers are also capable of storing small pieces of infor-*
       mation from web sites about the user, such as a password or usage his-
       tory. These bits of information are known as "cookies." The same
       security preferences dialog box mentioned earlier also allows those con-
       cerned about cookies to disable them or have the browser announce the
       delivery of a cookie from the web site. Turning off cookies will disable
       password memory and history features of some web sites. The decision
       to turn off cookies depends on the user's concerns about her privacy and
       the web pages she views most often.

WWW Servers

Installing and configuring a web server is a much more involved process. A web server is a very complex daemon with numerous features controlled by a couple of configuration files. Web servers not only access files containing web pages, graphics, and other media types for distribution to clients, but assemble pages from more than one file, run CGI applications, and negotiate secure communications. Basic server configuration issues are discussed in the following sections.

➥   *NOTE: Solaris 8 ships with the Apache web server. Apache is available for older versions of Solaris from the web site listed at the end of this chapter.*

AnswerBook2 Is a Web Server

If you have *AnswerBook2* installed on your system, you should be aware that you have already installed a web server. *AnswerBook2* is web based and starts a web server (*dwhttpd*) by default at boot time. By default, the *AnswerBook2* web server listens on port 8888.

Because *AnswerBook2* is a web server, it does not need to be installed on every system; a central server can be used. In addition, it represents another avenue of access to a system and should not be enabled unless needed. The administrator can stop and start the *AnswerBook2* web server with the following commands.

```
/usr/lib/ab2/lib/ab2admin -o stop
/usr/lib/ab2/lib/ab2admin -o start
```

To disable the *AnswerBook2* web server from starting at boot time, the *ab2mgr init* script needs to be removed from the */etc/rc2.d* directory. The following command will do this.

```
rm /etc/rc2.d/S96ab2mgr
```

Choosing a Server

Choice of a general-purpose web server is dependent on several factors, including the list of desired features, cost, and availability. Solaris systems support several different web servers. The following are two common choices.

- *iPlanet*: A commercial web server with many web-based commerce features built in. This server is updated and supported by the Sun-Netscape Alliance and is intended for production-quality web service.

- *Apache*: A free web server developed by a community of Internet programmers, it is available for many UNIX systems, as well as Solaris in source code form. According to the NetCraft survey, Apache is the most widely used web server package, and for good reason. Apache includes the latest features and provides for a broad range of customization and configuration. It is a good choice for sites that require the latest web server features and possess the required software development tools to compile and maintain the program. The Apache web server is available at the web site listed at end of this chapter.

A web server installation begins by installing the web server software itself, followed by careful configuration of numerous server options and features. CGI programs run by the server for building web pages on demand or for processing information received in forms will also be installed and integrated with the web server.

> ✗   *WARNING: As with other network services, security is a primary con-*
> *cern when configuring a web server. A great number of system attacks*
> *focus on the web server and the associated CGI programs. See the section*
> *that follows on server security for more details.*

Server Configuration

How a web server functions is controlled by its configuration file.
Configuration options of the major web servers are similar. In the
following discussion, the configuration file locations and options
used by the Apache web server will be used. The Apache web
server's function is controlled via the configuration files described
in table 22-1, which follows.

Table 22-1: Apache Configuration Files

File	Function
httpd.conf	Server configuration file containing the base configuration information.
srm.conf	Resource configuration file that controls local resource information such as the location of icons and image maps. This file is empty in current versions of Apache. The directives previously placed here are now found in the *httpd.conf* file.
access.conf	Access configuration file that specifies the actions allowed by specific clients on specific files. This file is empty in current versions of Apache. The directives previously placed here are now found in the *httpd.conf* file.
mime.types	Specifies the media types and respective file suffixes.

The Apache server distribution includes a set of samples of the
previously listed files, which the system administrator can modify.
Over 100 configuration options can be applied to control the
behavior of the Apache web server. A few of the most basic
options to be examined upon setting up a new web server are
examined in this section.

➡ *NOTE: As of Apache version 1.3, all of the configuration information from the* srm.conf, access.conf, *and* httpd.conf *files has been merged into a single file, the* httpd.conf *file.*

httpd.conf *File*

The *httpd.conf* file is the base on which all other configuration files rest. Among the options that can be controlled by this file are the locations of the *httpd* and associated files, server log files, and other configuration files.

At a minimum, the system administrator will want to modify the *User, Group, ServerAdmin, ServerRoot*, and *ServerName* lines to reflect the local site. The User and Group lines specify the *user id* and *group id* the web server will operate under once started. The *ServerAdmin* is an e-mail address to which the server can send problem reports. The *ServerRoot* specifies the installation directory for the server. The *ServerName* is the name of the server returned to clients. The following is an example of an *httpd.conf* file.

```
User nobody
Group nobody
ServerAdmin dwight@astro-corp.com
ServerRoot /usr/local/apache
ServerName www.astro-corp.com
```

➡ *NOTE: The User and Group directives in the* httpd.conf *file have significant security implications. The nobody user is used to severely limit the access privileges the web server has in order to limit what an attacker might be able to access via the web server. These directives also specify the default user under which any CGI program is run. Limiting the privileges a CGI program has is an important step in making the CGI program secure.*

The *Alias* lines may also require updating to reflect the location of icons and other local files. The *Alias* lines allow web page designers to use shortened names for resources such as icons, instead of specifying full paths. An example follows.

```
DocumentRoot /usr/local/http/htdocs
UserDir WWW
Alias /icons/ /usr/local/http/icons/
ScriptAlias /cgi-bin/ /usr/local/http/cgi-bin/
```

Besides making web page construction easier by providing short names for icons and CGI programs, these directives allow access to users' web pages. The *UserDir* line specifies the subdirectory each user can create in his home directory to hold web pages. This directory, *WWW* in the example, is mapped to the user's log-in name as follows. A user whose log-in name is *bob* has his WWW directory mapped to *http://www.astro-corp.com/~bob/*. By default, the Apache web server will display the *index.html* file in that directory, or a directory listing if the *index.html* file is not found.

A new installation of Apache will also require changing the *<Directory>* directives to indicate where the server should look for documents to serve and for CGI programs. For example, if the server is installed in */usr/local/apache*, with the documents and CGI programs in directories under that directory, the following *<Directory >* line may be necessary.

```
<Directory /usr/local/apache/htdocs>
```

Modules

One of the more useful features found in the Apache web server is the use of modules for extending base server functionality. These modules provide such services as web server status monitoring, URL rewriting, and addition of native versions of CGI tools such as Perl. For modules built as part of the standard

Apache build, activating them is a matter of calling the directive associated with the module. For example, the following are the lines required to activate the *mod_status* module, which allows the administrator to query the web server for status information.

```
<Location /server-status>
  SetHandler server-status
  Order deny,allow
  Deny from all
  Allow from .astro.com
</Location>
```

The *Location* directive describes the "page" used to view that status information, whereas *SetHandler* specifies the server-status entry to the *mod_status* module. The triple of *Order*, *Deny*, and *Allow* directives controls access to this "page," limiting it to only hosts within the specified domain. If the server's name were *www.astro.com*, the URL used to access this page would be *http://www.astro.com/server-status/*. The following illustration shows the status page for an active web server.

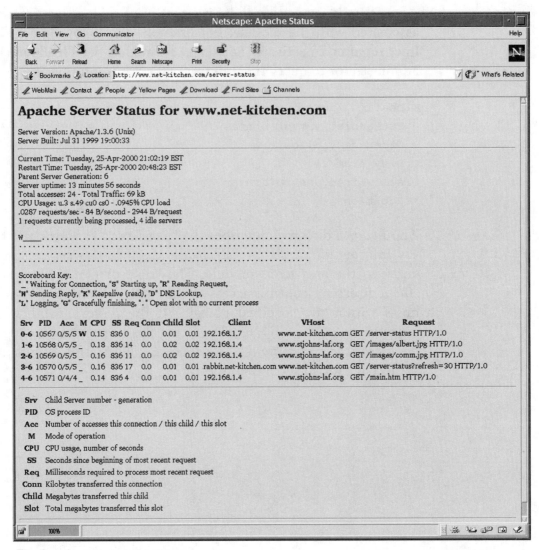

The Apache server status page.

Another commonly used module is *mod_spelling*. This module corrects misspelled URLs by fixing common typos such as transposition of characters. The *CheckSpelling* directive controls this module, and it can be made to work over a specific set of web pages in a manner similar to that shown for the server-status directive. For more information on modules, see the documentation on the Apache web site listed at the end of this chapter.

mime.types *File*

The *mime.types* file includes the mapping from a mime type to a file extension. The most common types are provided in the sample file provided with the Apache distribution. A more detailed description of the myriad available configuration options can be found in the books and web sites listed at the end of the chapter.

Starting and Stopping the Web Server

The web server is started and stopped in the same manner as other daemons. The run-control script names will vary, depending on the server used. The Apache web server provides a control program named *apachectl* for starting, stopping, and restarting the web server. This is used to start and stop the server manually, as shown in the following.

```
/usr/local/apache/bin/apachectl start
/usr/local/apache/bin/apachectl stop
```

Automating the startup of the server is a matter of placing links to *apachectl* in the */etc/rc2.d* directory, as described in Chapter 5, "Refining System Boot and Shutdown Procedures."

Server Security Issues

Web servers present a difficult security challenge. They must be widely accessible to be useful, but tightly controlled to prevent security breaches. They must be tolerant of any requests submitted to them, including requests specifically constructed to gain unauthorized access to files or to exercise bugs in CGI programs. Basic precautions that should be taken to help prevent problems are discussed in the following sections. Additional information on securing a web server can be found in the Garfinkel and Spafford books listed at the end of the chapter.

Port 80

By default a web server listens on port 80. This is a well-known port and will be examined by attackers. The port can be changed via the server configuration file. However, this will cause web browsers to be unable to connect to the server unless the port number is included in the URL specification. For example, if the web server on *www.astro.com* were set to listen on port 8000, the URL for the server's default page would be the following.

http://www.astro.com:8000/

File Access Control

The control files that determine the web server's function, as well as the log files it produces, should not be accessible to the user ID under which the web server runs. Individuals attempting to gain unauthorized access are thwarted to the extent that they cannot obtain information about the web server's configuration and function. In addition, web server options under which web pages include other files, and optionally execute programs, should be carefully scrutinized for potential access to files not intended for distribution.

Many web servers provide access control for individual directories by means of control files found in those directories. Apache uses a file named *.htaccess*, which contains directives specifying access. The access directives can include IP address ranges and references to password databases if desired. The details of access control under Apache can be found in the documentation on the Apache web site.

Dangers of CGI

CGI programs are among the greatest potential dangers to web server security. These programs are run based on a URL passed to the web server by a client. In normal operations, this URL comes from a form or page. However, the URL provided to a CGI program can be given to the web server by other means, and can be carefully constructed to exercise bugs in the CGI program itself.

✗ *WARNING: Perhaps the most common attack against a web server is via the* phf *CGI program. The* phf *program is not included with recent versions of Apache, but was present in earlier versions. To disable this CGI program, remove it from the* cgi-bin *directory specified in the web server configuration file. As a general rule, any unused CGI program should be removed from the* cgi-bin *directory.*

CGI programs must be carefully constructed to avert potential problems resulting from the input passed to them. One successful method is to use the *tainted* variable facility found in the Perl scripting language. If other languages are used, care must be taken to ensure that *all* possible input characters are properly handled, including shell metacharacters, quotes, asterisks, and braces. Administrators must also be alert to the well-known problem of very large input strings designed to overwrite small input buffers. Security-conscious sites should carefully audit CGI programs before putting them into operation.

✗ *WARNING: The* mod_perl *module for the Apache web server does not provide any security advantages over a standalone CGI program written in Perl. Although* mod_perl *does offer a substantial performance improvement, CGI programs making use of* mod_perl *need to be as carefully audited as standalone CGI programs.*

Some alternative approaches to standalone CGI programs are dynamic page construction tools such as PHP and ZOPE. These tools provide a standardized CGI interface designed specifically to avoid some of the problems found in input from web pages. These tools also provide for rapid development of dynamic pages, used in a growing number of web applications. PHP is also available as an Apache module, giving better performance than that of a standalone CGI program.

> ✗   *WARNING: Both PHP and ZOPE are powerful scripting languages and have security concerns specific to the features of each. If a site chooses to use one of these tools, the script developers will need to review the specific security issues for the chosen tool. Pay particular attention to the file access concerns raised by these tools.*

Log Files

Web servers maintain several log files that can aid in monitoring the security of the web server. Table 22-2, which follows, describes several such log files for the Apache web server.

Table 22-2: Apache Log Files

File Name	Content
access_log	Listing of each request fielded by the web server.
agent_log	Listing of every program run by the web server. This log is optional in the default Apache installation, and can be enabled by editing the *httpd.conf* file.
error_log	Listing of the errors the server encountered. Errors from CGI programs, as well as the server itself, are logged to this file.
refer_log	Listing of the previous URL accessed by a given browser. This log is optional in the default Apache installation, and can be enabled by editing the *httpd.conf* file.

Of principal interest from a security standpoint are *error_log*, *agent_log*, and *access_log*. These logs should be reviewed periodically for purposes of identifying CGI program problems and attempts to access files not intended for distribution.

Another aspect of web server log files is the wealth of information they hold regarding use of the web site. Log analysis tools such as *http-analyze (http://www.netstore.de/Supply/http-analyze/)* can provide the web site administrator with a variety of useful statistics on the use of the web site.

Web Servers and Firewalls

Another common error in deploying web servers is to place the web server behind the firewall and allow requests to the web server to pass through the firewall. Although this might seem like a good way of protecting the web server, it in fact more often leads to the web server becoming a conduit for attackers to pass through the firewall and gain access to the secured network behind it.

A better approach is to place the web server outside the firewall. In this configuration, the web server is dedicated to web serving only, and all other services except for a secure communications facility such as *ssh* are removed for the system. Chapter 18 both describes the removal of services and provides details on obtaining and installing *ssh*. Placing the web server outside the firewall acts to prevent an attack on the web server from proceeding to the systems protected by the firewall. The following illustration shows an example of such a network configuration.

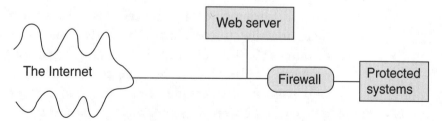

A web server located outside the firewall helps prevent attacks on the web server from proceeding on to the other systems protected by the firewall.

A still better approach for larger networks is to establish a so-called "Demilitarized Zone" (or DMZ) area between the firewall-protected internal network and the Internet using a second firewall. The advantage of this approach is that the firewall between the Internet and the DMZ offers some protection to the web server, while allowing web requests to pass into and out of the DMZ. The firewall between the DMZ and the internal network then acts to prevent an attack on the web server from proceeding on to systems on the internal network. The following illustration shows an example of such a network configuration.

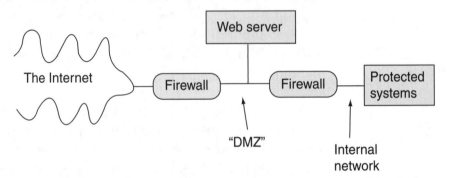

A DMZ area containing a web server is protected from the Internet by a firewall, whereas the internal network is protected from the DMZ (and the Internet) by a second firewall.

Web Performance Issues

The performance of a web server is a mixture of several factors, including the style of data served (dynamic versus static), system resources (CPU, I/O), and the available network bandwidth. Web requests can be viewed as requests for various objects. A typical web page might include some text and one or more graphical images. A web browser will make separate requests, often in parallel, for each element of the page. The web server fills each request as a separate item. Web server load is measured in the size of individual requests and the number of requests it can fill per unit of time. Requests are referred to as "hits."

Apache Web Server

The Apache web server deals with requests by using a pool of slave processes. The number of processes in the pool is managed dynamically by the parent web process, within the bounds set in the *httpd.conf* file. The following are the parameters that control the pool.

```
MinSpareServers    5
MaxSpareServers    10
StartServers       5
```

The *MinSpareServers* parameter specifies the minimum number of server processes in the pool. The *MaxSpareServers* parameter specifies the maximum number of server processes in the pool. *StartServers* specifies how many servers to start when Apache is started. The values listed for each of these parameters is the default, and in general should not be changed. Sites that see very large numbers of hits may consider increasing the number of servers but will need to pay careful attention to system resources, especially memory.

Server processing of data before a request is filled by page processing tools such as PHP, or by CGI programs, adds additional load on the server. Servers with dynamic page content may require additional memory or faster processors to provide reasonable speed in responding to requests.

Likewise, the speed of the network connection between the web server and web clients will limit the maximum number of hits per unit time that can be processed. Sites concerned about web performance should consult the detailed discussion of server and network sizing for web service found in Brian Wong's *Configura-*

tion and Capacity Planning for Solaris Servers, listed in the "References" section of this chapter.

Web Caches

Another method of improving web performance is the use of an external cache system. Most web browsers have a cache of recently viewed pages, graphics, and other page elements. This allows the browser to rapidly view the page again by loading elements from the local cache instead of re-requesting them from a web server. A similar technique can be applied to both the serving of web pages and the local network. Squid, a commonly used web cache program is listed in the reference section of this chapter.

For a local network with a slow connection to the Internet, a proxy web cache can be used to improve performance and conserve bandwidth on the slow-speed link. A proxy web cache acts as a local reference for all web requests. Web browsers on the local network are configured to use the proxy cache, and the proxy cache in turn makes requests for web pages not in its cache, or simply replies with the page elements already in the cache. The following illustration shows how this works.

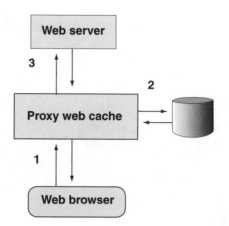

A proxy cache improves web performance for browsers by providing recently viewed page elements directly from a local cache.

1. The web browser requests a page element from the proxy web cache instead of directly from the web server.
2. If the requested element is in the cache, the proxy replies directly to the web browser using its local copy of the page element.
3. If the requested element is not in the cache, the proxy requests the element from the web server, stores it in its cache, and then passes the element along to the web server.

Some web sites use a web cache as the "front end" to their web server. This improves the performance for page serving by allowing the web cache to reply to frequently requested pages from its cache, off-loading that work from the web server itself. One situation where this is helpful is a web site with a mixture of static and dynamic web pages. The web cache can take on the load of serving the static pages while requests for dynamic pages are passed on the web server itself. The following illustration shows an example of this configuration.

A web cache handles requests for frequently viewed page elements off-loading requests from a busy web server.

1. Requests for web pages coming in from the Internet are directed to the web cache.
2. If the requested element is in the cache, the web cache replies directly using its local copy of the page element.
3. If the requested element is not in the cache, the proxy requests the element from the web server, stores it in its cache, and then passes the element along to requesting web browser.

References

The discussion of web client and server configuration in this chapter is intended as an overview. If you require more detailed information, consider the sources listed below in tables 22-3 and 22-4.

Table 22-3: Books

Source	Comments
Simson Garfinkel and Gene Spafford, *Web Security and Commerce*. O'Reilly & Associates, 1997 (ISBN 1-56592-269-7).	Specifically targeted at web server security and securing transactions carried out over the Web.
Stephen Spainhour and Valerie Querci, *Webmaster in a Nutshell*. O'Reilly & Associates, 1996 (ISBN 1-56592-229-8).	Covers a wide range of topics, including HTML, frames, JavaScript, and server configuration.
Brian L. Wong, *Configuration and Capacity Planning for Solaris Servers*. Prentice Hall, 1997 (ISBN 0-13-349952-9).	Covers server "sizing" for a variety of servers, including web servers.

Table 22-4: Web Sites

Contact Information	Comments
http://www.apache.org	Official home page for the Apache web server. This page includes complete on-line documentation and access to the web server source code.
http://www-genome.wi.mit.edu/WWW/faqs/www-security-faq.html	Lincoln D. Stein's *WWW Security FAQ*. A complete and current discussion of web server security problems and solutions.
http://hoohoo.ncsa.uiuc.edu/cgi/security.html	CGI security document of the National Center for Supercomputer Applications. Provides guidelines for writing more secure CGI programs.
http://www.php.net	Home web site for the PHP hypertext preprocessor.
http://www.zope.org	Home web site for the Z object publishing environment.
http://www.netstore.de/Supply/http-analyze/	*Http-analyze* is a web-based web server log file analysis tool that extracts several useful statistics from a web server's log files.

Summary

Web servers are becoming a common service that nearly every site will offer in some fashion. In this chapter, the basics of web service have been covered, including an introduction to configuring the Apache web server, and a discussion of web server security and performance issues. Administering WWW services is an ongoing task. The rapid development of web-based media means that continual updating of web browsers and web servers will be required. Administrators running servers should pay close attention to the configuration and security of their systems as they are updated and grow in complexity.

Throughout this book, the goal has been to prepare the system administrator to run systems in a robust, secure, and efficient manner. The emphasis on understanding the basic concepts of various system elements has led to a discussion of the basic tools every system administrator needs to know. But concepts and tools are not enough.

A system administrator needs to know some of the proven techniques other system administrators use that can be applied as needed to the system management challenges at hand. It is hoped that the discussion presented here will help the system administrator effectively manage Solaris systems, and thereby meet the needs of the organization he or she serves.

APPENDIX A
Solaris Version Information

With each release of Solaris, Sun Microsystems adds new features and support for new hardware. At the same time, it is not unusual for End of Life (EOL) statements to accompany a new Solaris release. These statements inform the user about the hardware and software that will no longer be supported in future releases.

The information in this appendix summarizes the major new features of, and EOL statements that accompanied the release of, Solaris 2.6, Solaris 7, and Solaris 8. This information is a compilation of information garnered from several Sun Microsystems documents.

This appendix also provides information on minimum SPARC operating system requirements for, and minimum Solaris firmware requirements for, Solaris 2.6, Solaris 7, and Solaris 8. Readers desiring more information (or more details) should read through the installation instructions, hardware compatibility information, and release notes for specific Solaris releases.

Solaris 2.6

The sections that follow present features introduced in Solaris 2.6, end-of-life statements announced with Solaris 2.6, and minimum SPARC operating system requirements for Solaris 2.6.

Features Introduced in Solaris 2.6

Table A-1, which follows, lists and describes features introduced in Solaris 2.6.

Table A-1: Features Introduced in Solaris 2.6

Feature	Description
Java Virtual Machine	The Java Virtual Machine 1.1 includes Java runtime support, and basic applet and application development tools.
HotJava Browser	Solaris 2.6 includes the HotJava web browser.
WebNFS support	Solaris 2.6 includes support for the WebNFS extensions, which allow file system access via a web browser interface.
UFS direct I/O	Solaris 2.6 provides routines that allow applications to read/write directly from/to file systems, bypassing the virtual memory buffer cache.
Raw I/O	I/O routines allow raw read/write to disks, without a file system (typically used by database software).
Kernel Sockets	Kernel sockets interface improved to provide better compatibility with BSD-style sockets.
TCP large windows	TCP large window support is RFC1323 compliant, and provides improved performance over high-bandwidth network links (ATM, satellite). TCP window sizes in excess of 64 Kbyte are supported.
Zero copy TCP/ hardware checksum	TCP modified such that the kernel does not have to do user-space to kernel-space memory copies. Support for hardware checksums added (for adapters that support this feature, such as the SunATM card).
Solaris WebStart based installation	Solaris Web Start is a browser-based utility that may be used to install the operating system and bundled application software.

Feature	Description
X86 device configuration	Configuration assistant simplifies Solaris on x86 installations by automating the hardware discovery phase of the install.
X86 peripheral configuration	The *kdmconfig* command provides the ability to configure the display, mouse, and graphics adapter on a Solaris x86 system.
Power Management for SPARC	Power management software allows the user to configure the system to go into power-saving mode during periods of inactivity.
Unicode 2.0 support	Solaris 2.6 is ISO 10646 compliant.
Y2K support	Solaris 2.6 is Y2K compliant.
POSIX 1003.1b support	Solaris 2.6 is POSIX 1003.1b compliant, allowing real-time functionality, including support for POSIX AIO extensions.
Large-file support	Solaris 2.6 provides large-file support for UFS, NFS, and CacheFS file systems per the Large File Summit. Applications can create and access files up to one Tbyte (UFS) (larger for NFS and CacheFS).
Pluggable Authentication Module (PAM)	Solaris 2.6 supports the PAM authentication standards.
Bind 4.9.4-P1	Solaris 2.6 provides the Berkeley Internet Name Daemon version 4.9.4-P1.
Network Time Protocol (NTP)	Solaris 2.6 includes the NTP time synchronization package.
Dynamic Host Configuration Protocol (DHCP)	Solaris 2.6 provides support for the DHCP protocol.
Variable Length Subnet Masks	Solaris 2.6 supports VLSM by enabling Classless Inter-Domain Routing (CIDR).
Processor Sets	Solaris 2.6 allows the administrator to allocate processes to a specific processor within a multiprocessor system.
Patch tools	Solaris 2.6 includes new patch management tools (*patchadd*, *patchrm*) to replace, respectively, the *installpatch* and *backoutpatch* commands.
dtlogin	The default log-in screen manager is now *dtlogin* (a portion of the CDE environment).

End-of-Life Statements Announced with Solaris 2.6

Solaris 2.6 does *not* provide support for the following products.

- SparcSystem 600 series (630MP, 670MP, 690MP)
- Direct Xlib software
- *ssacli* command (used with the SparcStorage array)
- NeWSprint and NeWSprint based printers
- SX Compute Acceleration in XIL
- Sbus Expansion Subsystem (SES/C – X1072A)
- Power PC (Ppc) Systems

The following frame buffer and network interface products *may not* be supported in future Solaris releases.

Frame Buffers:

- MG1 (bwtwo)
- MG2 (bwtwo)
- CG2 (cgfour)
- CG4 (cgfour)
- TC (cgeight)
- ZX (leo)
- TZX (leo)

Network Interface:

- BEC

Solaris 2.6 Minimum SPARC Operating System Requirements

In order to run Solaris 2.6 on a SPARC system, you will need the following. Table A-2, which follows, presents recommended space allotments for various SPARC platform software configurations.

- A boot disk with capacity of 1.05 Gb or larger

- A 2.1-Gb or larger disk recommended

- At least 48 Mbytes of RAM memory

- A sun4c, sun4d, sun4m, sun4u, or sun4u (PCI) SPARC system

Table A-2: Recommended SPARC Platform Software Configuration Sizes for Solaris 2.6

Software Group	Recommended Space (Mb)
Entire distribution with OEM support	616
Entire distribution	608
Developer support	537
End user support	281

Solaris 7

The sections that follow present features introduced in Solaris 7, end-of-life statements announced with Solaris 7, minimum SPARC operating system requirements for Solaris 7, and minimum Solaris 7 firmware requirements.

Features Introduced in Solaris 7

Table A-3, which follows, lists and describes features introduced in Solaris 7.

Table A-3: Features Introduced in Solaris 7

Feature	Description
64-bit operating environment (SPARC only)	The 64-bit environment supports 32- or 64-bit application and development environments. This environment permits maximum compatibility and interoperability for existing applications. The 64-bit environment removes many limitations of 32-bit environments, including providing 64-bit virtual address space.
Netscape Communicator	Solaris 7 includes Netscape Communicator.
UFS logging	UFS logging stores file system transactions in a log before applying changes to the file system. UFS logging prevents file system inconsistencies, and reduces the time required for boot-time file system checks.
-o noatime option	Informs system not to update access times when mounting a UFS file system, causing substantial reduction of disk activity on file systems where access times are unimportant (e.g., Usenet News).
LDAP	Solaris 7 includes library support for the Lightweight Directory Access Protocol.
Dynamic reconfiguration	Dynamic reconfiguration allows hot-swapping of system boards in certain SPARC systems.
New commands	*pgrep*: Look at active processes on the system and display information on processes that match command line arguments. *pkill*: Issue a kill signal to processes that match command line arguments.
Sendmail 8.9	Solaris 7 includes Sendmail 8.9, which allows anti-spam filters, and virtual hosting.
traceroute	Solaris 7 includes *traceroute*.
System crash dump facility	The *dumpadm* command allows administrator to configure crash dump parameters. Dump information is stored in compressed format. System dump information is saved in the background if a non-*swap* partition is specified as the dump device.

Feature	Description
TCP with SACK	RFC 2018 compliant TCP selective acknowledge (TCP SACK) is supported under Solaris 7.
RPCSEC_GSS	The RPC system has been modified to be GSS-API compliant, improving NFS security.
NIS+ extended with Diffie-Hellman	NIS+ security has been improved by incorporating Diffie-Hellman keys (key length increased from 192 bits to 640 bits).
BIND 8.1.2	Solaris 7 ships with version 8.1.2 of the Berkeley Internet Name Daemon (BIND) name service package.
Enhanced language support	Solaris 7 includes six new UTF-8 locales.
Expanded locale support	Solaris 7 includes support for the new Euro currency.
SGML support	The Solaris 7 manual pages are available in *Answerbook2* SGML format.
Intelligent I/O support (I2O)	Solaris 7 supports the I2O framework on Intel systems.
CDE environment	CDE supports Motif 2.1 extensions.

End-of-Life Statements Announced with Solaris 7

The following frame buffer, network interface, and software products are *not* supported in Solaris 7.

Frame Buffers:

- MG1 (bwtwo)
- MG2 (bwtwo)
- CG2 (cgfour)
- CG4 (cgfour)
- TC (cgeight)

- ZX (leo)
- TZX (leo)

Network Interface:

- BEC

Software:

- Solstice SyMON
- Sun MediaCenter One
- Dynamic reconfiguration for Enterprise 10000
- Alternate pathing for Enterprise 3x00, 4x00, 5x00, 6x00 servers
- PEX graphics libraries

The following products *may not* be supported in future Solaris releases.

- HotJava Browser
- The sun4c architecture
  - SPARCstation SLC
  - SPARCstation ELC
  - SPARCstation IPC
  - SPARCstation IPX
  - SPARCstation 1
  - SPARCstation 1+
  - SPARCstation 2
- SPARCstation Voyager
- SPARC Xterminal 1

Solaris 7 Minimum SPARC Operating System Requirements

In order to run Solaris 7 on a SPARC system, you will need the following.

- A boot disk with capacity of 2.0 Gb or larger

- At least 64 Mbytes of RAM memory

- A sun4c, sun4d, sun4m, sun4u, or sun4u (PCI) SPARC system (see following Note)

➡ *NOTE: sun4c and sun4d systems run Solaris 7 in 32-bit mode only.*

Solaris 7 Minimum Firmware Requirements

In order to run Solaris 7, the system firmware may need to be upgraded. Table A-4, which follows, provides details for systems that may require firmware upgrades, including the minimum firmware version required to support Solaris 7 on those systems.

Table A-4: Minimum Firmware Requirements for Solaris 7

System Type (from uname –i)	Minimum Firmware Version (from prtconf –V)
SUNW,Ultra-1-Engine	3.10.0
SUNW,Ultra-1	3.11.1
SUNW,Ultra-2	3.11.2
SUNW,Ultra-4	3.7.107
SUNW,Ultra-Enterprise	3.2.16

Table A-5, which follows, lists recommended space allotments for 32-bit and 64-bit support for various Solaris software configurations.

Table A-5: Recommended Software Configuration Sizes for Solaris 7

Software Group	Recommended Space at 32–bit Support (Mb)	Recommended Space at 64–bit Support (Mb)
Entire distribution with OEM support	801	909
Entire distribution	787	895
Developer support	716	837
End user support	438	532

Solaris 8

The sections that follow present features introduced in Solaris 8, end-of-life statements announced with Solaris 8, minimum SPARC operating system requirements for Solaris 8, and minimum Solaris 8 firmware requirements.

Features Introduced in Solaris 8

Table A-6, which follows, lists and describes features introduced in Solaris 8.

Table A-6: Features Introduced in Solaris 8

Feature	Description
IPv6 support	Solaris 8 provides support for the next-generation Internet Protocol (IPv6). The Solaris 8 TCP/IP implementation allows for IPv4, IPv6, or dual-stack (IPv4 + IPv6) network operations. IPv6 adds increased address space, improves Internet functionality using simplified header format, adds support for authentication and privacy, adds auto-configuration of address assignments, and enables quality of service capabilities. The IPv6 support includes NIS/NIS+ and NFS/ RPC support.
Lightweight Directory Access Protocol (LDAP)	Native LDAP provides the Naming service switch with back-end support for LDAP-based directory services.
Improved installation and management	Solaris WebStart, a graphical wizard-based Java application, installs the Solaris operating environment, and other software.
DHCP install support	Network installs can now use DHCP to acquire boot and network configuration information.
IA: Boot partition changes	The Intel edition now requires a separate boot partition when using WebStart installation software.
StarOffice 5.1	Solaris 8 ships with a copy of the StarOffice 5.1 office suite. This package includes applications that can read/write files compatible with the Microsoft Office suite.
IA: CD-ROM boot	The Intel edition now supports "El-Torito" standards, which enable the system to be booted directly from the distribution CD.
Java 2 Software Development Kit (SDK)	Solaris 8 includes an updated Java Development Kit (version 1.2.1_04), which provides improved scalability, improved class libraries (including two new APIs), enhanced memory management, high-performance scalable Java virtual machine (JVM), Just In Time (JIT) compiler optimization, and faster Java thread synchronization.
DHCP manager	The DHCP manager now provides a Java-based graphical front end.
IA: Large-disk support	The Intel edition now supports disks larger than 8 gigabytes.
Solaris Web Based Enterprise Management (WBEM) support	Solaris WBEM Services software implements Web Based Enterprise Management (WBEM) standards and technologies. WBEM is intended for developers and administrators of WBEM-enabled environments.

Feature	Description
DNS support in system identification utilities	Solaris 8 supports DNS as a name service. The installation process "identification" utilities now allow the user to enter DNS information.
Unlimited number of pseudo terminals	Solaris 8 software allows an unlimited number of pseudo terminals for use with applications such as *rlogin*, *telnet*, and *ssh*.
Product Registry	The Product Registry tool manages the software installed using Solaris WebStart or *pkgadd* commands. The registry allows the administrator to view a list of installed software, install new software, uninstall software, and browse/launch an installer.
IPsec support	Solaris 8 provides support for the IPsec (IP Security) standards.
Sendmail 8.9.3	Solaris 8 includes Sendmail version 8.9.3.
Service Location Protocol (SLP)	SLP is an Internet Engineering Task Force protocol for discovering shared resources on the network.
Streams enhancements	Solaris 8 includes numerous streams enhancements.
Network Time Protocol (NTP)	Solaris 8 includes NTP version 3-5.93e.
Universal Disk Format (UDF)	Solaris 8 includes drivers to support the UDF format on CD-ROM, disks, diskettes, and DVD media.
NFS server logging	NFS server logging allows the server to provide a record of file operations performed on file systems.
IA: Extended memory support	The Intel edition now supports the XMEM standard, increasing memory space to 4 gigabytes. The improved memory support includes PAE mode support, which allows address spaces up to 32 gigabytes of memory.
Solaris SmartCard support	The Solaris 8 SmartCard support implements the Open Card Framework (OCF) 1.1 standard.
Personal Digital Assistant (PDA) support	Solaris 8 includes PDASync, a Java-based application that allows users to synchronize calendar, mail, address book, and memos with their PDA.
Hot Key Editor	The Hot Key Editor allows users to predefine function keys to perform key sequence operations.
X11R6.4 support	Solaris 8 is X11R6.4 compliant.

Feature	Description
Netscape Communicator 4.7	Solaris 8 includes Netscape Communicator version 4.7.
Apache web server	Solaris 8 includes version 1.3.9 of the popular Apache web server.
Perl5	Solaris 8 includes Perl 5.005-3.
Role Based Access Control (RBAC)	RBAC allows administrators to assign limited administrative capabilities to normal users.

End-of-Life Statements Announced with Solaris 8

The following features are *not* supported in Solaris 8.

- The sun4c architecture
  - SPARCstation SLC
  - SPARCstation ELC
  - SPARCstation IPC
  - SPARCstation IPX
  - SPARCstation 1
  - SPARCstation 1+
  - SPARCstation 2
- SPARCstation Voyager
- SPARC Xterminal 1

The following features *may not* be supported in future Solaris releases.

- Sun4d Servers (32-bit only)
  - SparcServer 1000
  - SparcServer 2000
- Ethernet Quad Drivers

- *qe* device
- *qec* device

Solaris 8 Minimum SPARC Operating System Requirements

In order to run Solaris 8 on a SPARC system, you will need the following.

- Boot disk with capacity of 2.0 Gb or larger

- 4.2-Gb (or larger) disk drive recommended

- At least 64 Mbytes of RAM memory

- A sun4m, sun4u, or sun4u (PCI) SPARC system

Solaris 8 Minimum Firmware Requirements

In order to run Solaris 8, the system firmware may need to be upgraded. Table A-7, which follows, provides details on systems that may require firmware upgrades, including the minimum firmware version required to support Solaris 8 on those systems.

Table A-7: Minimum Firmware Requirements for Solaris 8

System Type (from uname –i)	Minimum Firmware Version (from prtconf –V)
SUNW,Ultra-1-Engine	3.10.0
SUNW,Ultra-1	3.11.1
SUNW,Ultra-2	3.11.2
SUNW,Ultra-4	3.7.107
SUNW,Ultra-Enterprise	3.2.16

Table A-8, which follows, presents recommended space allotments for various Solaris 8 software configurations.

Table A-8: Recommended Software Configuration Sizes for Solaris 8

Software Group	Recommended Space (Gb)
Entire distribution with OEM support	2.4
Entire distribution	2.3
Developer support	1.9
End user support	1.6

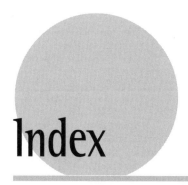

Index

Also Available from OnWord Press or Your Favorite Technical Bookstore

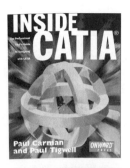

INSIDE CATIA
Paul Carman and
Paul Tigwell
Includes CD-ROM
Order No. 1-56690-153-7

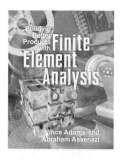

*Building Better Products
with Finite Element Analysis*
Vince Adams and Abraham
Askenazi
Order No. 1-56690-160-X

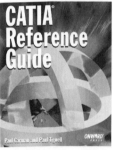

CATIA Reference Guide
Paul Carman and
Paul Tigwell
Order No. 1-56690-155-3

INSIDE Pro/SURFACE
Norman Ladouceur
Order No. 1-56690-134-0

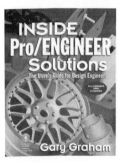

*INSIDE Pro/ENGINEER
Solutions*
Gary Graham
Includes CD-ROM
Order No. 1-56690-168-5

*Pro/ENGINEER Solutions
and Plastic Design*
Norman Ladouceur and
John McKeen
Order No. 1-56690-188-X

*Pro/ENGINEER Solutions
Advanced Techniques and
Workarounds*
Bob Townsend and Gary
Schmidt
Order No. 1-56690-163-4

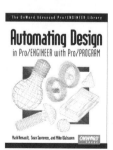

*Automating Design in Pro/
ENGINEER with Pro/
PROGRAM*
M. Henault, S. Sevrence, M.
Walraven, F. Karam, G.
Mlotkowski
Order No. 1-56690-117-0

INSIDE ArcInfo
Jay Flynn and Teresa Pitts
Includes CD-ROM
Order No. 1-56690-194-4

GIS Online: Information Retrieval, Mapping, and the Internet
Gordon Plewe
Order No. 1-56690-137-5

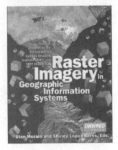

Raster Imagery in Geographic Information Systems
Eds. Stan Morain and Shirley López Baros
Order No. 1-56690-097-2

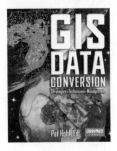

GIS Data Conversion: Strategies, Techniques, Management
Ed. Pat Hohl
Order No. 1-56690-75-8

Focus on GIS Component Software: Featuring ESRI's MapObjects
Robert Hartman
Includes CD-ROM
Order No. 1-56690-136-7

GIS Solutions in Natural Resource Management
Ed. Stan Morain
Order No. 1-56690-146-4

For a complete list of OnWord Press titles, visit our Web site at:
www.onwordpress.com
TO ORDER CALL: 800-347-7707

Your opinion matters! If you have a question, comment, or suggestion about OnWord Press or any of our books, please send email to *info@delmar.com*. Your feedback is important as we strive to produce the best how-to and reference books possible.